T0329521

**Sensing Technologies
for Real Time Monitoring
of Water Quality**

Sensing Technologies for Real Time Monitoring of Water Quality

Libu Manjakkal
University of Glasgow, Glasgow, UK
Edinburgh Napier University, Edinburgh, UK

Leandro Lorenzelli
Fondazione Bruno Kessler, Trento, Italy

Magnus Willander
Linköping University, Linköping, Sweden

IEEE Press Series on Sensors
Vladimir Lumelsky, Series Editor

IEEE PRESS

WILEY

Published by John Wiley & Sons, Inc., Hoboken, New Jersey.
Published simultaneously in Canada.

For general information on our other products and services or for technical support, please contact our Customer Care Department within the United States at (800) 762-2974, outside the United States at (317) 572-3993 or fax (317) 572-4002.

Wiley also publishes its books in a variety of electronic formats. Some content that appears in print may not be available in electronic formats. For more information about Wiley products, visit our web site at www.wiley.com.

Library of Congress Cataloging-in-Publication Data:
Names: Manjakkal, Libu, author. | Lorenzelli, Leandro, author. | Willander, Magnus, author.
Title: Sensing technologies for real time monitoring of water quality / Libu Manjakkal, Leandro Lorenzelli, Magnus Willander.
Description: Hoboken, New Jersey : Wiley-IEEE Press, [2023] | Includes index.
Identifiers: LCCN 2023012128 (print) | LCCN 2023012129 (ebook) | ISBN 9781119775812 (cloth) | ISBN 9781119775829 (adobe pdf) | ISBN 9781119775836 (epub)
Subjects: LCSH: Water quality–Measurement. | Water quality–Remote sensing. | Intelligent sensors.
Classification: LCC TD367 .M3175 2023 (print) | LCC TD367 (ebook) | DDC 628.1/61–dc23/eng/20230422
LC record available at https://lccn.loc.gov/2023012128
LC ebook record available at https://lccn.loc.gov/2023012129

Cover Design: Wiley
Cover Image: © webphotographeer/Getty Images

Set in 9.5/12.5pt STIXTwoText by Straive, Pondicherry, India

Contents

About the Editors

Libu Manjakkal (Phd, MRSC) received B.Sc. and M.Sc. degrees in physics from Calicut University and Mahatma Gandhi University, India, in 2006 and 2008, respectively. From 2009 to 2012, he was with CMET, Thrissur, India, and in 2012, for a short period, he worked at CENIMAT, FCT-NOVA, Portugal. He completed his PhD in electronic engineering from the Institute of Electron Technology, Poland (2012–2015) (Marie Curie ITN Program), and after this one year, he continues as a post-doctoral fellow in the same institute. From 2016 to 2022, he has been a post-doctoral fellow at the University of Glasgow (UoG) and also worked in the role of Scientific Project Manager in a Marie Curie ITN project (AQUASENSE) at UoG. Currently, he is working as a Lecturer at Edinburgh Napier University. He has authored/co-authored more than 65 peer-reviewed papers (47 journals) and 1 patent. His research interests include material synthesis, wearable energy storage, electrochemical sensors, pH sensors, water quality monitoring, health monitoring, supercapacitors, and energy-autonomous sensing systems.

Leandro Lorenzelli, Head of the Microsystems Technology Research Unit at FBK-Center for Materials and Microsystems (Trento, Italy), received the Laurea degree in Electronic Engineering from the University of Genova in 1994 and a PhD in 1998 in Electronics Materials and Technologies from the University of Trento. His objective has been to strengthen the reliability of the microfabrication technologies in the sectors of MEMS and BioMEMS (lab on a chip and microfluidics), microsensors, flexible electronics, and to set up initiatives for technological transfer. His current main research interests are in the fields of technologies for organs on a chip and bendable electromagnetic metasurfaces. He is the author or co-author of about 90 scientific publications and four patents. He has been a coordinator of European projects in the areas of innovative technologies for flexible electronics, microsystems for food analysis, quality control, and water monitoring, and sensors in prosthetic applications. He has been a program chair and speaker in many international conferences.

Magnus Willander has been a chair professor in Physics at Göteborg University and Linköping University in Sweden. He has also been a visiting scientist and visiting professor in different parts of the world. His research was concentrated on material synthesis, characterization, and devices, and he used both experimental and theoretical approaches. For the last 15 years, Professor Willander has concentrated his work on chemical and mechanical problems for sensing and energy conversions. In these fields, Professor Willander has published around 1000 scientific/technical papers and 13 international scientific books, which are cited around 29 000 times. In addition, Willander has supervised 55 PhD students and worked as a specialist in electronics in the industry. He has been a PhD examiner for numerous PhD theses around the world.

List of Contributors

Andrea Adami
Fondazione Bruno Kessler Center
for Sensors and Devices (FBK-SD)
Trento, Italy

Fowzia Akhter
Faculty of Science and Engineering
Macquarie University, Sydney
NSW Australia

Md. E. E. Alahi
Shenzhen Institute of Advanced
Technology, Chinese Academy of
Science, Shenzhen, China

Akshaya Kumar Aliyana
Department of Electronics, Mangalore
University, Konaje, India

Aiswarya Baburaj
Department of Electronics, Mangalore
University, Konaje, India

Robert J. W. Brewin
Centre for Geography and
Environmental Science, University of
Exeter, Penryn, UK

Ravinder Dahiya
Bendable Electronics and Sensing
Technologies Group, Department of
Electrical and Computer Engineering
Northeastern University, Boston, USA

Fabiane Fantinelli Franco
Infrastructure and Environment
Research Division, James Watt School
of Engineering, University of Glasgow
Glasgow, UK

Renny Edwin Fernandez
Department of Engineering, Norfolk
State University, Norfolk, VA, USA

Priyanka Ganguly
Chemical and Pharmaceutical Sciences,
School of Human Sciences, London
Metropolitan University London, UK

Noushin Ghaderi
Fondazione Bruno Kessler Center
for Sensors and Devices (FBK-SD)
Trento, Italy

Naveen Kumar S. K.
Department of Electronics, Mangalore
University, Konaje, India

Peter E. Land
Plymouth Marine Laboratory
Plymouth, UK

Leandro Lorenzelli
Fondazione Bruno Kessler Center
for Sensors and Devices (FBK-SD)
Trento, Italy

Pierre Lovera
Nanotechnology Group, Tyndall
National Institute, University College
Cork, Cork, Ireland

Muhammad Hassan Malik
Silicon Austria Labs GmbH
Heterogeneous Integration
Technologies, Villach, Austria

Bappa Mitra
Fondazione Bruno Kessler Center
for Sensors and Devices (FBK-SD)
Trento, Italy

S. C. Mukhopadhyay
Faculty of Science and Engineering
Macquarie University, Sydney
NSW Australia

Tarun Narayan
Nanotechnology Group, Tyndall
National Institute, University College
Cork, Cork, Ireland

Omer Nur
Department of Sciences and
Technology, Physics and
Electronics, Linköping University
Norrköping, Sweden

Alan O'Riordan
Nanotechnology Group, Tyndall
National Institute, University College
Cork, Cork, Ireland

Nasrin Razmi
Department of Sciences and Technology,
Physics and Electronics, Linköping
University Norrköping, Sweden

Ali Roshanghias
Silicon Austria Labs GmbH
Heterogeneous Integration
Technologies, Villach, Austria

Sohail Sarang
Faculty of Technical Sciences
University of Novi Sad
Novi Sad, Serbia

H. R. Siddiquei
Faculty of Science and Engineering
Macquarie University
Sydney, NSW, Australia

Joao L. E. Simon
Centre for Geography and
Environmental Science, University of
Exeter, Penryn, UK

Jamie D. Shutler
Centre for Geography and
Environmental Science University
of Exeter, Penryn, UK

Stevan Stankovski
Faculty of Technical Sciences
University of Novi Sad, Novi Sad, Serbia

Goran M. Stojanović
Faculty of Technical Sciences
University of Novi Sad
Novi Sad, Serbia

Kiranmai Uppuluri
Lukasiewicz Research Network – Institute
of Microelectronics and Photonics
Department of LTCC Technology and
Printed Electronics, Krakow, Poland

Magnus Willander
Department of Sciences and Technology
Physics and Electronics, Linköping
University, Norrköping, Sweden

Yuqing Yang
Nanotechnology Group, Tyndall
National Institute, University College
Cork, Cork, Ireland

Preface

Water quality (WQ) degradation is caused due to multiple reasons that directly impact public health and the economy. Sensor technology and various programs are implemented for monitoring and assessing the status and the causes of WQ degradation. This is reflected through programs such as National Rural Drinking Water Quality Monitoring and Surveillance in India, the Water Framework Directive in the European Union, and part of the Water Quality Framework of Environmental Protection Agencies (EPA) in the United States. Over the past decade, WQ observing technology has risen to the challenge of scientists to identify and mitigate poor WQ by providing them with tools that can take measurements of essential biogeochemical variables autonomously. Commercial sensors for in situ monitoring using buoys and boats are being deployed to broaden data coverage in space and time. Yet, despite these options becoming more readily available, there is a gap between the technology and the end user and a disconnect between data quality, data gathering by autonomous sensors, and data analysis. Further, real-time monitoring of various physical-chemical-biological (PCB) parameters remains a challenge, and methods that allow holistic water management approach (also considering the catchment management) need greater attention and innovation. The need for real-time water quality monitoring (WQM) is highlighted in recent white papers from the European Innovation Partnerships on Water (EIP water), an initiative of the European Commission (EC). Given these, a fundamental rethinking of monitoring approaches could yield substantial savings and increased benefits of monitoring.

This book will cover a complete set of sensing technologies for WQM, particularly in relation with real-time monitoring. A few review articles and books reported in this field have reviewed some of the sensing technologies only and that too is not directly related to real-time monitoring. This book will cover smart sensing technologies for WQM, and it will highlight the current progress in this area. As mentioned above, the major obstacle in WQM is related to (i) data quality, (ii) data gathering, and (iii) data analysis. To address these challenges, this book will present a detailed overview of these topics through several chapters related to each. The above topics also define the three broad sections of this book. In fact, this will be one of the distinguishing features of this book as compared to previously reported review articles or books. Potential capabilities and critical limitations of each sensing technology and wireless system will be highlighted and possible solutions or alternatives will be explored. In terms of sensors, the book will evaluate various sensing materials, substrates, and designs of sensors including flexible or non-flexible and multi-sensory patches. Overall, this book will be the first choice for researchers to explore the potential of different sensing technologies, electronics/communication designs, and algorithms for data analysis and select the best one closely matching to their resources for desired application including water and food quality in harsh environments.

This book will be organized into four major sections. *Section I (Materials and Sensors Development Including Case Study)* will be devoted to the introduction and development of various materials and sensors for WQM. To describe all these features categorically, Section I is divided further into six chapters as described later in more detail. *Section II (Readout Electronic and Packaging)* will describe in full detail the various design of electronics, communication systems, and packaging. Section II is divided into two chapters as described later in more detail. *Section III (Sensing Data Assessment and Deployment Including Extreme Environment and Advanced Pollutants)* will present innovative deployment strategies used for remote monitoring of WQ in various atmospheres. This includes chapters related to microplastics, the deployment of sensors, and sensors for extreme environmental conditions. Section III is divided into three chapters as described later in more detail. *Section IV (Sensing Data Analysis and Internet of Things with a Case Study)* discusses real-time WQM using IoT and wireless sensor networks. The section will be divided into two chapters. The unique combination as proposed here has not been provided so far by any other book on WQM. Editors acknowledged the support provided by the European Commission through the AQUASENSE (H2020-MSCA-ITN-2018-813680) project.

July, 2023

Libu Manjakkal
Edinburgh, UK

Leandro Lorenzelli
Trento, Italy

Magnus Willander
Linköping, Sweden

Section I

Materials and Sensors Development Including Case Study

1

Smart Sensors for Monitoring pH, Dissolved Oxygen, Electrical Conductivity, and Temperature in Water

Kiranmai Uppuluri

Lukasiewicz Research Network – Institute of Microelectronics and Photonics, Department of LTCC Technology and Printed Electronics, Krakow, Poland

1.1 Introduction

All life on planet Earth is supported by water and it has not been found in liquid form anywhere else in the universe. However, this precious resource suffers from issues such as pollution. To solve these problems and to avoid them in the future, it is critical to continuously monitor the quality of water. Water quality monitoring is defined as the collection of data at regular intervals from set locations across and along the water body to establish accurate values of various parameters such that trends, variations, and conditions can be observed. Water can be contaminated by pollutants due to human activities, especially industrial and agricultural practices [1]. Polluted water is unfit for use and poses a threat to the health of humans and other organisms including plants and animals that depend on water for the sustenance of life. Use of harmful chemical and the disposal of untreated waste into the environment pollute the groundwater as well as the surface water. A few examples of such activities are mining of materials, manufacturing of products, addition of chemical fertilizers to the soil, production of energy, improper treatment and disposal of sewage, and transportation via waterways.

The traditional laboratory-based methods of water quality testing are time-consuming and expensive with low value for money because they require specific infrastructure and equipment, skilled personnel, sample collection, transportation, and storage. When water samples are collected from the site and transported to the laboratory, the final result of the quality test may be biased because different

Sensing Technologies for Real Time Monitoring of Water Quality, First Edition.
Libu Manjakkal, Leandro Lorenzelli, and Magnus Willander.
© 2023 The Institute of Electrical and Electronics Engineers, Inc.
Published 2023 by John Wiley & Sons, Inc.

errors are introduced during collection and transportation, such as contamination and changes in data [2]. Additionally, some oxidation–reduction processes and parameters such as temperature must be measured on-site. Laboratory-based testing also does not provide real-time data. This is dangerous because the faster we know about a problem in the water, the faster we can act on it. Real-time water quality data is the paramount requirement for early warning systems (EWS) and contamination warning systems (CWS) which help to immediately notify the responsible authorities of undesirable changes in the water. Sensors are crucial in this process because for engineers and operators handling such systems, accurate sensors aid in tracking changes in water quality, predicting the generation of regulated compounds, and ensuring their elimination after a treatment or remediation process. Online water quality monitoring is designed for faster data collection and communication using smart sensor systems that eliminate the need for sample collection and transportation using wireless communication and Internet of Things (IoT). Consequently, this reduces the overall time, expense, space, workforce, and energy required for water quality monitoring.

1.2 Water Quality Parameters and Their Importance

Water quality testing indicates the health of water and informs us whether it is fit or unfit for a particular application such as drinking, agriculture, recreation, or disposal to open waters and thereby acts as the guiding factor in decisions related to national and international environmental regulations for water quality. Additionally, the variation in a particular parameter may indicate the presence or absence of a threat such as a microbial community or ecosystem warming. Therefore, real-time water quality monitoring simultaneously acts as an early warning system and thus the importance of low cost, miniaturized, and effective wireless sensors is well acknowledged by researchers, industries, and environmental authorities worldwide.

Compared to microbial parameters, physicochemical parameters such as electrical conductivity, temperature, pH, and chlorine are cheaper, simpler to detect, and have the ability to be measured using online instrumentation [3]. Moreover, variation in a parameter can also act as an indicator of changes in another parameter.

1.2.1 Impact of pH on Water Quality

There have been many studies showing evidence of the impact of pH on water quality and consequently aquatic life [4–6]. Marine ecological systems are so fragile yet complex that changes in one water quality parameter can lead to a domino effect causing changes in the hydrologic system and put an entire species at risk, eventually affecting other species including humans, their health, and livelihood. For example, Bradley and Sprague [6] found that zinc toxicity to rainbow trout (*Salmo gairdneri*) was directly related to pH and water hardness levels. Zinc toxicity rose by factors of two to five when pH changed from 5.5 to 7.0 and at pH 9.0 it was 10 times more toxic. However, at higher pH, zinc precipitated and has much lesser lethality to fish. Another study by Schubaur-Berigan et al. [4] observed that total ammonia was more toxic at higher pH to a species of nonbiting midge larvae (*Chironomus tentans*) and blackwork (*Lumbriculus variegatus*). At high pH, more unionized ammonia prevails which is important for determination of toxicity of ammonia to the two species.

1.2.2 Impact of Dissolved Oxygen on Water Quality

Dissolved oxygen (DO) is a very crucial parameter while monitoring water quality, especially to ensure the well-being of aquatic creatures [7]. In wastewater treatment, if DO level is too low, the important bacteria that decompose the solids will die whereas if DO content is too high, aeration consumes much more energy. In aquaculture, DO acts as the lifeline of fish and if it is too low, the fish will suffocate and perish.

1.2.3 Impact of Electrical Conductivity on Water Quality

In water quality monitoring, electrical conductivity is nonselective and does not consider the individual concentrations of various ions but instead their collective concentration. Regardless of that, changes in conductivity act as a warning signal of contamination and environmental changes. High levels of electrical conductivity can be due to industrial or urban runoff, low-flow conditions, or long periods of dry weather whereas low levels might occur due to organic compounds such as oils [8]. The measurement of electrical conductivity of water is therefore a good method to investigate the dissolved substances, chemicals, and minerals in it.

1.2.4 Impact of Temperature on Water Quality

Temperature has garnered more attention from researchers than any other environmental variable for aquatic organisms, possibly because of the simplicity in measuring it for both field and laboratory experiments [9]. Temperature alone may be lethal. Most aquatic organisms are ectothermic which means that they have a body temperature that is similar to their environment. The worrisome phenomenon of coral bleaching has very often been attributed to elevated temperatures [10]. A study by Gock et al. in 2003 highlighted that temperature with respect to water activity strongly impacted the germination of xerophilic fungi, which is responsible for spoilage of bakery and very important in food quality testing [11].

Therefore, in order to safeguard delicate natural ecosystems, it is vital for environmental authorities to have continuous real-time data of water quality parameters on a high spatial resolution at various depths. For example, rapid detection of hydrologic variability is crucial for EWS and subsequent swift response [12].

1.3 Water Quality Sensors

A sensor is a machine or a device that detects a change or an event in its environment and sends a signal to another electronic device to display the message. It is usually used in combination with another electronic device that translates its signal into a readable output. A sensor may be quantitative or qualitative. Quantitative sensors give numerical output in respective units whereas qualitative sensors suggest the descriptive value or a specific feature of a sample. A sensor may furthermore measure chemical, physical, or biological properties of water and examples of their quantitative counterparts are pH, temperature, and bacterial density, respectively. On the other hand, odor and taste are qualitative physical properties of water.

The progress in online environmental monitoring goes hand in hand with the technological development of solid-state sensors. With the possibility of miniaturization and controlled by a signal processing unit, they have no moving parts and require a single parameter measurement such as impedance, current, or voltage alongside a linear calculation of the signal. Such features additionally allow for increased portability of sensors, on-site testing, remote sensing, applicability on flexible substrates, and integration of multiple sensors within a single platform. The interest in thin and thick film solid-state sensors over traditionally used water quality sensors also arises from the development of new materials such as metal oxides (RuO_2, IrO_2, TiO_2, etc.) and the variety of technologies that can be used to manufacture them, such as printing (screen, ink-jet, 3D, etc.) and sputter deposition (magnetron, radio frequency, reactive, etc.). Additionally, they have a longer shelf life as they are less prone to breakability and can tolerate extreme environmental conditions.

In a water quality monitoring setup, it is advisable not to interfere with the environment in which the parameter needs to be tested. Therefore, it significantly depends on the sensor to work well with its environment and deliver an accurate reading. This is why the most important component of the solid-state sensor which decides the fate of its performance is the material used to fabricate it and the design that is most well aligned with the principle in operation. The relationship between the material and the electrolyte is defined by the electrochemistry that takes place upon their contact. When there is a change in the morphological and structural properties of the sensing electrode of the sensor, there is a significant impact on key sensor characteristics such as response time, sensitivity, and selectivity [13]. Therefore, manipulation of sensor material and design is a popular strategy among researchers to improve the quality of solid-state sensors.

Described here are the principles, designs, and materials used for making sensors that detect four different types of water quality parameters: pH, DO, temperature, and conductivity.

1.3.1 pH

Introduced as the hydrogen ion exponent with notation pH [8], the Danish chemist S. P. L. Sørensen was the first to present the concept of pH as a measure of acidity and alkalinity in 1909 at Carlsberg Laboratories in Copenhagen [14]. In 1922, W.S. Hughes invented the famous pH glass electrode, recorded as the first chemical sensor [12]. According to the International Union of Pure and Applied Chemistry (IUPAC), the base-10 logarithm of the inverse, of the hydrogen ion activity in a solution is defined as the pH of that solution [15]. The equation to represent this relation is:

$$pH = -\log[H^+]$$

Instead of hydrogen ion concentration, it is more advantageous to define pH with respect to hydrogen ion activity because ionic activity can be measured directly using potentiometry [10].

1.3.1.1 pH Sensors: Principles, Materials, and Designs

pH measurement can be classified in various groups such as optical sensing, acoustic, electrochemical, and nuclear magnetic resonance (NMR) method. Most common among these groups are electrochemical methods and they are the most developed class of sensors fabricated to detect pH due to their faster response, simplicity, and low cost. pH sensors may depend on one of the principles of potentiometric, chemiresistive, chemi-transistor, conductimetric, ion-sensitive field effect transistor (ISFET), extended gate field effect transistor (EGFET), etc., and the primary materials to fabricate them are glass, metal oxides, mixed metal oxides, polymers, and carbon [16].

1.3.1.2 Glass Electrode

The typical glass electrode has established itself as a comfortable favorite over the past few decades not just as a laboratory pH measurement device but also a household tool for measuring pH in aquarium ponds, culinary arts, soil for gardening, etc. The glass electrode relies on the principle of an electrochemical cell.

The materials comprising a glass electrode are usually an external body and sensing bulb made from specific glass, an internal electrode (calomel or silver chloride electrode), and an internal solution which is usually a buffer solution of pH 7. However, the glass electrode is very dependent on position when used or stored [17]. They are challenging to minimize, mechanically delicate, instable in aggressive electrolytes [18], and require wet storage with the bulb immersed in electrolyte [19]. A critical quality of solid-state sensors that sets them apart from the glass electrode is the absence of an internal electrolyte solution. This is the motivation for development of all-solid-state electrodes with large pH range and low sensitivity to redox agents.

1.3.1.3 Solid-State Ion-Selective Electrodes

Solid-state ion-selective electrodes (ISE) are associated with smaller size, easier fabrication, and simpler operation in comparison to conventional ISE such as the glass pH electrode which has an internal filling solution. The first solid-state ISE design to be invented was a coated-wire electrode but the blocked interface between the ionically conducting ion-selective membrane and the electronic conductor resulted in an unstable potential. Hydrogel-based electrolytes, instead of the liquid internal electrolyte, help to overcome the issue of blocked interface by providing a distinct and clear pathway for the ion-to-electron transduction. Another modification in solid-state ISEs is the application of an intermediate layer of redox active compounds between the ion-selective membrane and the electronic conductor. Due to this requirement of high redox capacitance in the intermediate layer, conducting polymers are applicable as ion-to-electron transducers. Conducting polymers are suitable for use in solid-state ISEs because they are electroactive, soluble, electronically conducting, and can be easily deposited on the conductor by the electro-polymerization of big and diverse monomers [20].

1.3.1.4 Metal Oxide pH Sensors

pH measurement techniques that have gained popularity in recent times using oxides as pH-sensitive layers are conductimetric/capacitive, potentiometric, and ISFETs, based on standard metal oxide field effect transistor (MOSFET). Among these methods, the potentiometric was the most commonly found technique in commercial devices due to its simplicity in fabrication and advancement in technology over the years [21].

Several possible mechanisms of pH sensitivity in metal oxides were suggested by Fog and Buck [22]. They include oxygen or hydrogen intercalation, ion exchange in the surface layer, corrosion of the material, and a possible equilibrium between the two oxidizing forms of the metal. They strongly supported the possibility of oxygen intercalation because there is nonstoichiometric oxygen content in the oxides. This implies that the activity of oxygen in the solid phase should also be considered while calculating the electrode potential.

Among the many metal oxides that have been researched, ruthenium dioxide (RuO_2) and iridium dioxide have been most promising [18] with fast response due to their resistance to space charge accumulation which is credited to high conductivity and chemical stability. RuO_2 is especially a favorite among researchers as it shows close to Nernstian response even in the presence of strong oxidizing agents [22], organic sediments [23], and contaminants [24].

Different methods of fabricating RuO_2 electrodes are screen printing [16], radio frequency magnetron sputtering (RFMS) [25], molecular beam epitaxy [26], physical or chemical vapor deposition [27], thermal decomposition [28], sol–gel [29, 30], and ultrasonic spray pyrolysis [31]. The most preferred method to fabricate RuO_2 electrodes is screen printing because it has been reported to deliver the best sensitivity and response time [16]. Additionally, it gives the flexibility to use various sizes and compositions along with being fast and cost effective [32].

Binary metal oxides have been used for pH sensing most commonly as dimensionally stable anodes (DSA) in various applications [33]. In these systems, chemically inert oxides are mixed with an active transition oxide which helps to enhance electrochemical properties, stability, and longevity [16]. For example, doping has been found to improve antifouling properties in RuO_2 electrodes. Both un-doped nanostructured RuO_2 [34] and doped RuO_2 or mixed with other metal oxide have been recommended for pH sensing [16, 33, 34]. Some mixtures have been described below.

Utilization of tin oxide (IV) for electrode fabrication allows increase of the lifetime of the electrodes and shortens response time to five to nine seconds. RuO_2–SnO_2 mixed oxide-based electrodes show the Nernstian response with the sensitivity of -56.5 mV/pH for the RuO_2:SnO_2 ratio of 70 : 30 wt.% [35]. Since RuO_2 is expensive, combination with TiO_2 is used to reduce the cost. Furthermore, substituting part of RuO_2 with TiO_2 allows higher electric conductivity [36]. The sensitivity of RuO_2:TiO_2 70 : 30 mol% ratio was found to be -56.11 mV/pH when fabricated by the Pechini method [18] and -56.03 mV/pH when fabricated by screen-printing [35]. The RuO_2-TiO_2 electrode fabricated by Pocrifka et al. [18] experienced very low interference from anions such as Li^+, Na^+, and K^+ which is a desirable quality in pH sensors. Combination of RuO_2 with Ta_2O_5, which is known for its use as ISFET [37], allows to lower the cost, minimize potential drift, and time response [38–40]. Doping of RuO_2 with up to 20 mol% of Cu_2O allowed to produce electrodes with the nonporous surface that is favorable from the point of antifouling and stable performance even after three months of implementation [41, 42]. Sensitivity of 10 mol% Cu_2O-doped RuO_2 electrode was -47.4 mV/pH and did not change with the increase of Cu_2O concentration [42].

1.3.2 Dissolved Oxygen

The amount of oxygen dissolved in a unit volume of water or other liquids is termed as DO and is expressed in the unit milligram/litre. For pure water at 25 °C and 1 atm, the saturation level of DO is known to be 8.11 mg/l. DO means the free, non-compounded oxygen (O_2) molecules in the water which are not bound to any other element. Therefore, when measuring DO levels, the bound oxygen molecule in water (H_2O) which is in a compound is not considered. By natural processes, DO may enter water through slow diffusion from air across the water surface or aeration of water by various forms of running water such as waterfalls, waves, groundwater discharge, etc., and by aquatic plants, seaweed, algae, and phytoplankton as a byproduct of photosynthesis.

1.3.2.1 DO Sensors: Principles, Materials, and Designs

There are three ways to measure DO based on three different principles: chemical, electrochemical, and photochemical. These principles and the types of methods based on them are shown in Figure 1.1.

1.3.2.2 Chemical Sensors

The classical titration also known as volumetric analysis or the indicator method is the oldest known method of measuring DO with a chemical test and it is considered the benchmark of DO measurement [43]. The most popular indicator method for DO detection and measurement is Winkler's method or the iodometric method. Its principle is to produce manganese hydroxide (II) by adding manganese sulfate and alkaline potassium iodide to water. Due to unstable nature, manganese hydroxide produces manganic acid by combining with DO. Potassium

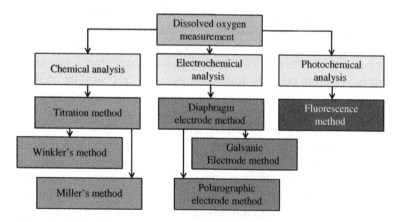

Figure 1.1 Principles and methods of measuring dissolved oxygen.

iodide and concentrated sulfuric acid are added to separate the iodine and react with DO, respectively. Using starch as an indicator, the amount of DO can be calculated with the titration of released iodine with sodium thiosulfate. Another method of DO measurement by titration is called Miller's method in which the oxidation of ferrous ions is measured in an alkaline medium [44]. Titration methods are only prevalent in laboratories and are not used for online measurement of DO because the design is simply too complicated to serve the demands of a continuous water quality monitoring system [43].

1.3.2.3 Electrochemical Sensors

Even though all methods are good and sensitive, in wireless sensor networks, electrochemical methods are most preferred due to miniaturization, robustness, and high resistance to fouling. The most commonly used type of electrochemical sensor is Clarke cell [13], shown in Figure 1.2. This sensor is controlled by a flux of oxygen that diffuses through the gas-permeable membrane (usually a polytetrafluoroethane film) separating the cell and the solution. By Fick's law, this flux which is indicated by the current produced is proportional to the DO concentration. The oxidation occurs at the anode (negative electrode) and the reduction occurs at the cathode (positive electrode) immediately when the oxygen diffuses through the film. Commonly used electrode material for Galvanic DO sensors are lead, iron, or zinc for anode and silver or platinum for cathode.

The two primary electrochemical techniques in practice for DO measurement are galvanic and polarographic. They both are electrode systems but with a slight difference between them. In a polarographic DO sensing system, an external voltage is applied. However, in the galvanic DO sensing system, an external potential is not required because the electrode materials are selected such that there is a

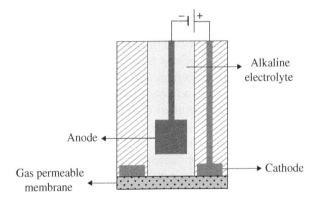

Figure 1.2 Schematic representation of as Clarke cell-type electrochemical DO sensor.

−0.5 V or greater potential between the cathode and the anode. Electrodes in the diaphragm method, on the other hand, measure the amount of oxygen passing through the gas-permeable diaphragm.

Semiconductor-Based Potentiometric DO Sensors Solid-state sensors based on semiconductors have helped to overcome the limitations of size and maintenance while providing researchers with the flexibility to explore various designs and materials. For example, a DO sensor was developed using platinum thin film electrodes which were coated with a solid-state proton conductive matrix (PCM) and the surface of the planar membrane electrode was covered with solid proton conductive material using spin coating process [44]. Therefore, instead of the traditional Clarke sensor which has a built-in electrolyte, this sensor had a solid polymer electrolyte. DO sensing systems made with RuO_2 as sensitive electrode and Ag/AgCl, Cl^- as reference electrode exhibited linear response with detection limits of 0.5–8.0 ppm [45].

Doped semiconductors are a popular choice of electrode material for electrochemical DO sensors. For example, nanostructured RuO_2 has been found to exhibit one of the best performances for DO sensing [13]. When doped with ZnO and Cu_2O oxides, it shows an increase in DO sensitivity by −48.6 mV/decade for 10 mol% ZnO [46] and −47.4 mV/decade for 10 mol% Cu_2O [42], respectively. When a semiconductor is doped, there is a formation of donor levels located on the upper band of the semiconductor due to the deviation in electronic structure of the doped mixed oxide. These donor levels contribute to the enhancement of sensor properties. As a charge donor, the dopant increases the conductivity of the electrode by transforming the structure of the nano-oxide and supplying excess carriers to the conductivity band [47].

1.3.2.4 Optical or Photochemical Sensors

There are different optical principles which can be used to measure DO such as the phosphorus quenching principle, near-infrared principle, the absorption principle, and the fluorescence principle. Among these principles, currently the most commonly used one is the fluorescence quenching principle and a DO sensor based on it consists of three constituents: excitation light sources, an optoelectronic detection element, and a substrate film attached to fluorescence-sensitive substances. In fluorescence quenching, the sensitive material absorbs the ultraviolet light of specific wavelength and the electrons gain energy which they release in order to return to ground state by emitting fluorescence. The amount of DO in the water can be estimated by the intensity of fluorescence or the fluorescence lifetime generated at the sensitive interface because the collisions between oxygen molecules and excited fluorescent substances interfere with the excitation process of fluorescent substance.

Sensitive materials that are fluorescent substances and used for the film are pyrene butyric acid, fluoroanthene, and polycyclic aromatic compounds [43]. Fluorescence indicators used in optical DO sensors to improve DO detection sensitivity include platinum phosphor porphyrins such as platinum octaethyl porphyrins (PtOEP), platinum tetrafluorophenyl porphyrins (PtTFPP), and ruthenium–chromium complexes. The choice of substrate material while fabricating optical DO sensors is critical because the DO sensors are principally based on the oxygen quenching effect of fluorescence of the luminescent body fixed in the matrix [43].

1.3.3 Electrical Conductivity

A measure of the ability of a solution to conduct electrical current indicated by the amount of free-flowing electrons and/or ions is termed as conductivity or specific conductance since conductivity is directly proportional to conductance ($1/R$, where R is resistance of the circuit). Expressed in Siemens per meter (S/m) or micro-Siemens per centimeter (μS/cm), conductivity is the inverse of the electrical parameter of resistance (ohm).

Measurement of conductivity is used for the determination of the following [13]:

- Number of free ions in the water influences some physiological processes in living organism.
- Sudden changes in waste and natural water systems.
- Amount of reagent required or sample sizes in chemical analysis of water.
- Concentration of total dissolved solids.

The conductivity of deionized and pure distilled water is about 0.05 μS/cm, drinking water is 200–800 μS/cm, and sea water is 50 mS/cm.

1.3.3.1 Conductivity Sensors: Principles, Materials, and Designs

Conductivity sensors are usually based on Ohm's law wherein the resistance of the water sample can be calculated from known values of resistor, voltage, current, and cell constants, i.e. area and length of the sample [3]. Such sensors are called conductive, contacting, or electrode sensors. In a conductive or electrode sensor which may have two, three, or four electrodes [48], the geometrical design of the sensor is based on the target conductivity range and determines the proportionality coefficient (KC) [8]. In three- or four-electrode system, if there is any damage, fouling, or polarization on the electrodes, the reference voltage allows compensation for it. Therefore, the reference electrode helps in achieving better accuracy over wide ranges unaffected by minor changes in electrode condition. In this system, an electric field is generated within the electrolyte when an electric current is applied on one of the electrodes causing the negatively charged ions to the cathode and the positively charged ions to the anode. Therefore, the current in the electrolyte is carried by the ions, meaning that an increase in their mobility and concentration is expected to increase conductivity [13].

In traditional dip or flow-through style two-electrode conductivity sensors, electrodes are usually made of platinum, graphite, titanium, or gold-plated nickel. A comparative study [23] of four-electrode systems found electrodes which have high resistance to be more sensitive than electrodes made from Ru nanostructures. Metallic Pt/Ag/Pd alloys exhibited ion sensitivity within the range of 5–200 μS/cm with a response time of approximately one to three seconds whereas Ru-based electrodes system had a lower limit of 500 μS/cm and response time of two minutes [13]. Figure 1.3 presents the schematic representation of two-electrode and four-electrode types of conductivity sensors.

Another way to measure electrical conductivity of water is by using the inductive method based on the principle of mutual inductance [49]. Conductivity sensors based on this method are called toroidal or inductive sensors. This type of sensor has two conducting coils enclosed in a casing made of nonconducting material. By Faraday's law of induction, when an electrical current is induced by the first coil in the water, the second coil measures the magnitude of the induced current [8]. This measured value is directly proportional to the conductivity of the water sample.

Between conductive and inductive sensor, there is a vast divide between their prevalence in practice. Even though electrode conductivity sensors run the risk of delivering inaccurate measurements due to corrosion or deposition of fine

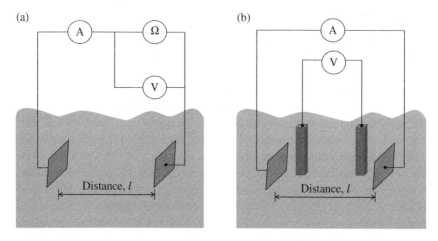

Figure 1.3 Schematic representations of (a) two-electrode and (b) four-electrode types of conductivity sensors where I is the current source, R is the resistance, and V is the voltage.

particles or bacteria on the electrode [49], their popularity far outweighs that of the inductive conductivity sensor due to simpler maintenance, lower cost [3], reduced fringe effect, sensitivity to contact resistance, and extendable linearity which can be achieved by varying the excitation voltage of the cell [48].

However, some researchers are of the opinion that from the perspective of long-term monitoring, the requirement of the electrodes to be repeatedly cleaned and replaced is somewhat contradictory to the objectives of wireless sensor networks [49]. The reason for their skepticism toward the precision of conductive sensors is that several factors such as volume of water, salinity of water, and size of coils [50] that influence the magnitude of induced current are overlooked in the conductive methodology but not in inductive methodology. Nevertheless, the lack of exploration in the potential of inductive sensors has left them with a commercially failed fate in industry. Researchers have not been as vocal in their criticism of inductive sensors as they have been in their appreciation of conductive sensors. Hence, it is most probable that the coil was simply too complicated in comparison to the two-dimensional electrode as suggested by a group of researchers [48] who prefer conductive sensors over inductive sensors due to simple and practical advantages such as ease in fabrication, directness in operational principle, and miniature size. However, the same researchers admit that the inductive sensor has a circuitry such that the input and output points do not come in contact with the water, thereby reducing the probability of fouling but there can still be signal loss and electrical interference.

Commercially, such a bias does not exist as both inductive and electrode conductivity sensors are easily available on the market. An example of commercial inductive conductivity sensor is AANDERAA Conductivity Sensor 4319 (range: 0–75 mS/ cm, material of coil: titanium) and an example of commercial electrode sensor is YOKOGAWA Model SC4A (two-electrode system, range: 0.1–50 mS/cm, material of electrode: Stainless steel AISI 316L). In solid-state sensors, conductivity can be measured using either two electrodes or four electrodes.

1.3.4 Temperature

One of the most common data logging applications, temperature measurement is required across a multitude of industries and in almost every practical application. When two objects are at the same temperature and do not exchange any thermal energy, they are said to be in thermal equilibrium with each other. A scalar unit, temperature can be expressed using various units according to the thermometric scale used. Most popular temperature measurement scales are Kelvin (K), Fahrenheit (°F), and Celsius (°C). Based on the three physical states of matter, fundamental fixed points defining the temperature scales are listed in Table 1.1.

Table 1.1 Fixed points defined by temperature scales.

Material physical state	K	°F	°C
Absolute zero[a]	0	−459.67	−273
Ice point of water (1 atm)	273	32	0
Boiling point of water (1 atm)	373	212	100

[a] Temperature at which adiabatic and isothermic processes occur at the same time.

The environmental temperature is proportional to the current or voltage output generated using voltage reference or a diode with a well-established voltage versus temperature characteristic. For example, Analog Device AD 592 transducer has an output current of $1\,\mu A/K$ with output of $248\,\mu A$ at $-25\,°C$. Many such IC temperature sensors are available commercially. They have an operation range of $-55\,°C$ to $+150\,°C$ and an accuracy of $0.3–0.5\,°C$.

1.3.4.1 Temperature Sensors: Principles, Materials, and Designs
Temperature sensors can be broadly divided into two design types: contact and non-contact. Figure 1.4 presents a classification of different types of temperature sensors. Contact temperature transducers implement transduction using heat transmission to create energy balance in the system when it comes in contact with the medium. Different types of contact temperature transducers

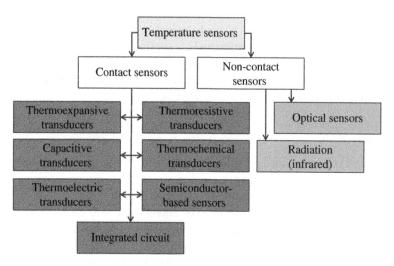

Figure 1.4 Different types of temperature sensors.

work using different types of principles such as capacitive transduction, thermochemical and thermoelectric transduction, thermoexpansive and thermoresistive transduction, semiconductor elements, and transduction with diodes. Contactless or non-contact temperature sensors use electromagnetic radiation of an object with temperature above 0 K. This radiation has an intensity which depends on the temperature of the object. Hence, these temperature sensors, also known as pyrometers, do not need any physical contact with the object. Although contactless temperature sensors can be used to measure temperature of water surface, they are mostly used for solid surfaces. In water quality monitoring, contact sensors are more common and the most popular ones have been discussed later in this chapter.

The following four types of temperature sensors are currently available in the market [43]:

1.3.4.2 Thermocouples
These sensors operate on the principle of Seedbeck effect. In this mechanism, two distinct electrodes produce a voltage difference that is proportional to the temperature difference between them [51].

1.3.4.3 Resistance Temperature Detector
Resistance Temperature Detectors or RTDs measure the change in electrical resistance of the material. Most commonly used materials for fabricating RTDs are pure metals because their temperature coefficients are small, positive, and accurate. RTDs made using platinum wire are known to be linear from temperatures ranging from -259 to $600\,°C$, well characterized and most precise.

1.3.4.4 Thermistor
The sensor design is similar to RTD and the only exception is that instead of a metal inside the device, there is a ceramic or a polymer [52]. The temperature-sensitive resistor has a negative temperature coefficient that is large and nonlinear which means that as the temperature increases, the resistance nonlinearly decreases. Their maximum usable temperatures range between 150 and 300 °C.

Since a thermistor is a nonlinear and very sensitive detector, it must be regularly calibrated based on desired temperature ranges. However, it responds linearly over small temperature ranges. Thermistor suffers from poor repeatability and is susceptible to self-heating errors caused by the power ($P = I^2R$) produced in it.

Therefore, magnitude of passing current is suggested to be kept at minimum in a thermistor. Most commonly used composition for making thermistors is mixture of semiconducting materials.

1.3.4.5 Integrated Circuit

Also known as monolithic or semiconductor temperature transducers, integrated circuits (IC) with two terminals produce a voltage signal that could be digital or analog or a current. This voltage signal is directly proportional to the absolute temperature. Typically, an IC operates over a temperature range of 0–100 °C in which it delivers an output of 10 mV/K.

1.4 Smart Sensors

Smart or intelligent sensors are the electronic systems which include in addition to detection and measurement, a complete and wireless system for the conditioning and collection of analog signals. They are more popular than traditional sensors because they have the advantage of combining multiple features and processes in one device, including the connection to a wider network of sensors over the internet. A smart sensor system consists of a detection/transduction system which converts the analog signal into a digital signal, a measurement and signal processing interface, a calibration system to properly process the data, and an autonomous power source to ensure uninterrupted operation of the sensor device [53]. In a smart sensor, a microcontroller is embedded in the device for the purpose of internal calculations and processing with a bidirectional communication channel which receives instructions and transmits the results. Overall, a smart sensor is simply a device containing a sensor with an attached facility for data computing and wireless transmission in one common enclosure.

1.5 Conclusion

The demand for real-time data in water quality monitoring grows simultaneously with the demand for advanced materials and sensor technologies, especially for application in CWS. Temperature, pH, DO, and conductivity are some of the most critical indicators of the health of a water body. Sensors are devices which aid in the measurement of these parameters. Recent research has paved the way for alternative methods for water quality testing based on thick film solid-state sensors which provide an array of advantages over traditional methods such as lower cost, ease of manufacturing, applicability in online monitoring system, deposition on flexible materials, robustness, and longer operational time. Knowledge of different principles, designs, and materials of these water quality sensors enables manufacturers, researchers, and operators to apply them effectively for the purpose of remote and real-time water quality testing.

Acknowledgment

This work is supported by the European Union's Horizon 2020 Research and Innovation program under the Marie Skłodowska-Curie grant agreement number 813680 and CHIST-ERA IV grant agreement number 857925.

References

1 Owa, F.D. (2013). Water pollution: sources, effects, control and management. *Mediterr. J. Soc. Sci.* 4: 65–68. https://doi.org/10.5901/mjss.2013.v4n8p65.
2 Acharya, K., Blackburn, A., Mohammed, J. et al. (2020). Metagenomic water quality monitoring with a portable laboratory. *Water Res.* 184: 116112. https://doi.org/10.1016/j.watres.2020.116112.
3 Cloete, N.A., Malekian, R., and Nair, L. (2016). Design of smart sensors for real-time water quality monitoring. *IEEE Access* 4: 3975–3990. https://doi.org/10.1109/ACCESS.2016.2592958.
4 Schubaur-Berigan, M.K., Monson, P.D., West, C.W., and Ankley, G.T. (1995). Influence of pH on the toxicity of ammonia to *Chironomus tentans* and *Lumbriculus variegatus*. *Environ. Toxicol. Chem.* 14: 713–717. https://doi.org/10.1002/etc.5620140419.
5 Miskimmin, B.M., Rudd, J.W.M., and Kelly, C.A. (1998). Influence of dissolved organic carbon, pH, and microbial respiration rates on mercury methylation and demethylation in lake water. *Can. J. Fish. Aquat. Sci.* 49: 17–22. https://doi.org/10.1139/f92-002.
6 Bradley, R.W. and Sprague, J.B. (1985). The influence of pH, water hardness, and alkalinity on the acute lethality of zinc to rainbow trout (*Salmo gairdneri*). *Can. J. Fish. Aquat. Sci.* 42: 731–736. https://doi.org/10.1139/f85-094.
7 Lee, H.j., Kim, H.M., Park, J.H., and Lee, S.K. (2017). Fabrication and characterization of micro dissolved oxygen sensor activated on demand using electrolysis. *Sensors Actuators B Chem.* 241: 923–930. https://doi.org/10.1016/j.snb.2016.10.145.
8 Ramos, P.M., Pereira, J.M.D., Member, S. et al. (2008). A four-terminal water-quality-monitoring conductivity sensor. *IEEE Trans. Instrum. Meas.* 57: 577–583.
9 Cairns, J., Heath, A.G., and Parker, B.C. (1975). Temperature influence on chemical toxicity to aquatic organisms. *J. Water Pollut. Control Fed.* 47: 267–280.

10 Brown, B.E. (1997). Coral bleaching causes and consequences. *Coral Reefs* 2100: 129–138.

11 Gock, M.A., Hocking, A.D., Pitt, J.I., and Poulos, P.G. (2003). Influence of temperature, water activity and pH on growth of some xerophilic fungi. *Int. J. Food Microbiol.* 81: 11–19. https://doi.org/10.1016/S0168-1605(02)00166-6.

12 Glasgow, H.B., Burkholder, J.A.M., Reed, R.E. et al. (2004). Real-time remote monitoring of water quality: a review of current applications, and advancements in sensor, telemetry, and computing technologies. *J. Exp. Mar. Biol. Ecol.* 300: 409–448. https://doi.org/10.1016/j.jembe.2004.02.022.

13 Zhuiykov, S. (2012). Solid-state sensors monitoring parameters of water quality for the next generation of wireless sensor networks. *Sensors Actuators B Chem.* 161: 1–20. https://doi.org/10.1016/j.snb.2011.10.078.

14 Myers, R.J. (2010). One-hundred years of pH. *J. Chem. Educ.* 87: 30–32. https://doi.org/10.1021/ed800002c.

15 Lonsdale, W. (2018). *Development, Manufacture and Application of a Solid-State pH Sensor Using Ruthenium Oxide*. Edith Cowan University. https://ro.ecu.edu.au/theses/2095 (accessed 21 June 2018).

16 Manjakkal, L., Szwagierczak, D., and Dahiya, R. (2020). Progress in materials science metal oxides based electrochemical pH sensors: current progress and future perspectives. *Prog. Mater. Sci.* 109: 100635. https://doi.org/10.1016/j.pmatsci.2019.100635.

17 Vonau, W. and Guth, U. (2006). pH monitoring: a review. *J. Solid State Electrochem.* 10: 746–752. https://doi.org/10.1007/s10008-006-0120-4.

18 Pocrifka, L.A., Gonçalves, C., Grossi, P. et al. (2006). Development of RuO_2–TiO_2 (70–30) mol% for pH measurements. *Sensors Actuators B Chem* 113: 1012–1016. https://doi.org/10.1016/j.snb.2005.03.087.

19 Salvo, P., Melai, B., Calisi, N. et al. (2018). Graphene-based devices for measuring pH. *Sensors Actuators B Chem.* 256: 976–991. https://doi.org/10.1016/j.snb.2017.10.037.

20 Bobacka, J. (2006). Conducting polymer-based solid-state ion-selective electrodes. *Electroanalysis* 18: 7–18. https://doi.org/10.1002/elan.200503384.

21 Gill, E., Arshak, K., Arshak, A., and Korostynska, O. (2008). Mixed metal oxide films as pH sensing materials. *Microsyst. Technol.* 14: 499–507. https://doi.org/10.1007/s00542-007-0435-9.

22 Fog, R. and Buck, A. (1984). Electronic semiconducting oxides as pH sensors. *Sensors Actuators* 5: 137–146. https://doi.org/10.1016/0250-6874(84)80004-9.

23 Zhuiykov, S. (2009). Morphology of Pt-doped nanofabricated RuO_2 sensing electrodes and their properties in water quality monitoring sensors. *Sensors Actuators B Chem.* 136: 248–256. https://doi.org/10.1016/j.snb.2008.10.030.

24 Uppuluri, K., Lazouskaya, M., Szwagierczak, D. et al. (2021). Fabrication, potentiometric characterization, and application of screen-printed RuO_2 pH electrodes for water quality testing. *Sensors* 21: https://doi.org/10.3390/s21165399.

25 Lonsdale, W., Wajrak, M., and Alameh, K. (2017). Effect of conditioning protocol, redox species and material thickness on the pH sensitivity and hysteresis of sputtered RuO_2 electrodes. *Sensors Actuators B Chem.* 252: 251–256. https://doi.org/10.1016/j.snb.2017.05.171.

26 Jia, Q.X., Wu, X.D., Foltyn, S.R. et al. (1995). Heteroepitaxial growth of highly conductive metal oxide RuO_2 thin films by pulsed laser deposition. *Appl. Phys. Lett.* 67: 1677. https://doi.org/10.1063/1.115054.

27 Miao, G.X., Gupta, A., Xiao, G., and Anguelouch, A. (2005). Epitaxial growth of ruthenium dioxide films by chemical vapor deposition and its comparison with similarly grown chromium dioxide films. *Thin Solid Films* 478: 159–163. https://doi.org/10.1016/j.tsf.2004.10.032.

28 Siviglia, P., Daghetti, A., and Trasatti, S. (1983). Influence of the preparation temperature of ruthenium dioxide on its point of zero charge. *Colloids Surf.* 7: 15–27. https://doi.org/10.1016/0166-6622(83)80038-9.

29 Armelao, L., Barreca, D., and Moraru, B. (2003). A molecular approach to RuO_2-based thin films: sol–gel synthesis and characterisation. *J. Non-Cryst. Solids* 316: 364–371. https://doi.org/10.1016/S0022-3093(02)01636-8.

30 Osman, J.R., Crayston, J.A., Pratt, A., and Richens, D.T. (2008). RuO_2-TiO_2 mixed oxides prepared from the hydrolysis of the metal alkoxides. *Mater. Chem. Phys.* 110: 256–262. https://doi.org/10.1016/j.matchemphys.2008.02.003.

31 Fugare, B.Y. and Lokhande, B.J. (2017). Study on structural, morphological, electrochemical and corrosion properties of mesoporous RuO_2 thin films prepared by ultrasonic spray pyrolysis for supercapacitor electrode application. *Mater. Sci. Semicond. Process.* 71: 121–127. https://doi.org/10.1016/j.mssp.2017.07.016.

32 Li, M., Li, Y.T., Li, D.W., and Long, Y.T. (2012). Recent developments and applications of screen-printed electrodes in environmental assays – a review. *Anal. Chim. Acta* 734: 31–44. https://doi.org/10.1016/j.aca.2012.05.018.

33 Ribeiro, J., Moats, M.S., and De Andrade, A.R. (2008). Morphological and electrochemical investigation of RuO_2-Ta_2O_5 oxide films prepared by the Pechini-Adams method. *J. Appl. Electrochem.* 38: 767–775. https://doi.org/10.1007/s10800-008-9506-6.

34 Zhuiykov, S. (2011). Effect of heterogeneous oxidation on electrochemical properties of tailored $Cu_{0.4}Ru_{3.4}O_7 + RuO_2$ sensing electrode of potentiometric DO sensor. *Mater. Lett.* 65: 3219–3222. https://doi.org/10.1016/j.matlet.2011.07.020.

35 Manjakkal, L., Cvejin, K., Kulawik, J. et al. (2014). Fabrication of thick film sensitive RuO_2-TiO_2 and Ag/AgCl/KCl reference electrodes and their application for pH measurements. *Sensors Actuators B Chem.* 204: 57–67. https://doi.org/10.1016/j.snb.2014.07.067.

36 Yue, H., Xue, L., and Chen, F. (2017). Environmental efficiently electrochemical removal of nitrite contamination with stable RuO_2–TiO_2/Ti electrodes. *Appl. Catal. B Environ. J.* 206: 683–691. https://doi.org/10.1016/j.apcatb.2017.02.005.

37 Hara, H. and Ohta, T. (1996). Dynamic response of a Ta_2O_5-gate pH-sensitive field-effect transistor. *Sensors Actuators B Chem.* 32: 115–119. https://doi.org/10.1016/0925-4005(96)80119-5.

38 Manjakkal, L., Zaraska, K., Cvejin, K. et al. (2016). Potentiometric RuO_2–Ta_2O_5 pH sensors fabricated using thick film and LTCC technologies. *Talanta* 147: 233–240. https://doi.org/10.1016/j.talanta.2015.09.069.

39 Lonsdale, W., Wajrak, M., and Alameh, K. (2018). Manufacture and application of RuO_2 solid-state metal-oxide pH sensor to common beverages. *Talanta* 180: 277–281. https://doi.org/10.1016/j.talanta.2017.12.070.

40 Kwona, D.H., Cho, B.W., Kim, C.S., and Sohn, B.K. (1996). Effects of heat treatment on Ta_2O_5 sensing membrane for low drift and high sensitivity pH-ISFET. *Sensors Actuators B Chem.* 34: 441–445.

41 Zhuiykov, S., Kats, E., Marney, D., and Kalantar-Zadeh, K. (2011). Improved antifouling resistance of electrochemical water quality sensors based on Cu_2O-doped RuO_2 sensing electrode. *Prog. Org. Coat.* 70: 67–73. https://doi.org/10.1016/j.porgcoat.2010.10.003.

42 Zhuiykov, S., Kats, E., and Marney, D. (2010). Potentiometric sensor using sub-micron Cu_2O-doped RuO_2 sensing electrode with improved antifouling resistance. *Talanta* 82: 502–507. https://doi.org/10.1016/j.talanta.2010.04.066.

43 Wei, Y., Jiao, Y., An, D. et al. (2019). Review of dissolved oxygen detection technology: from laboratory analysis to online intelligent detection. *Sensors* 19: https://doi.org/10.3390/s19183995.

44 Walker, K.F., Williams, W.D., and Hammer, U.T. (2003). The miller method for oxygen determination applied to saline lakes. *Limnol. Oceanogr.* 15: 814–815. https://doi.org/10.4319/lo.1970.15.5.0814.

45 Martínez-Máñez, R., Soto, J., García-Breijo, E. et al. (2005). A multisensor in thick-film technology for water quality control. *Sensors Actuator A Phys.* 120: 589–595. https://doi.org/10.1016/j.sna.2005.03.006.

46 Zhuiykov, S., Kats, E., Plashnitsa, V., and Miura, N. (2011). Toward selective electrochemical "e-tongue": potentiometric DO sensor based on sub-micron ZnO–RuO_2 sensing electrode. *Electrochim. Acta* 56: 5435–5442. https://doi.org/10.1016/j.electacta.2011.01.062.

47 Li, L.M., Du, Z.F., and Wang, T.H. (2010). Enhanced sensing properties of defect-controlled ZnO nanotetrapods arising from aluminum doping. *Sensors Actuators B Chem.* 147: 165–169. https://doi.org/10.1016/j.snb.2009.12.058.

48 Banna, M.H., Najjaran, H., Sadiq, R. et al. (2014). Miniaturized water quality monitoring pH and conductivity sensors. *Sensors Actuators B Chem.* 193: 434–441. https://doi.org/10.1016/j.snb.2013.12.002.

49 Parra, L., Sendra, S., Lloret, J., and Bosch, I. (2015). Development of a conductivity sensor for monitoring groundwater resources to optimize water management in smart city environments. *Sensors* 15: 20990–21015. https://doi.org/10.3390/s150920990.

50 Wood, R.T., Bannazadeh, A., Nguyen, N.Q., and Bushnell, L.G. (2010). A salinity sensor for long-term data collection in estuary studies. *Oceans 2010 MTS/IEEE Seattle*, Seattle, WA (20–23 September 2010).

51 Childs, P.R.N., Greenwood, J.R., and Long, C.A. (2000). Review of temperature measurement. *Rev. Sci. Instrum.* 71: 2959–2978. https://doi.org/10.1063/1.1305516.

52 Hayat, H., Griffiths, T., Brennan, D. et al. (2019). The state-of-the-art of sensors and environmental monitoring technologies in buildings. *Sensors* 19: https://doi.org/10.3390/s19173648.

53 Garrido-Momparler, V. and Peris, M. (2022). Smart sensors in environmental/water quality monitoring using IoT and cloud services. *Trends Environ. Anal. Chem.* 35: e00173. https://doi.org/10.1016/j.teac.2022.e00173.

2

Dissolved Heavy Metal Ions Monitoring Sensors for Water Quality Analysis

Tarun Narayan, Pierre Lovera, and Alan O'Riordan

Nanotechnology Group, Tyndall National Institute, University College Cork, Cork, Ireland

2.1 Introduction

Drinking water is one of the world's most precious resources and has been essential for sustainable and socioeconomic development. Access to clean drinking water is a major challenge that most of the developing countries face [1, 2]. Rapid industrialization and the growing human population have led to an increase in the pollution of freshwater resources. This degradation in the quality of water is leading to environmental and economic problems. The organic and inorganic effluents from industries and domestic wastes are the sources of contamination of freshwater resources such as lakes, rivers, and ground water [3]. These effluents can increase heavy metals concentration. Some heavy metals, like manganese, molybdenum, copper, iron, vanadium, strontium, and zinc, are essential elements occurring in trace amount in humans [4]. Other heavy metals such as arsenic (As), cadmium (Cd), chromium (Cr), lead (Pb), and mercury (Hg) are priority pollutant metals as they are a threat to public health. These heavy metals could accumulate through the food chain, and because of their highly reactive nature, they can be toxic and carcinogenic even in small concentrations [5, 6]. If the concentration of any of these heavy metals increases in the body, i.e. through bioaccumulation, it can lead to various morphological defects, genetic abnormalities, depletion of antioxidants, oxidation of sulfhydryl groups of proteins, and several other effects [7]. Consequently, detection of heavy metals is essential to preserve the environmental quality and health of living beings. Thus, organizations like the US Environmental Protection Agency (USEPA), World Health Organization (WHO),

Sensing Technologies for Real Time Monitoring of Water Quality, First Edition.
Libu Manjakkal, Leandro Lorenzelli, and Magnus Willander.
© 2023 The Institute of Electrical and Electronics Engineers, Inc.
Published 2023 by John Wiley & Sons, Inc.

Centers for Disease Control (CDC), and European Union (EU) have introduced stringent rules for regulating and monitoring the levels of heavy metals in water sources. Consequently, there is an urgent requirement for the development of analytical monitoring techniques to keep a check on heavy metals concentration.

2.2 Sources and Effects of Heavy Metals

Most of the environmental heavy metals contamination results from industries (petroleum refining, textiles, and plastics), mining, fertilizers, pesticides, and automobiles. The heavy metal uptake in humans occurs by consuming vegetable and dairy products from affected land, water, and air inhalation. The recommended levels of toxic heavy metals by the WHO and EU with their sources and effects on humans are summarized in Table 2.1.

2.3 Detection Techniques

2.3.1 Analytical Detection: Conventional Detection Techniques of Heavy Metals

The conventional laboratory-based analytical techniques that are used for heavy metal detection are atomic fluorescence spectroscopy (AFS) [26], inductive coupled plasma mass spectroscopy (ICP-MS) [27, 28], atomic absorption spectroscopy (AAS) [29], and inductive coupled plasma atomic emission spectrometry (ICP-AES) [30]. However, these techniques are expensive, time-consuming, require a trained analyst, and an elaborate sample preparation. Spectrophotometry may also be used for the detection of heavy metals, but the method requires complex equipment with high power and high precision, which is not suited for on-field application [31]. Therefore, there is a need for the development of a rapid, portable, and simple technique for the detection of heavy metals. Electrochemical techniques can overcome most of these limitations as they are more user friendly, economical, and these systems can easily be integrated with wireless networks for on-field monitoring [32, 33]. An electrochemical sensor allows simple fabrication for portable devices and easy procedure for online monitoring of contaminated samples with a short time to result. Figure 2.1 shows the comparison of different techniques used for electrochemical detection and their limits of detection [34].

2.3.2 Electrochemical Detection Techniques of Heavy Metals

In an electrochemical sensor, the transducing layer is an electrode onto which a receptor (metal, polymer [35], DNA/RNA, peptides, aptamers, etc.) is immobilized. This working electrode is then subsequently constituted in an

Table 2.1 Permissible limit as recommended by WHO and EU of some toxic heavy metal with their sources and effects on the living organisms [8, 9].

Heavy metal	WHO limit (mg/l)	EU limits (mg/l)	Sources	Effects	References
Lead (Pb)	0.05	0.01	E-waste, smelting operations, ceramics, bangle industry, lead-acid batteries, and paints	Kidney damage, ataxia, mental retardation, paralysis, stupor	[10, 11]
Cadmium (Cd)	0.003	0.005	E-waste, zinc smelting, batteries, paint sludge, electroplated parts, incinerations, engraving process, and photovoltaic cells	Renal dysfunction, hypertension, cancer, immune disorders, microcytic hypochromic anemia, lymphocytosis, osteocalcin, and osteoporosis	[12, 13]
Mercury (Hg)	0.006	0.001	Municipal solid waste incineration, electrical appliances, thermal power plants, fluorescent lamps, and hospital waste (damaged thermometers, barometers)	Poisoning of protoplasm, acrodynia, digestive system, Minamata, immune system, lungs, hypotonia, and hypertension	[14–16]
Arsenic (As)	0.01	0.01	Preservatives, pesticides, fertilizers, smelting operations, thermal power plants, and fuel	Skin and respiratory problems, peripheral nervous system (PNS), cardiovascular, pulmonary diseases, gastrointestinal tract (GI), teratogenic diseases, and anorexia	[17–19]
Chromium (Cr)	0.05	0.05	Leather industry, tanning, mining, industrial coolants, and chromium salts manufacturing	Skin ulcers mutagenicity, carcinogenicity, lung cancer, dermatitis, dermatitis, reproductive toxicity, and allergies	[20, 21]
Zinc (Zn)	3	—	Smelting, electroplating, cosmetics, and pigments	Anemia, arteriosclerosis, bronchiolar leukocytes, neuronal disorder, prostate cancer risks, macular degeneration	[22, 23]
Copper (Cu)	2	2	Mining, electroplating, fertilizers, and tanning	Wilson's disease, kidney problems, stomach aches, alopecia, arthritis, autism, cystic fibrosis, diabetes, hemorrhaging, vomiting	[24, 25]

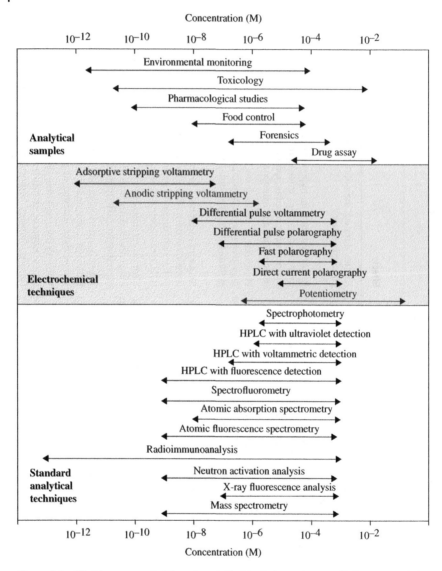

Figure 2.1 Working range of different analytical techniques to quantify heavy metals in different sample types. *Source:* Reproduced with permission from Ref. [34]. © 2021, American Chemical Society.

electrochemical cell-like three-electrode configuration, where one electrode acts as a counter electrode (usually Pt) and the other acts as a reference electrode (usually Ag/AgCl or calomel). Nanomaterial modification on the transducing layer can lead to an increase in surface area and sensitivity with enhanced electronic and quantum mechanical properties. A wide range of nanomaterials is used in sensing such as metallic, carbon, metal-oxide, and polymer-based materials. Different structures of nanomaterials like nanowire and nanorods help to increase the reactivity of electrochemical reaction and provide high spatial resolution for localized detection [36, 37]. Biosensors are another class of electrochemical sensors comprising of biomolecules on the working electrode. Many biorecognition elements can be used for these sensors, such as nucleic acid, proteins, aptamer, enzymes, and whole cells [38].

The various electrochemical techniques that can be used are Cyclic Voltammetry (CV), Square Wave Voltammetry (SQW), Differential Pulse Voltammetry (DPV), and Linear Square Voltammetry (LSV). The difference between these techniques is the time waveform produced with respect of its functional application. Stripping voltammetry involves two steps, preconcentration and stripping. The preconcentration step involves reduction and accumulation of zero-valent metals ions on the working electrode using Faraday's reaction. In the second step, the ion is reoxidized, and the signal is obtained by using anodic stripping [39, 40]. The current peak potential changes correspond to different concentrations of heavy metal ion. Impedance measurements that are based on the change in double-layer capacitance, charge transfer resistance, and solvent resistance are also employed [41, 42]. Figure 2.2 shows a summary of various electrochemical detection techniques to detect heavy metals with a standard three-electrode electrochemical setup and commercial screen-printed electrodes.

2.3.2.1 Nanomaterial-Modified Electrodes

The main factor for a reliable electrochemical sensor is the material used in the detection platform (working electrode). Different nanomaterials and nanostructures have brought about some significant progress and advantages in the field of electrochemistry. The unique properties of nanomaterials make electrochemical sensors attractive compared to conventional techniques. They offer fast electron transfer, a high degree of functionalization, increased mass-transport rate, and high surface area, making them highly sensitive to achieve low detection limit (LOD). Table 2.2 shows different modifications using nanomaterials for heavy metal detection. In this section, we have discussed different nanomaterials, their physicochemical properties, and recent works on nanomaterials-based heavy metal detection.

2.3.2.2 Metal Nanoparticle-Based Modification

Metallic nanomaterials offer unique catalytic and electrical properties allowing high specific detection of the heavy metal ion. Electrochemical detection systems based on mercury as working electrode were previously used because of their high selectivity toward certain heavy metals. However, due to the toxicity of mercury to the environment, other alternative metals like bismuth, gold, and silver came into use. Particularly, bismuth-modified electrodes have been heavily

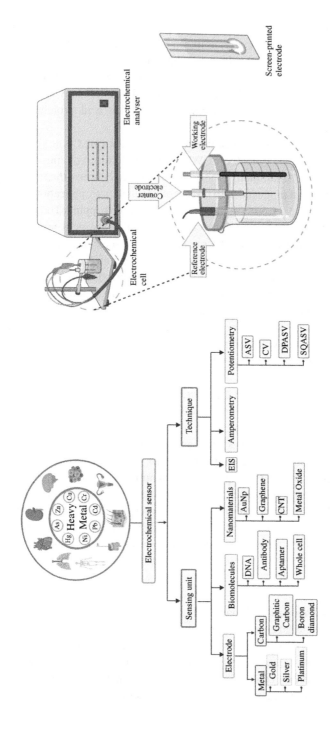

Figure 2.2 Schematic showing the various electrochemical methods for the detection of heavy metals and standard three-electrode setup for electrochemical detection. Created with http://BioRender.com.

Table 2.2 Different nanomaterial modifications of electrochemical sensor for heavy metal detection.

Heavy metal ion	Interface material	Technique	Range of detection	Limit of detection (LOD)	References
Metal-based materials					
As^{3+}	Porous gold electrode	SWASV	0.1–14 µg/l	0.1 µg/l	[43]
Hg	Hexagonal-shaped bismuth NPs/ITO	SWASV	0.74–148 µg/l	0.74 µg/l	[44]
Cd^{2+}, Pb^{2+}	Bi/OMC-MW/GCPE	SWASV	1.0–70.0 µg/l	0.07 µg/l, 0.08 µg/l	[45]
As^{3+}, Hg^{2+}, Pb^{2+}	EG-Bi	SWASV	1–250 mg/l	0.053 µg/l, 0.014 µg/l, 0.081 µg/l	[46]
Pb^{2+}, Cd^{2+}	Bi-NCNF/GCE	SWASV	1–120 µg/l	0.02 µg/l, 0.03 µg/l	[47]
As^{3+}, Hg^{2+}, Pb^{2+}	AuNS/CPSPEs	SWASV	0–200 µg/l	0.8 µg/l, 0.5 µg/l, and 4.3 µg/l	[48]
Pb^{2+}, Cu^{2+}	AuNPs/CFs	DPASV	20.7–414.2 µg/l, 12.7–127 µg/l	1.07 µg/l, 0.57 µg/l	[49]
Cd^{2+}	Au–Ph–AuNP–GSH	OSWV	11.2–11250 ng/l	11.2 ng/l	[50]
Hg^{2+}, Cu^{2+}	GQDs/AuNPs	LSV	4.01–300.7 ng/l, 12.7–95.2 ng/l	10.02 ng/l, 3.17 ng/l	[51]
As^{3+}	Nanostructured Au arrays	LSV	0.125–5 µg/l	0.0423 µg/l	[52]
As^{3+}	FePd NPs/Si	SWASV	1–5 µg/l	0.8 µg/l	[53]
Hg^{2+}	AgNP-FA-PGE	CV	2005–5012.5 µg/l	1690.2 µg/l	[54]
Metal oxide-based materials					
Cd^{2+}, Pb^{2+} Hg^{2+}	GO-Fe₃O₄/GCE	SWASV	22.4–337.2 µg/l, 41.5–623.1 µg/l, 40.1–601.5 µg/l	0.89 µg/l, 1.24 µg/l, 0.8 µg/l	[55]
Zn^{2+}, Cd^{2+}, and Pb^{2+}	Fe₂O₃/G	DPASV	1–100 µg/l	0.11 µg/l, 0.08 µg/l, and 0.07 µg/l	[56]

(Continued)

Table 2.2 (Continued)

Heavy metal ion	Interface material	Technique	Range of detection	Limit of detection (LOD)	References
Carbon-based materials					
Cd^{2+}, Pb^{2+}	Bi/TRGO/Au	SWASV	1.0–120.0 µg/l	0.4 µg/l, 0.1 µg/l	[57]
Hg^{2+}, Cu^{2+}	GQDs/AuNPs	LSV	4.01–300.75 ng/l, 1.27–95.2 ng/l	4.01 ng/l, 3.17 ng/l	[51]
Cd^{2+}, Pb^{2+}	Au/rGOCNT/Bi	SWASV	20–200 µg/l.	0.6 µg/l, 0.2 µg/l	[58]
Cd^{2+}, Pb^{2+}, As^{3+}, Hg^{2+}	[Ru(bpy)$_3$]$^{2+}$-GO/Au	DPV	5.62–33.7 µg/l, 10.3–51.8 µg/l, 3.74–134/8 µg/l, and 20.05–240.6 µg/l	314.7 ng/l, 236.2 ng/l, 172.2 ng/l, and 320.8 ng/l	[59]
Pb^{2+}, Cd^{2+}, Cu^{2+}, Hg^{2+}	AgNPs/RGO	SWASV	10.3–518 µg/l, 5.62–393.4 µg/l, 3.17–222.2 µg/l, and 100.2–601.5 µg/l	29.2 µg/l, 28.5 µg/l, 10.7 µg/l, and 56.1 µg/l	[60]
Pb^{2+}	Carbon nanofiber	SWASV	9987.04–99 994.7 ng/l	358.4 ng/l	[61]
Pb^{2+}, Cd^{2+}	4-Aminothiophenol 1-N doped MWCNT	SWASV	2–50 µg/l	0.4 µg/l and 0.3 µg/l	[62]
Hg^{2+}	Bi NPs@Gr-CNTs	DPV	200 ng/l to 43 588.7 µg/l	20.05 ng/l	[63]
Cd^{2+}, Pb^{2+}	PyTS-CNTs/Nafion/PGE	DPASV	1.0–90 µg/l, 1.0–110 µg/l	0.8 µg/l, 0.02 µg/l	[64]
Zn^{2+}, Cd^{2+}, Pb^{2+}	AG/NA/Bi	DPASV	5–100 µg/l	0.57 µg/l, 0.07 µg/l, 0.05 µg/l	[65]
Pb^{2+}, Cd^{2+}, Cu^{2+}, Zn^2	CNT tower	ASV	—	414.4 ng/l, 2810 ng/l, 2794 ng/l, and 5018.3 ng/l	[66]

ASV, Anodic Stripping Voltammetry; CV: Cyclic Voltammetry; DPV, Differential Pulse Voltammetry; LSV, Linear Square Voltammetry; OSV, Osteryoung Square Wave Voltammetry; SWASV, Square-Wave Anodic Stripping Voltammetry.

utilized for anodic stripping voltammetry because of simple preparation, and their environment friendliness [44, 46, 67, 68]. Heterostructure semiconductor nanomaterials like bismuth vanadate ($BiVO_4$), which is used for its photocatalytic properties, has also been used for electrochemical detection of Cr^{6+} [69]. It can also be made into a composite with various materials for heavy metal detection. Bismuth film was made into a composite with ordered mesoporous carbon (OMC) and diphenylacetylene on a graphitic carbon paste electrode [70]. The electrode exhibited dual detection of Cd^{2+} and Pb^{2+} metal ions with detection limits of 0.07 μg/l and 0.08 μg/l, respectively [45].

Gold and silver nanoparticles have proven to be promising materials for heavy metals detection [43, 54, 71]. Further, the electron transfer rate is improved significantly by compositing them with carbon materials [49, 60]. Many different structures of these metal nanoparticles have been reported and can be easily synthesized by chemical and physical deposition methods. Gold nanostars (AuNS) were prepared using Good's buffer method to modify carbon paste screen-printed electrodes for the detection of As^{3+}, Hg^{2+}, and Pb^{2+} ions. AuNS were found to have more repeatable sensing results as compared to gold nanoparticles (AuNPs) [48]. AuNPs can also be easily functionalized to increase selectivity and stability [50]. There are also a number of reports on bimetallic nanoparticles because of several advantages over single metallic materials [53, 69, 72–74]. However, due to the high cost and surface fouling issue, gold is not much preferred as an electrode modifier in comparison to metal oxides.

2.3.2.3 Metal Oxide Nanoparticle-Based Modification

Metal oxide nanostructures exhibit biocompatibility, large surface area, and effective catalytic property, making them excellent candidates for electrochemical detection of heavy metal ions. Many metal oxides such as CuO, ZnO, Fe_3O_4, ZrO_4, TiO_2, and MgO have been reported for heavy metals detection. Fe_3O_4 nanoparticles are highly chemically stable, nontoxic, and have a high affinity toward heavy metal detection. Deshmukh et al. capped Fe_3O_4 with terephthalic acid for the individual detection of Hg^{2+}, Pb^{2+}, and Cd^{2+}. Each heavy metal ion was first detected individually with the LOD of 0.1 μM, 0.01 μM, and 0.05 μM, respectively, and then simultaneously with LOD of 0.3 μM, 0.04 μM, and 0.2 μM, respectively [75]. This study revealed how the capping agent can increase the selectivity and electrochemical property of the sensor. Iron oxide nanoparticles have also been incorporated with materials like graphene/graphene oxide for metal detection [55, 76]. The nanocomposite of Fe_2O_3/graphene (G) was plated on Bi electrode for electrochemical detection of Zn^{2+}, Cd^{2+}, and Pb^{2+} using differential pulse anodic stripping voltammetry (DPASV). These metals were detected for a combined target range of 1–100 μg/l, with LOD of 0.11 μg/l (Zn^{2+}), 0.08 μg/l (Cd^{2+}), and 0.07 μg/l (Pb^{2+}) [56]. ZnO is also a widely used metal oxide nanoparticle for various applications. In a recent report, ZnO was doped with Ag for improved performance. It was tested for various applications including heavy metal detection, biodiesel production, photocatalytic, and antibacterial activities. Electrocatalytic behavior was tested for Ag–ZnO NPs modified glassy carbon electrode (GCE) to detect Pb^{2+} and Cd^{2+} ions. The linear range of detection was obtained within the concentration range of 50–300 nM and the LOD of 3.5 nM and 3.8 nM, respectively [77].

2.3.2.4 Carbon Nanomaterials-Based Modification

Carbon nanostructures such as graphene, carbon nanotubes (CNT), carbon dots, and fullerenes have attracted a lot of attention in various fields, including heavy metal sensing [78]. Graphene and CNT have attracted much attention owing to their mechanical property, high electron mobility, and easy tunability of properties by chemical or physical doping [79, 80]. Graphene due to its two-dimensional structure provides good surface coverage with a wide potential window. These properties significantly change when its size is reduced to quantum level [81]. Graphene quantum dots with AuNPs were drop cast onto a GCE for the detection of Hg^{2+} and Cu^{2+} with the LOD of 0.02 nM and 0.05 nM, respectively. Doping of graphene/GO with the highly electronegative atom fluorine enhances its electrocatalytic properties and temperature resistance. These properties make fluorinated graphene oxide (FGO) highly suitable for batteries, capacitors, catalysts, and biomedical devices. FGO-modified electrode was reported to simultaneously detect heavy metal ions Cd^{2+}, Pb^{2+}, Cu^{2+}, and Hg^{2+} and found to have significantly better sensitivity than GO [82]. GO can also be doped with redox inorganic complexes such as $[Ru(bpy)_3]^{2+}$ which further enhances its mass transport efficiency and charge transfer rate. A recently reported $[Ru(bpy)_3]^{2+}$-GO/Au electrode was able to detect the presence of Cd^{2+}, Pb^{2+}, As^{3+}, and Hg^{2+} ions in different sources such as river and tap water [59]. Recently, a simple and reliable method was used for the development of micropatterned rGO film without affecting its electrochemical property. Microfabrication technique using negative photoresist SU-8 was used to pattern the electrode on the Au surface. The working electrode was screen-printed with TRGO (solvothermal-assisted reduced graphene oxide), and Bi was electrochemically deposited for heavy metals detection. The SWASV was used to detect lead and cadmium with detection limits of 0.4 µg/l and 1.0 µg/l, respectively. Lu et al. made a highly porous graphene aerogel (GA) with large surface area and low density. Metal-organic framework (UiO-66-NH$_2$) was incorporated with the GA for the detection of Cd^{2+}, Pb^{2+}, Cu^{2+}, and Hg^{2+} in water, soil, and vegetables samples. The composite was drop casted on GCE and was able to simultaneously detect the heavy metals with LOD of 8 nM for Cu^{2+}, 9 nM for Cd^{2+}, 0.9 nM for Hg^{2+}, and 1 nM for Pb^{2+} [83].

CNT can be of two different types, SWCNT and MWCNT (single or multi-walled CNT). SWCNT has a simple structure with single-layer graphene rolled up whereas MWCNT has several tubes rolled up together. Grimaldi et al. modified screen-printed electrode and compared the properties of SWCNTs, MWCNTs, and graphene [84]. The graphene-modified electrode showed typical thin layer properties, whereas MWCNT showed the highest electrochemical signal. However, the capacitance current of CNT changes lab to lab depending on the synthesis but has higher current as compared to GCE because of large microscopic surface area. Moreover, in some reports, high negative potentials can reduce other chemicals leading to interference with metal ion detection. Guo et al. fabricated an array of CNT with millions of ordered CNT and called it tower CNT. The fabricated tower CNT was shown to detect Pb^{2+}, Cd^{2+}, Cu^{2+}, and Zn^{2+} with the detection limit of 12 nM, 25 nM, 44 nM, and 67 nM, respectively [66]. Carbon-based nanocomposite of CNT and rGO was used to make flexible microelectrode on gold. It was further modified with a thin layer of Bi to increase the sensitivity for the detection of Pb and Cd ions. The resulting flexible Au/rGO/CNT/Bi microelectrode significantly improved the response time of the detection and can further be easily integrated with a circuit for on-field detection [58]. Li et al. showed that N doping in MWCNT can enhance its electrical property as it has one more valence electron as compared to carbon. This N-doped MWCNT was further functionalized with thiol using 4-aminothiophenol. This dual functionalized MWCNT was used for the detection of Pb^{2+} and Cd^{2+}. Bismuth was also mixed in an electrolyte to increase the detection sensitivity. The LOD was calculated to be 0.4 μg/l and 0.3 μg/l, respectively [62].

2.3.3 Biomolecules Modification for Heavy Metal Detection

Biosensors are considered a promising technique for heavy metal detection because of their specificity, sensitivity, and reliability. Biosensors can not only detect trace amounts of heavy metals, but can also provide insights about the toxicity level of heavy metals toward living beings. Biosensors are analytical devices that have enzymes [85], antibody [86, 87], DNA [88], whole cells [89], and/or aptamers [90] as a sensing element. When molecules interact with a specific heavy metal, they either act as inhibitors or catalysts; this results in a catalytic reaction. This section details works on the utilization of biomolecules (in biosensors) for heavy metal detection. Table 2.3 gives a summary of recent advances in biosensors for heavy metal ion detection.

2.3.3.1 Antibody-Based Detection

Immunosensors are antibody-based devices, wherein the antibody is specific to an "antigen" that may be chemical (e.g. a metal ion) or biological (pathogenic entities) in nature. When metal ions come in contact with a specific antibody, a metal

Table 2.3 Biosensor modification for heavy metal detection.

Heavy metal ion	Interface material	Technique	Range of detection	Limit of detection (LOD)	References
Pb^{2+}	Au-PWE/(Fe–P)$_n$-MOF-Au-GR	EIS	0.03–1000 nmol/l	0.02 nmol/l	[91]
Cu^{2+}	Cu^{2+}-EDTA McAb/CdSe/ZnS QDs	FL	10–1000 ng/ml	0.33 ng/ml	[92]
Hg^{2+}	DNA MWCNT	LFS	0.05–1 µg/l	0.05 µg/l	[93]
Pb^{2+}	GR–5/(Fe–P)$_n$-MOF probe	Chrono	10.3–41 440 ng/l	7.04 ng/l	[94]
Ag^+, Hg^{2+}	DNA/Fe$_3$O$_4$@Au	SWV	1078–16 170 ng/l, 2005–20 050 ng/l	366.5 ng/l, 183.2 ng/l	[95]
Pb^{2+}, Ag^+, Hg^{2+}	G-rich DNA; C–C mismatch; T–T mismatch	EIS	2072 µg/l to 2072 pg/l, 10 780–86 240 ng/l; 0 µg/l to 20.05 ng/l	2072 pg/l, 1078 ng/l, 20.05 ng/l	[96]
Hg^{2+}	T/CA/AuNPs/rGO	DPV	10 ng/l to 1.0 µg/l	1.5 ng/l	[97]
Hg^{2+}	T/Cu$_2$O@NCs	EIS	200.5–20 050 ng/l	30.07 ng/l	[98]
Pb^{2+}	Hemin/G-quadruplex DNAzyme	SWV	10.3–10 360 ng/l	6630.4 pg/l	[99]
Hg^2	T/3D-rGO@PANI	EIS	20.05–20 050 ng/l	7.01 ng/l	[100]
Ag^+	EG/C/GN-CPE	DPV	97.02 pg/l to 107.08 ng/l	21.5 pg/l	[101]
Pb^{2+}, As^{3+}	Fe-MOF@mFe$_3$O$_4$@mC	EIS	2.07–2072 ng/l, 1.07–1078 ng/l	470.3 pg/l and 725.4 pg/l	[102]
Pb^{2+}, Hg^{2+}	DNA/DNAzyme/substrate DNA/NH$_2$-rGO	EIS	2.07–20 720 ng/l, 2–20 050 ng/l	1616.1 and 1082.7 pg/l	[98]
Hg^{2+}	CS@3D-rGO@DNA	EIS	20.05–2005 ng/l	3.20 ng/l	[103]

Chrono, Chronoamperometry; EIS, Electrochemical Impedance Spectroscopy; LFS, lateral flow strip.

chelate antibody complex is formed generating a detectable signal. Wylie et al. developed a commercial immunosensor for the detection of mercury [104]. It was based on the formation of a mercury-sulfhydryl complex; the sensor could detect in the concentration range of 0.02–300 ppm Some of the antibodies are highly specific to metal-ethylenediamine tetra acetic acid (EDTA) – a principle that has been utilized for the detection of metals like cadmium, lead, nickel, mercury, and copper. For example, monitoring of Co^{2+}, Cd^{2+}, Pb^{2+}, and U^{6+} ions were performed using four different types of antibody complexes. These ions were specific to metal-EDTA complex of DTPA–Co(II), EDTA–Cd(II), 2,9-dicarboxyl-1,10-phenanthroline–U(VI), and cyclohexyl–DTPA–Pb(II), respectively. This also was utilized for the detection of Cu^{2+} ion using mouse anti-Cu^{2+}-EDTA antibody (McAb). A fluorescent-based immunosensor was developed with CdSe/ZnS quantum dots quenching effect in response to Cu^{2+}. Wang et al. developed an immunochromatographic test strip (EFITS) for the detection of Hg^{2+}, using an anti-Hg^{2+} monoclonal antibody (mAb). Hg^{2+} competes with MNA–ovalbumin antibody-AuNPs probe causing a color change, whose intensity correlates to the concentration of Hg^{2+} [87].

2.3.3.2 Nucleic Acid-Based Detection

The interaction of nucleic acids with heavy metal ions induces different types of effects on their structure. Heavy metal ions can influence different functions like replication, mutation, transcription, and may even induce damage. A nucleic acid-based sensor utilizes the following principles for heavy metal detection: (i) binding with DNA bases to form stable DNA-duplex, (ii) cleavage of DNA, and (iii) interaction with G-quadruplex structure. DNA probes are tagged with different "labels"; these label molecules enhance electrochemical response, sensitivity, and selectivity of detection. A biosensor based on DNA-modified Fe_3O_4/Au magnetic nanoparticles was used for simultaneous detection of Ag^+ and Hg^{2+}. They used two different electrochemical dyes (ferrocene [Fc] and methylene blue [MB]) tagged on DNA probes to enhance the sensitivity of detection [95]. Metal-organic frameworks (MOFs) are porous crystalline metal-organic hybrid materials. They have gained attention in the past few years owing to their high porosity and mechanical properties. Among various types of MOFs, Fe(III)-based MOF is reported to have high stability and specificity in the detection of heavy metals. Aptamer (GR-5)-based iron porphyrinic MOF was developed for detecting Pb^{2+} [94]. Aptamers are small DNA molecules that are specific to a particular analyte and upon binding, change their secondary structure resulting in a detectable response. AuNP paper with DNA-modified iron porphyrinic MOF was fabricated to detect Pb^{2+} with a detection limit of 0.02 nmol/l [105].

DNAzyme is also explored widely as a probe for the detection of heavy metal. They are oligonucleotide sequences that cleave DNA explicitly in the presence of a metal ion. DNAzymes have been used in colorimetric, fluorescent, and electrochemical biosensors for the detection of various metal ions such as Pb^{2+}, Mg^{2+}, Hg^{2+}, and Zn^{2+}. A Pb^{2+} specific sensor with DNAzymes was developed and integrated with a glucometer. Upon interaction of Pb^{2+}, DNA was cleaved at a specific site to give an electrochemical response with a detection limit of 1.0 pM [106]. However, it can have a long reaction time for DNA-mediated catalyst polymerization [99].

Thymine mismatch is also reported to show high selectivity toward a number of metals. A selective electrode for the detection of Hg^{2+} was fabricated using thymine-rich cuprous oxide (Cu_2O) and nano-chitosan (NCs) composite [98]. The use of NCs provided a biocompatible surface for immobilization, and Cu_2O provided an increase in surface area and enhanced electrochemical properties. However, Wang et al. fabricated a more sensitive electrode while using the same sensing principle with AuNPs/rGO [97]. The sensor was able to detect in the range of 10 ng/l to 1.0 µg/l with the detection limit of 1.5 ng/l. The sensor surface could also be regenerated and reused after washing with EDTA solution. Cytosine interaction with silver is also studied widely ($C–Ag^+–C$) as it makes stable metal-mediated DNA duplexes. A sensitive electrochemical sensor was developed using AuNPs and polyaniline for the detection of silver ions. A multistage signal amplification strategy was used to detect Hg^{2+} [107]. A triple helix DNA with thymine-rich DNA was immobilized on AuNP-decorated bovine serum albumin/rGO composite. In the presence of Hg^{2+}, the triple helix DNA unwinds to form stable $T–Hg^{2+}–T$, leaving sDNA on the electrode. Further, cDNA-AuNP was used in the presence of Ag^+ to form a $C–Ag^+–C$ complex. This creates a silver nanocluster, and eventually stripping voltammetry was utilized to record the detection of Hg^{2+}. Thus, the sensitive detection of Hg^{2+} was achieved with a detection limit of 0.03 nM. Yao et al. created a lateral flow assay for the detection of Hg^{2+}. They used sandwich assay using $T–Hg^{2+}–T$ recognition mechanism and DNA label using carboxylated MWCNT to increase the specificity [93].

2.3.3.3 Cell-Based Sensor

Whole cell-based (CB) sensors have been developed for the detection of various environmental contaminants including heavy metals [108]. These biosensors comprise both transduction and bioreceptor components. These sensors not only detect heavy metals but also provide deeper insights into the interaction of heavy metals with biological systems at the cellular level, e.g. their inhibitory concentration, physicochemical effects, etc. Both prokaryotic and eukaryotic cells such as fibroblasts, algae, fibroblasts, and bacteria have been used for CB sensor. Yagi et al. differentiated CB sensor into effect-specific response stimulated by metal ions and physicochemical changes on cell [109]. A recent study shows a double mediator system with *Saccharomyces cereviseae* cells [110]. The cells were immobilized on chitosan hydrogel layer, which provides a biocompatible layer with boron-doped nanocrystalline diamond (BND) particles. This sensor was used for the determination of heavy metals (Cu^{2+}, Cd^{2+}, Ni^{2+}, and Pb^{2+}) and phenolic compounds. In other study, chitosan with redox mediator thionine was used for the detection of Cu^{2+}, Cd^{2+}, Zn^{2+}, and Pb^{2+}. There is a lot of work reported on fluorescence-based detection when it comes to CB sensor. In this context, *Torulosa* cells in a cellulose membrane were used for fluorescence-based detection of Cu^{2+}, Pb^{2+}, and Cd^{2+} with the detection limit of 1.195 µg/l, 0.100 µg/l, and 0.027 µg/l, respectively [111]. Self-powered biosensors, based on microbial fuel cell (MFCs), have shown potential in continuous heavy metal monitoring [112, 113]. MFC is a system that uses microbes growing anaerobically on an electrode surface to oxidize organic media and produces electricity [114]. Adekunle et al. designed a membrane-less, air-breathing cathode MFC for online monitoring of heavy metals [115].

However, the main drawback of using CB sensor is the stability of the sensor for a longer duration, low sensitivity, and in some reports delayed response time of detection. This can be further improved by combining them with DNA which will increase the response time significantly.

Table 2.4 summarizes electrode modifications for electrochemical detection of heavy metals.

Table 2.4 Comparison of different modification methods in electrochemical sensor.

Modification material	Advantage	Disadvantage
Metal nanoparticle	ppb detection	Reproducibility
	Miniaturization	Formation of intermetallic compound
	On-field detection	
Metal oxide	High adsorption	Interference
	Low cost	
	Nontoxic	
Carbon materials	Miniaturization	Selectivity
	On-field detection	Poor reusability
	Stability	
	Sensitivity	
Biomolecules	Specificity	High cost
		Long-term stability
		On-field detection not available
Cell-based sensor	Cellular response of different heavy metals, can be used for drug and toxicity screening	Specialized equipment for cell culture
		Minor contamination can affect the result

2.4 Future Direction

The pollution caused by industrialization and urbanization has led to decrease in safe, clean, and drinkable water sources. The challenge is to provide high-quality water which correlates with the Drinking Water Directive (EU) and WHO guidelines. The detection technologies developed must address these challenges such as (i) identification of metal ion from complex samples, (ii) achieve fast response time, (iii) overcoming the interference and fouling of electrode, and (iv) continuous monitoring of heavy metal ions in water resources.

The integration of heavy metal sensor with smartphones, wireless sensor network (WSNs), photometers, and portable electrochemical devices would create an end-to-end system [116]. The ion-specific electrodes solve a lot of these issues by making it highly selective toward the desired analytes. These electrodes are coated with different polymer membranes that allow only selective analyte to pass through the electrode. Besides these, microfluidic systems can be used as it allows the low intake of sample volume. However, pretreatment is not included in most of the microfluidic analysis system, which hinders the practicality when dealing with complex samples. The integration of nanomaterials in such devices could enhance the practicality and performance of the sensor [117]. Likewise, the disposable electrode modified with nanomaterials offer simple and low-cost way for on-field detection [118, 119].

There is a lot of improvement needed for on-field sensing and commercialization. Even though the WSN [120] is being deployed for water monitoring, there is lack of hybrid analysis for continuous monitoring of water bodies. These WSNs should be employed in the site of water consumption to better understand the quality of water that reaches the end user. It is important to reduce the complexity of the system in large-scale deployment of WSN to prevent transmission interruptions and loss of data points. The hybrid integration of different systems, reliable, user-friendly interface, and hand-held devices will make the sensor more dependable for minimizing the contaminants that are present in water.

2.5 Conclusions

This chapter presents an overview of the recent developments in electrochemical detection of different heavy metal ions. The need for pure and safe drinking water has led many researchers to work on an electrochemical sensor that provides rapid analysis and trace level detection. Different electrochemical techniques such as voltammetry, potentiometric, and impedance have been reported for selective detection of heavy metals. Among the techniques, anodic stripping is widely used and is most potent in sensitive detection [121]. The use of nanomaterials has significantly improved the performance of the sensing platform. Various types of materials, such as metal nanoparticles, CNT, and graphene, have been explored for modifying the electrode surface. The properties like high electron transfer and ease of surface functionalization have enabled sensor to detect with enhanced sensitivity and selectivity. However, suitability in complex matrix and reproducibility of the sensor in the real sample need further improvement. Moreover, the toxicity of some nanomaterials like quantum dots also should be considered before their use in rivers and lakes. The use of bimetallic materials and improved modification strategies can enhance the specificity and stability of the sensor.

Many biosensors based on whole-cell, antibodies, enzymes, and DNA have been reported for different heavy metal detection. The use of biomolecules with nanoparticles has been used effectively in the development of a robust sensor platform. The performance and stability of biosensor depend on the type of biomolecules used as the transducer. The whole CB sensor has been reported with slow response time and with a nonspecific response. One of the disadvantages of electrochemical biosensors could also be a suppressed signal due to background current and nonspecific adsorption. However, the use of multiple probes and electrochemical redox indicators could amplify the current response. Different organic compounds in the real sample can interfere with the sensor performance. Aptamers and DNAzyme can be used to make them highly specific, but they are costly to fabricate.

Sensing methods such as portable electrochemical analyzer, disposable electrodes, colorimetric paper strips, and lateral flow chambers have a potential for on-field testing and mass manufacturing. The integration of nanomaterials in such devices could enhance the practicality and performance of the sensor. In future, we believe more portable and smart devices with nanomaterial integration would be in use for real-time monitoring of heavy metals in water bodies.

Acknowledgment

Authors are thankful for research supported by grant from European Union's Horizon 2020 research and innovation programme under the Marie Skłodowska-Curie grant agreement (H2020-MSCA-ITN-2018-813680) and emanated in part from the financial support of Science Foundation Ireland (SFI) and the Department of Agriculture, Food and Marine on behalf of the Government of Ireland under Grant Number [16/RC/3835] VistaMilk.

References

1 World Health Organization (2015). *Guidelines for Drinking-Water Quality*, 4e. World Health Organization.

2 Cotruvo, J.A. (2017). 2017 WHO guidelines for drinking water quality: first addendum to the fourth edition. *J. Am. Water Works Assoc.* 109 (7): 44–51.

3 Calderon, R.L. (2000). The epidemiology of chemical contaminants of drinking water. *Food Chem. Toxicol.* 38 (1 Suppl): S13–S20.

4 Khan, A., Khan, S., Khan, M.A. et al. (2015). The uptake and bioaccumulation of heavy metals by food plants, their effects on plants nutrients, and associated health risk: a review. *Environ. Sci. Pollut. Res.* 22 (18): 13772–13799.

5 Li, M., Gou, H., Al-Ogaidi, I., and Wu, N. (2013). Nanostructured sensors for detection of heavy metals: a review. *ACS Sustain. Chem. Eng.* 1: 713–723.

6 Kallis, G. and Butler, D. (2001). The EU water framework directive: measures and implications. *Water Policy* 3 (2): 125–142.

7 Valko, M., Morris, H., and Cronin, M.T. (2005). Metals, toxicity and oxidative stress. *Curr. Med. Chem.* 12 (10): 1161–1208.

8 European Commission (1998). Council directive 98/83/EC of 3 November 1998 on the quality of water intended for human consumption. *Off. J. Eur. Communities L* 330: 32–54.

9 World Health Organization (1993). *Guidelines for Drinking-Water Quality*. World Health Organization.

10 Patrick, L. (2006). Lead toxicity part II: the role of free radical damage and the use of antioxidants in the pathology and treatment of lead toxicity. *Altern. Med. Rev.* 11 (2): 114–127.

11 Ettinger, A.S. and Wengrovitz, A. M. (2010). Guidelines for the identification and management of lead exposure in pregnant and lactating women. https://www.cdc.gov/nceh/lead/publications/leadandpregnancy2010.pdf

12 Ercal, N., Gurer-Orhan, H., and Aykin-Burns, N. (2001). Toxic metals and oxidative stress part I: mechanisms involved in metal-induced oxidative damage. *Curr. Top. Med. Chem.* 1 (6): 529–539.

13 Verma, N., Kumar, S., and Kaur, H. (2010). Fiber optic biosensor for the detection of Cd in milk. *J. Biosens. Bioelectron* 1: 102.

14 Clarkson, T.W. and Magos, L. (2006). The toxicology of mercury and its chemical compounds. *Crit. Rev. Toxicol.* 36 (8): 609–662.

15 Clarkson, T.W. (1997). The toxicology of mercury. *Crit. Rev. Clin. Lab. Sci.* 34 (4): 369–403.

16 Kobal, A.B., Horvat, M., Prezelj, M. et al. (2004). The impact of long-term past exposure to elemental mercury on antioxidative capacity and lipid peroxidation in mercury miners. *J. Trace Elem. Med. Biol.* 17 (4): 261–274.

17 Abernathy, C.O., Thomas, D.J., and Calderon, R.L. (2003). Health effects and risk assessment of arsenic. *J. Nutr.* 133 (5 Suppl 1): 1536S–1538S.

18 Kapaj, S., Peterson, H., Liber, K., and Bhattacharya, P. (2006). Human health effects from chronic arsenic poisoning: a review. *J. Environ. Sci. Health A Tox. Hazard. Subst. Environ. Eng.* 41 (10): 2399–2428.

19 Luong, J.H., Lam, E., and Male, K.B. (2014). Recent advances in electrochemical detection of arsenic in drinking and ground waters. *Anal. Methods* 6 (16): 6157–6169.

20 Rodriguez, V.M., Jimenez-Capdeville, M.E., and Giordano, M. (2003). The effects of arsenic exposure on the nervous system. *Toxicol. Lett.* 145 (1): 1–18.

21 Cieślak-Golonka, M. (1996). Toxic and mutagenic effects of chromium(VI). A review. *Polyhedron* 15 (21): 3667–3689.

22 Walsh, C.T., Sandstead, H.H., Prasad, A.S. et al. (1994). Zinc: health effects and research priorities for the 1990s. *Environ. Health Perspect.* 102 (Suppl 2): 5–46.

23 Fosmire, G.J. (1990). Zinc toxicity. *Am. J. Clin. Nutr.* 51 (2): 225–227.

24 Gaetke, L.M. and Chow, C.K. (2003). Copper toxicity, oxidative stress, and antioxidant nutrients. *Toxicology* 189 (1–2): 147–163.

25 Selvaraj, S., Krishnaswamy, S., Devashya, V. et al. (2011). Investigations on membrane perturbation by chrysin and its copper complex using self-assembled lipid bilayers. *Langmuir* 27 (21): 13374–13382.

26 Sanchez-Rodas, D., Corns, W., Chen, B., and Stockwell, P. (2010). Atomic fluorescence spectrometry: a suitable detection technique in speciation studies for arsenic, selenium, antimony and mercury. *J. Anal. At. Spectrom.* 25 (7): 933–946.

27 Koelmel, J. and Amarasiriwardena, D. (2012). Imaging of metal bioaccumulation in hay-scented fern (*Dennstaedtia punctilobula*) rhizomes growing on contaminated soils by laser ablation ICP-MS. *Environ. Pollut.* 168: 62–70.

28 Corbisier, P., van der Lelie, D., Borremans, B. et al. (1999). Whole cell-and protein-based biosensors for the detection of bioavailable heavy metals in environmental samples. *Anal. Chim. Acta* 387 (3): 235–244.

29 Siraj, K. and Kitte, S.A. (2013). Analysis of copper, zinc and lead using atomic absorption spectrophotometer in ground water of Jimma town of Southwestern Ethiopia. *Int. J. Chem. Anal. Sci.* 4 (4): 201–204.

30 Rao, K.S., Balaji, T., Rao, T.P. et al. (2002). Determination of iron, cobalt, nickel, manganese, zinc, copper, cadmium and lead in human hair by inductively coupled plasma-atomic emission spectrometry. *Spectrochim. Acta B At. Spectrosc.* 57 (8): 1333–1338.

31 Aragay, G., Pons, J., and Merkoci, A. (2011). Recent trends in macro-, micro-, and nanomaterial-based tools and strategies for heavy-metal detection. *Chem. Rev.* 111 (5): 3433–3458.

32 Ring, G., O'Mullane, J., O'Riordan, A., and Furey, A. (2016). Trace metal determination as it relates to metallosis of orthopaedic implants: evolution and current status. *Clin. Biochem.* 49 (7–8): 617–635.

33 Seymour, I., Narayan, T., Creedon, N. et al. (2021). Advanced solid state nano-electrochemical sensors and system for Agri 4.0 applications. *Sensors* 21 (9): 3149.

34 Ding, R., Cheong, Y.H., Ahamed, A., and Lisak, G. (2021). Heavy metals detection with paper-based electrochemical sensors. *Anal. Chem.* 93: 1880–1888.

35 Luong, J.H., Narayan, T., Solanki, S., and Malhotra, B.D. (2020). Recent advances of conducting polymers and their composites for electrochemical biosensing applications. *J. Funct. Biomater.* 11 (4): 71.

36 Wahl, A.J., Seymour, I.P., Moore, M. et al. (2018). Diffusion profile simulations and enhanced iron sensing in generator-collector mode at interdigitated nanowire electrode arrays. *Electrochim. Acta* 277: 235–243.

37 O'Sullivan, B., O'Sullivan, S., Narayan, T. et al. (2022). A direct comparison of 2D versus 3D diffusion analysis at nanowire electrodes: a finite element analysis and experimental study. *Electrochim. Acta* 408: 139890.

38 Montrose, A., Creedon, N., Sayers, R. et al. (2015). Novel single gold nanowire-based electrochemical immunosensor for rapid detection of bovine viral diarrhoea antibodies in serum. *J. Biosens. Bioelectron* 6 (3): 1–7.

39 Daly, R., Narayan, T., Shao, H. et al. (2021). Platinum-based interdigitated micro-electrode arrays for reagent-free detection of copper. *Sensors* 21 (10): 3544.

40 Wasiewska, L.A., Seymour, I., Patella, B. et al. (2021). Reagent free electrochemical-based detection of silver ions at interdigitated microelectrodes using in-situ pH control. *Sensors Actuators B Chem.* 333: 129531.

41 Petovar, B., Xhanari, K., and Finšgar, M. (2018). A detailed electrochemical impedance spectroscopy study of a bismuth-film glassy carbon electrode for trace metal analysis. *Anal. Chim. Acta* 1004: 10–21.

42 Finšgar, M., Xhanari, K., and Petovar, B. (2019). Copper-film electrodes for Pb(II) trace analysis and a detailed electrochemical impedance spectroscopy study. *Microchem. J.* 147: 863–871.

43 Kim, J., Han, S., and Kim, Y. (2017). Electrochemical detection of arsenic(III) using porous gold via square wave voltammetry. *Korean J. Chem. Eng.* 34 (7): 2096–2098.

44 Gupta, S., Singh, R., Anoop, M.D. et al. (2018). Electrochemical sensor for detection of mercury(II) ions in water using nanostructured bismuth hexagons. *Appl. Phys. A* 124 (11): 737.

45 Zhao, G., Wang, H., Liu, G., and Wang, Z. (2017). Simultaneous and sensitive detection of cd(II) and Pb(II) using a novel bismuth film/ordered mesoporous carbon-molecular wire modified graphite carbon paste electrode. *Electroanalysis* 29 (2): 497–505.

46 Mafa, P.J., Idris, A.O., Mabuba, N., and Arotiba, O.A. (2016). Electrochemical co-detection of As(III), Hg(II) and Pb(II) on a bismuth modified exfoliated graphite electrode. *Talanta* 153: 99–106.

47 Lu, Z., Dai, W., Lin, X. et al. (2018). Facile one-step fabrication of a novel 3D honeycomb-like bismuth nanoparticles decorated N-doped carbon nanosheet frameworks: ultrasensitive electrochemical sensing of heavy metal ions. *Electrochim. Acta* 266: 94–102.

48 Dutta, S., Strack, G., and Kurup, P. (2019). Gold nanostar electrodes for heavy metal detection. *Sensors Actuators B Chem.* 281: 383–391.

49 Xiong, W., Zhou, L., and Liu, S. (2016). Development of gold-doped carbon foams as a sensitive electrochemical sensor for simultaneous determination of Pb(II) and Cu(II). *Chem. Eng. J.* 284: 650–656.

50 Liu, G., Zhang, Y., Qi, M., and Chen, F. (2015). Covalent anchoring of multifunctionized gold nanoparticles on electrodes towards an electrochemical sensor for the detection of cadmium ions. *Anal. Methods* 7 (13): 5619–5626.

51 Ting, S.L., Ee, S.J., Ananthanarayanan, A. et al. (2015). Graphene quantum dots functionalized gold nanoparticles for sensitive electrochemical detection of heavy metal ions. *Electrochim. Acta* 172: 7–11.

52 Podesva, P., Gablech, I., and Neuzil, P. (2018). Nanostructured gold microelectrode array for ultrasensitive detection of heavy metal contamination. *Anal. Chem.* 90 (2): 1161–1167.

53 Moghimi, N., Mohapatra, M., and Leung, K.T. (2015). Bimetallic nanoparticles for arsenic detection. *Anal. Chem.* 87 (11): 5546–5552.

54 Eksin, E., Erdem, A., Fafal, T., and Kıvçak, B. (2019). Eco-friendly sensors developed by herbal based silver nanoparticles for electrochemical detection of mercury(II) ion. *Electroanalysis* 31 (6): 1075–1082.

55 Xiong, S., Yang, B., Cai, D. et al. (2015). Individual and simultaneous stripping voltammetric and mutual interference analysis of Cd^{2+}, Pb^{2+} and Hg^{2+} with reduced graphene oxide-Fe_3O_4 nanocomposites. *Electrochim. Acta* 185: 52–61.

56 Lee, S., Oh, J., Kim, D., and Piao, Y. (2016). A sensitive electrochemical sensor using an iron oxide/graphene composite for the simultaneous detection of heavy metal ions. *Talanta* 160: 528–536.

57 Xuan, X., Hossain, M.F., and Park, J.Y. (2016). A fully integrated and miniaturized heavy-metal-detection sensor based on micro-patterned reduced graphene oxide. *Sci. Rep.* 6: 33125.

58 Xuan, X. and Park, J.Y. (2018). A miniaturized and flexible cadmium and lead ion detection sensor based on micro-patterned reduced graphene oxide/carbon nanotube/bismuth composite electrodes. *Sensors Actuators B Chem.* 255: 1220–1227.

59 Gumpu, M.B., Veerapandian, M., Krishnan, U.M., and Rayappan, J.B.B. (2017). Simultaneous electrochemical detection of Cd(II), Pb(II),

As(III) and Hg(II) ions using ruthenium(II)-textured graphene oxide nanocomposite. *Talanta* 162: 574–582.

60 Sang, S., Li, D., Zhang, H. et al. (2017). Facile synthesis of AgNPs on reduced graphene oxide for highly sensitive simultaneous detection of heavy metal ions. *RSC Adv.* 7 (35): 21618–21624.

61 Robinson, J.E., Heineman, W.R., Sagle, L.B. et al. (2016). Carbon nanofiber electrode array for the detection of lead. *Electrochem. Commun.* 73: 89–93.

62 Li, X., Zhou, H., Fu, C. et al. (2016). A novel design of engineered multi-walled carbon nanotubes material and its improved performance in simultaneous detection of Cd(II) and Pb(II) by square wave anodic stripping voltammetry. *Sensors Actuators B Chem.* 236: 144–152.

63 Jeromiyas, N., Elaiyappillai, E., Kumar, A.S. et al. (2019). Bismuth nanoparticles decorated graphenated carbon nanotubes modified screen-printed electrode for mercury detection. *J. Taiwan Inst. Chem. Eng.* 95: 466–474.

64 Jiang, R., Liu, N., Gao, S. et al. (2018). A facile electrochemical sensor based on PyTS–CNTs for simultaneous determination of cadmium and lead ions. *Sensors* 18 (5): 1567.

65 Lee, S., Bong, S., Ha, J. et al. (2015). Electrochemical deposition of bismuth on activated graphene-nafion composite for anodic stripping voltammetric determination of trace heavy metals. *Sensors Actuators B Chem.* 215: 62–69.

66 Guo, X., Yun, Y., Shanov, V.N. et al. (2011). Determination of trace metals by anodic stripping voltammetry using a carbon nanotube tower electrode. *Electroanalysis* 23 (5): 1252–1259.

67 Zhao, G., Wang, H., Liu, G., and Wang, Z. (2016). Box–Behnken response surface design for the optimization of electrochemical detection of cadmium by square wave anodic stripping voltammetry on bismuth film/glassy carbon electrode. *Sensors Actuators B Chem.* 235: 67–73.

68 Finšgar, M. and Kovačec, L. (2020). Copper-bismuth-film in situ electrodes for heavy metal detection. *Microchem. J.* 104635.

69 Jaihindh, D.P., Thirumalraj, B., Chen, S.M. et al. (2019). Facile synthesis of hierarchically nanostructured bismuth vanadate: an efficient photocatalyst for degradation and detection of hexavalent chromium. *J. Hazard. Mater.* 367: 647–657.

70 Gupta, N., Todi, K., Narayan, T., and Malhotra, B. (2022). Graphitic carbon nitride-based nanoplatforms for biosensors: design strategies and applications. *Mater. Today Chem.* 24: 100770.

71 Schopf, C., Wahl, A., Martin, A. et al. (2016). Direct observation of mercury amalgamation on individual gold nanorods using spectroelectrochemistry. *J. Phys. Chem. C* 120 (34): 19295–19301.

72 Van der Horst, C., Silwana, B., Iwuoha, E., and Somerset, V. (2017). Voltammetric analysis of platinum group metals using a bismuth-silver bimetallic nanoparticles sensor. In: *Recent Progress in Organometallic Chemistry* (ed. M.M. Rahman and A.M. Asiri), 123–137. IntechOpen.

73 Gong, J., Zhou, T., Song, D. et al. (2010). Stripping voltammetric detection of mercury(II) based on a bimetallic Au–Pt inorganic–organic hybrid nanocomposite modified glassy carbon electrode. *Anal. Chem.* 82 (2): 567–573.

74 Alam, M.M., Rashed, M.A., Rahman, M.M. et al. (2018). Electrochemical oxidation of As(III) on Pd immobilized Pt surface: kinetics and sensing performance. *RSC Adv.* 8 (15): 8071–8079.

75 Deshmukh, S., Kandasamy, G., Upadhyay, R.K. et al. (2017). Terephthalic acid capped iron oxide nanoparticles for sensitive electrochemical detection of heavy metal ions in water. *J. Electroanal. Chem.* 788: 91–98.

76 Chimezie, A.B., Hajian, R., Yusof, N.A. et al. (2017). Fabrication of reduced graphene oxide-magnetic nanocomposite (rGO-Fe$_3$O$_4$) as an electrochemical sensor for trace determination of As(III) in water resources. *J. Electroanal. Chem.* 796: 33–42.

77 Nagaraju, G., Prashanth, S., Shastri, M. et al. (2017). Electrochemical heavy metal detection, photocatalytic, photoluminescence, biodiesel production and antibacterial activities of Ag–ZnO nanomaterial. *Mater. Res. Bull.* 94: 54–63.

78 Wanekaya, A.K. (2011). Applications of nanoscale carbon-based materials in heavy metal sensing and detection. *Analyst* 136 (21): 4383–4391.

79 Zhang, L., Peng, D., Liang, R.-P., and Qiu, J.-D. (2018). Graphene-based optical nanosensors for detection of heavy metal ions. *TrAC Trends Anal. Chem.* 102: 280–289.

80 Lovera, P., Creedon, N., Alatawi, H. et al. (2014). Low-cost silver capped polystyrene nanotube arrays as super-hydrophobic substrates for SERS applications. *Nanotechnology* 25 (17): 175502.

81 Qian, Z.S., Shan, X.Y., Chai, L.J. et al. (2015). A fluorescent nanosensor based on graphene quantum dots-aptamer probe and graphene oxide platform for detection of lead(II) ion. *Biosens. Bioelectron.* 68: 225–231.

82 Thiruppathi, A.R., Sidhureddy, B., Keeler, W., and Chen, A. (2017). Facile one-pot synthesis of fluorinated graphene oxide for electrochemical sensing of heavy metal ions. *Electrochem. Commun.* 76: 42–46.

83 Lu, M., Deng, Y., Luo, Y. et al. (2019). Graphene aerogel-metal-organic framework-based electrochemical method for simultaneous detection of multiple heavy-metal ions. *Anal. Chem.* 91 (1): 888–895.

84 Grimaldi, A., Heijo, G., and Méndez, E. (2014). A multiple evaluation approach of commercially available screen-printed nanostructured carbon electrodes. *Electroanalysis* 26 (8): 1684–1693.

85 Chauhan, N., Tiwari, S., Narayan, T., and Jain, U. (2019). Bienzymatic assembly formed@ Pt nano sensing framework detecting acetylcholine in aqueous phase. *Appl. Surf. Sci.* 474: 154–160.

86 Narayan, T., Kumar, S., Kumar, S. et al. (2019). Protein functionalised self assembled monolayer based biosensor for colon cancer detection. *Talanta* 201: 465–473.

87 Wang, S., Wang, L., Chen, H. et al. (2016). Development of an eco-friendly immunochromatographic test strip and its application in detecting Hg^{2+} without chelators. *RSC Adv.* 6 (11): 8729–8735.

88 Prabhakar, N., Arora, K., Singh, H., and Malhotra, B.D. (2008). Polyaniline based nucleic acid sensor. *J. Phys. Chem. B* 112 (15): 4808–4816.

89 Gupta, N., Renugopalakrishnan, V., Liepmann, D. et al. (2019). Cell-based biosensors: recent trends, challenges and future perspectives. *Biosens. Bioelectron.* 141: 111435.

90 Farzin, L., Shamsipur, M., and Sheibani, S. (2017). A review: aptamer-based analytical strategies using the nanomaterials for environmental and human monitoring of toxic heavy metals. *Talanta* 174: 619–627.

91 Wang, X., Yang, C., Zhu, S. et al. (2017). 3D origami electrochemical device for sensitive Pb^{2+} testing based on DNA functionalized iron-porphyrinic metal-organic framework. *Biosens. Bioelectron.* 87: 108–115.

92 Shu, Q., Liu, M., Ouyang, H., and Fu, Z. (2017). Label-free fluorescent immunoassay for Cu^{2+} ion detection based on UV degradation of immunocomplex and metal ion chelates. *Nanoscale* 9 (34): 12302–12306.

93 Yao, L., Teng, J., Zhu, M. et al. (2016). MWCNTs based high sensitive lateral flow strip biosensor for rapid determination of aqueous mercury ions. *Biosens. Bioelectron.* 85: 331–336.

94 Cui, L., Wu, J., Li, J., and Ju, H. (2015). Electrochemical sensor for lead cation sensitized with a DNA functionalized porphyrinic metal-organic framework. *Anal. Chem.* 87 (20): 10635–10641.

95 Miao, P., Tang, Y., and Wang, L. (2017). DNA modified Fe_3O_4@Au magnetic nanoparticles as selective probes for simultaneous detection of heavy metal ions. *ACS Appl. Mater. Interfaces* 9 (4): 3940–3947.

96 Lin, Z., Li, X., and Kraatz, H.B. (2011). Impedimetric immobilized DNA-based sensor for simultaneous detection of Pb^{2+}, Ag^+, and Hg^{2+}. *Anal. Chem.* 83 (17): 6896–6901.

97 Wang, N., Lin, M., Dai, H., and Ma, H. (2016). Functionalized gold nanoparticles/ reduced graphene oxide nanocomposites for ultrasensitive electrochemical sensing of mercury ions based on thymine-mercury-thymine structure. *Biosens. Bioelectron.* 79: 320–326.

98 Liu, S., Kang, M., Yan, F. et al. (2015). Electrochemical DNA biosensor based on microspheres of cuprous oxide and nano-chitosan for Hg(II) detection. *Electrochim. Acta* 160: 64–73.

99 Zhang, B., Chen, J., Liu, B., and Tang, D. (2015). Amplified electrochemical sensing of lead ion based on DNA-mediated self-assembly-catalyzed polymerization. *Biosens. Bioelectron.* 69: 230–234.

100 Yang, Y., Kang, M., Fang, S. et al. (2015). Electrochemical biosensor based on three-dimensional reduced graphene oxide and polyaniline nanocomposite for selective detection of mercury ions. *Sens. Actuators B Chem.* 214: 63–69.

101 Ebrahimi, M., Raoof, J.B., and Ojani, R. (2015). Novel electrochemical DNA hybridization biosensors for selective determination of silver ions. *Talanta* 144: 619–626.

102 Zhang, Z., Ji, H., Song, Y. et al. (2017). Fe(III)-based metal-organic framework-derived core-shell nanostructure: sensitive electrochemical platform for high trace determination of heavy metal ions. *Biosens. Bioelectron.* 94: 358–364.

103 Wang, H., Wang, Y., Zhang, Y. et al. (2016). Photoelectrochemical immunosensor for detection of carcinoembryonic antigen based on 2D TiO_2 nanosheets and carboxylated graphitic carbon nitride. *Sci. Rep.* 6 (1): 1–7.

104 Wylie, D.E., Carlson, L.D., Carlson, R. et al. (1991). Detection of mercuric ions in water by ELISA with a mercury-specific antibody. *Anal. Biochem.* 194 (2): 381–387.

105 Dong, Y.-X., Cao, J.-T., Wang, B. et al. (2017). Exciton–plasmon interactions between CdS@g-C_3N_4 heterojunction and Au@Ag nanoparticles coupled with DNAase-triggered signal amplification: toward highly sensitive photoelectrochemical bioanalysis of MicroRNA. *ACS Sustain. Chem. Eng.* 5 (11): 10840–10848.

106 Fu, L., Zhuang, J., Lai, W. et al. (2013). Portable and quantitative monitoring of heavy metal ions using DNAzyme-capped mesoporous silica nanoparticles with a glucometer readout. *J. Mater. Chem. B* 1 (44): 6123–6128.

107 Wang, H., Zhang, Y., Ma, H. et al. (2016). Electrochemical DNA probe for Hg^{2+} detection based on a triple-helix DNA and multistage signal amplification strategy. *Biosens. Bioelectron.* 86: 907–912.

108 Gutiérrez, J.C., Martín-González, A., Díaz, S., and Ortega, R. (2003). Ciliates as a potential source of cellular and molecular biomarkers/biosensors for heavy metal pollution. *Eur. J. Protistol.* 39 (4): 461–467.

109 Yagi, K. (2007). Applications of whole-cell bacterial sensors in biotechnology and environmental science. *Appl. Microbiol. Biotechnol.* 73 (6): 1251–1258.

110 Gao, G., Fang, D., Yu, Y. et al. (2017). A double-mediator based whole cell electrochemical biosensor for acute biotoxicity assessment of wastewater. *Talanta* 167: 208–216.

111 Wong, L.S., Lee, Y.H., and Surif, S. (2013). Whole cell biosensor using *Anabaena torulosa* with optical transduction for environmental toxicity evaluation. *J. Sens.* 2013: 567272.

112 Veerubhotla, R. and Das, D. (2018). Application of microbial fuel cell as a biosensor. In: *Microbial Fuel Cell* (ed. D. Das), 389–402. Springer.

113 Liu, B., Lei, Y., and Li, B. (2014). A batch-mode cube microbial fuel cell based "shock" biosensor for wastewater quality monitoring. *Biosens. Bioelectron.* 62: 308–314.

114 Nara, S., Kandpal, R., Jaiswal, V. et al. (2020). Exploring *Providencia rettgeri* for application to eco-friendly paper based microbial fuel cell. *Biosens. Bioelectron.* 165: 112323.

115 Adekunle, A., Raghavan, V., and Tartakovsky, B. (2019). On-line monitoring of heavy metals-related toxicity with a microbial fuel cell biosensor. *Biosens. Bioelectron.* 132: 382–390.

116 Li, Y., Chen, Y., Yu, H. et al. (2018). Portable and smart devices for monitoring heavy metal ions integrated with nanomaterials. *TrAC Trends Anal. Chem.* 98: 190–200.

117 Kumar, P., Kim, K.-H., Bansal, V. et al. (2017). Progress in the sensing techniques for heavy metal ions using nanomaterials. *J. Ind. Eng. Chem.* 54: 30–43.

118 Lin, Y., Gritsenko, D., Feng, S. et al. (2016). Detection of heavy metal by paper-based microfluidics. *Biosens. Bioelectron.* 83: 256–266.

119 Liu, X., Yao, Y., Ying, Y., and Ping, J. (2019). Recent advances in nanomaterial-enabled screen-printed electrochemical sensors for heavy metal detection. *TrAC Trends Anal. Chem.* 115: 187–202.

120 Almazyad, A.S., Seddiq, Y.M., Alotaibi, A.M. et al. (2014). A proposed scalable design and simulation of wireless sensor network-based long-distance water pipeline leakage monitoring system. *Sensors* 14 (2): 3557–3577.

121 Patella, B., Narayan, T., O'Sullivan, B. et al. (2023). Simultaneous detection of copper and mercury in water samples using in-situ pH control with electrochemical stripping techniques. *Electrochimica Acta*, 439: 141668.

3

Ammonia, Nitrate, and Urea Sensors in Aquatic Environments

Fabiane Fantinelli Franco

Infrastructure and Environment Research Division, James Watt School of Engineering, University of Glasgow, Glasgow, UK

3.1 Introduction

Nitrogen is an important parameter in aquaculture and agriculture that can lead to environmental pollution if above a certain threshold, making it a key parameter to monitor. The nitrogen cycle (Figure 3.1a) gives an insight on which nitrogenous compounds to monitor and how they correlate to one another. Traditionally, the nitrogen cycle has been divided into three processes (dinitrate [N_2] fixation, nitrification, and denitrification). However, this has recently been expanded to include five nitrogen processes: ammonification, nitrification, denitrification, anammox, and nitrate–nitrate interconversion (Figure 3.1b) [1]. Mineralization and assimilation, which refer to the general processes of organic matter, complete the movement of the nitrogen cycle (Figure 3.1b). Ammonification is composed of two processes: – N_2 fixation, performed by bacteria and archae that have the nitrogenase enzyme, and anaerobic assimilatory and dissimilatory nitrite (NO_2^-) reduction to ammonium (NH_4^+), accomplished by bacteria and fungi. Nitrification is mediated by different groups of microorganisms that can perform nitritation, oxidation of ammonia to NO_2^-, nitratation, oxidation of NO_2^- to nitrate (NO_3^-), or completely oxidize ammonia to NO_3^-. Denitrification consists of the anaerobic respiration of NO_2^-, nitric oxide (NO), and nitrous oxide (N_2O) to N_2 from bacteria, archae, and several eukaryotes. Anammox, short for anaerobic ammonium oxidation, transforms NO_2^- and NH_4^+ to N_2 via the NO and hydrazine (N_2H_4) intermediates. The nitrate–nitrite process is achieved by nitrite-oxidizing or -reducing bacteria [1].

Sensing Technologies for Real Time Monitoring of Water Quality, First Edition.
Libu Manjakkal, Leandro Lorenzelli, and Magnus Willander.
© 2023 The Institute of Electrical and Electronics Engineers, Inc.
Published 2023 by John Wiley & Sons, Inc.

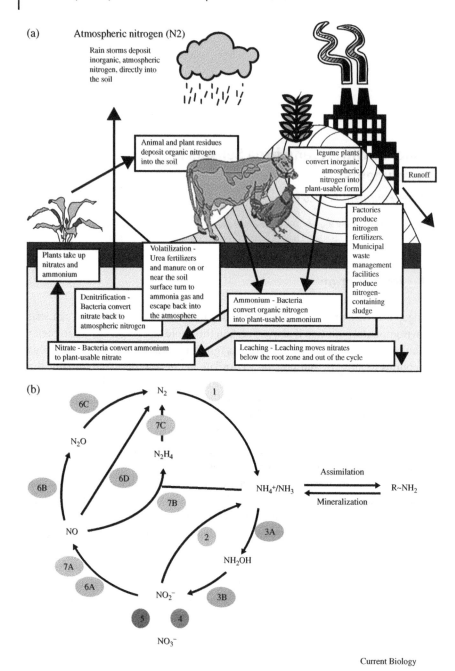

Current Biology

Figure 3.1 (a) Major processes of the nitrogen cycle, including nitrogen fixation, nitrification, and denitrification. *Source:* Copyright University of Missouri. (b) Detailed scheme of the processes in the nitrogen cycle. *Source:* From Ref. [1]/with permission of Elsevier. 1 – Nitrogen fixation; 2 – Dissimilatory nitrite reduction to ammonium; 3 – Nitritation; 4 – Nitratation; 5 – Reduction of nitrate to nitrite; 6 – Denitrification; 7 – Annamox.

In mineralization, the decomposition of organic nitrogen such as those from fertilizers or leaves into ammonia is performed by bacteria and fungi [2].

These nitrogenous compounds can have serious effects in the environment. In aquaculture, nitrogen is an important factor in fish growth. For example, fin fish, commonly grown in fish farms, are fed a protein-rich diet and, consequently, excrete dissolved nitrogen compounds (in particular, ammonia) [3]. Although this is a natural process, intensive fish farm may lead to an increased concentration of dissolved nitrogen compounds. This becomes an issue as some of these compounds, such as dissolved ammonia and nitrite, when not excreted, can lead to toxic buildup in tissues and blood in aquatic life [4]. Another problem is that a nitrogen-rich environment can lead to eutrophication, decreasing the amount of dissolved oxygen in water bodies which directly affects the aerobic organisms [5, 6]. Similar concerns arise from agriculture runoff into water bodies. As nitrogen plays a crucial role in plant lifecycle, nitrogen-rich fertilizers are widely used, sometimes promoting overfertilization [7]. If this nitrogen excess reaches the water bodies, it can disturb the aquatic life. Therefore, by continuously monitoring dissolved inorganic nitrogen (DIN), which consists of nitrate (NO_3^-), nitrite (NO_2^-), and ammonia (NH_3), and urea, a common substance in nitrogenous fertilizer, a safer aquatic environment can be achieved.

3.2 Detection Techniques for Ammonia, Nitrate, and Urea in Water

Different techniques have been implemented for the detection of ammonia, nitrate, and urea in water, with several materials and reagents used for the selective detection of these compounds. Although there are techniques that detect the total dissolved nitrogen (TDN), the dissolved organic nitrogen (DON), and DIN, this chapter will focus on the individual detection of these nitrogenous compounds. The main detection techniques can be divided into optical and electrochemical methods, with spectrophotometry, fluorometry, and electrochemical sensors being the most researched.

3.2.1 Spectrophotometry

In the spectrophotometry method, an electromagnetic wave from a light source irradiates the analyte and the light amount absorbed (or transmitted) is then measured. A spectrophotometer usually is comprised of a stable light source, a sample container, a radiation detector, a signal processor, and a readout unit [8]. A schematic representation of an UV–VIS spectrophotometer is given in Figure 3.2. The sample concentration is proportional to the decrease in radiation power, and it is usually given by the Beer–Lambert law:

$$A = \varepsilon C \times l \tag{3.1}$$

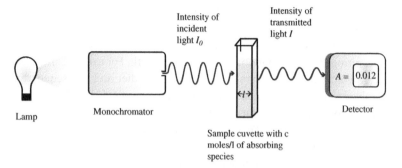

Figure 3.2 Schematic representation of an UV–Vis absorption measurement of a given sample. A light source irradiates the sample cell with an intensity I_0 and a transmitted light I is measured by the detector. *Source:* From Ref. [8]/with permission of Elsevier.

where A is the absorbance, ε is the molar absorption coefficient, C is the sample concentration, and l is width of the sample. ε is dependent on the incident wavelength radiation, the temperature, and the solvent employed. From the Beer–Lambert law, it is observed that the molar absorptivity is constant, and the absorbance is proportional to the concentration given the same parameters.

3.2.2 Fluorometry

Another method widely employed is fluorescence spectrophotometry. It is quite similar to the spectrophotometric technique; however, it instead analyses the fluorescence light emitted by the sample after excitation by a light source (Figure. 3.3). The fluorescence is detected with photomultiplier tubes and the spectrum can be analyzed by a signal processor and readout unit [9]. As a general rule, the same fluorescence emission spectra are observed irrespective of the excitation wavelength. Unlike in spectrophotometric detection, no sample pretreatment is needed, although the use of reagents is still necessary. Fluorometry allows for high sensitivity due to the nature of its detection. The technique can be automated with a high sample throughput and high precision by employing it together with an autonomous batch analyzer (ABA) or a flow injection analysis (FIA) [10].

3.2.3 Electrochemical Sensors

Electrochemical sensors present the possibility of miniaturized, low-cost, portable sensors that do not require a variety of reagents and specialized personnel to operate it. The sensing material can be varied to improve the sensing range, cross-sensitivity, and stability. Electrochemical sensors consist of a receptor, a transducer, and a signal processor (Figure 3.4a). These sensors are usually classified by the transduction type, which consists of conductometric, impedometric, potentiometric, amperometric, and voltammetric transduction modes

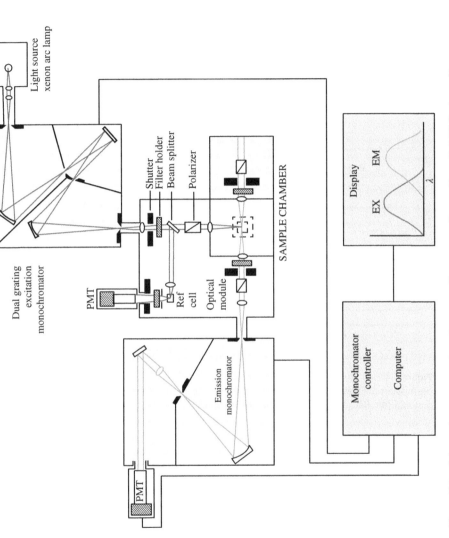

Figure 3.3 Schematic representation of a spectrofluorometer. *Source:* From Ref. [9]/Springer Nature.

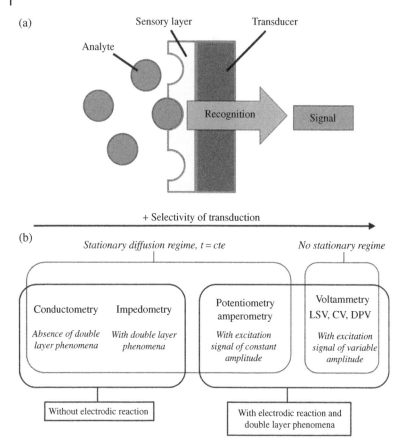

Figure 3.4 (a) Electrochemical sensors device scheme. *Source:* From Ref. [11]/with permission of Elsevier. (b) Transduction classification according to IUPAC classification. *Source:* From Ref. [12]/with permission of Elsevier.

(Figure 3.4b) [12]. Conductometric and impedometric sensors use the change in conductivity that arises when the sensing material interacts with the analyte. While conductometric sensors are based on the migration of ions of opposite charge in the presence of an electric field between two electrodes, impedometric sensors give more characterization details on the interaction of the sensing material with the analyte solution [12]. The latter can detect the double-layer formation as it is sensitive to surface phenomena and change of bulk properties [13, 14].

Potentiometric and amperometric sensors choose a constant amplitude signal to measure the correlate changing concentration with the output [15, 16]. For potentiometric sensors, no current passes through the system and the potential difference between the working electrode and the reference electrode is measured using a technique called open circuit potential (OCP). For amperometric sensors, a constant voltage is selected where the sensor will have a linear current response to a changing concentration of analyte [12]. Voltammetry sensors implement a voltage sweep that shows redox peaks of the analyte or substrate at a potential intrinsic to the material. Different voltammetry techniques exist, and their suitability depends on the intended outcome of the measurement. Some techniques include cyclic voltammetry (CV), linear sweep voltammetry (LSV), differential pulse voltammetry (DPV), and square wave voltammetry (SWV) [17, 18].

Different sensor designs are usually employed for these electrochemical techniques. Figure 3.5 exemplifies the different sensor designs and the output

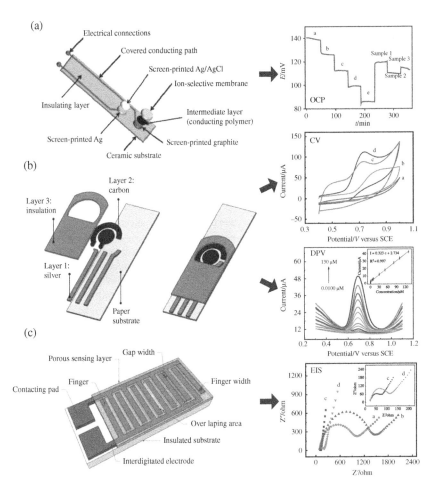

Figure 3.5 Common sensor designs used for the different electrochemical techniques and an illustration of the graphical output for standard techniques employed in electrochemical analyte detection. (a) Potentiometric sensor design and the signal output for OCP. *Source:* From Ref. [19]. (b) Voltammetric three-electrode design. *Source:* From Ref. [20]/Springer Nature. CV and DPV output. *Source:* From Ref. [21]/Springer Nature. (c) Impedimetric interdigitated sensor. *Source:* From Ref. [22]/MDPI/CC BY-3.0. The EIS output. *Source:* From Ref. [21]/Springer Nature.

technique. Potentiometric sensors (Figure 3.5a) usually have a sensing electrode covered with a selective membrane and a reference electrode to measure the electromotive force (emf) over time [19, 23]. In this way, a correlation can be done between different ion concentrations and the emf to test for pH or copper ions, for example. Differently than potentiometric sensors, where no current passes through the electrodes, voltammetric sensors require a counter electrode for current collection. In this way, they normally have three electrodes: the working electrode, the reference electrode, and the counter electrode (Figure 3.5b). The techniques used rely on faradaic reactions, where electron transfer between the sensitive material in the working electrode interacts with the analyte and produces a selective redox peak [21, 24]. Lastly, for impedimetric/conductimetric sensors, the spiral pattern or an interdigitated electrode (IDE) is normally used as it increases the effective surface-to-volume ratio of the electrode, leading to a higher conductivity and ion diffusion, which influences the signal intensity [25].

An overview of common and upcoming methods for dissolved NH_3, NO_3^-, NO_2^-, and urea is given in Table 3.1. Each technique is further discussed in this chapter.

Table 3.1 Overview of common and upcoming detection techniques for ammonia, nitrate, and urea.

Analyte	Method	Detection technique	Comments
NH_3/NH_4^+	Nessler	Spectrophotometer	Classical method that uses toxic reagents. Has been dropped as a standard method.
	Titration	Spectrophotometer	Needs preliminary distillation. Prone to high relative error for low NH_3-N concentrations.
	Indophenol	Spectrophotometer	Good linearity from 0.02 to 2 mg NH_3-N/l. Sample needs to be pretreated. Uses toxic reagents.
	OPA	Spectrofluorometer	Suitable for low concentrations of ammonia. High reagent consumption.
	ISE	Electrochemistry	Commercially available for continuous monitoring. Interference with K^+ ions.
	Metal oxides	Electrochemistry	Possibility of system miniaturization. Can present problem with selectivity and stability.

Table 3.1 (Continued)

Analyte	Method	Detection technique	Comments
NO_3^-/NO_2^-	Griess assay	Spectrophotometer	Most frequently used method. Can suffer interference from Cu^{2+}, Fe^{3+}, S^{2-}, and I^- ions.
	Nitrosation	Spectrophotometer	Low standard deviation. Susceptible to Fe^+ influence.
	Catalysis	Spectrophotometer	Needs controlled reaction conditions. Interference from several ions.
	Carbon-based materials	Electrochemistry	Possibility of miniaturization. Detection limit in μM or mM range. Can suffer interference in complex samples.
Urea	Diacetylmonoxime assay	Spectrophotometer	Low detection limit. Slow reaction time.
	Enzymatic	Electrochemistry	Fast response. Enzyme is inhibited in certain conditions. Long-term stability is an issue.

3.3 Ammonia

3.3.1 Ammonia in Aquatic Environments

NH_3 is a colorless gas that can readily dissolve in water due to its trigonal pyramidal shape and ability to form hydrogen bonds. It ionizes into ammonium ion (NH_4^+) in a relationship governed by temperature, pH values, and concentration of dissolved salts. The chemical ionization equation is shown in Eq. (3.2) [26], while Eq. (3.3) shows the acid dissociation constant, K_a, for the NH_4^+ ion [27].

$$NH_3 + H_2O \rightleftarrows NH_4^+ + OH^- \tag{3.2}$$

$$K_a = \frac{\left[NH_3\right]\left[H^+\right]}{\left[NH_4^+\right]} \tag{3.3}$$

The pK_a is temperature dependent and can be calculated using the following revised equation, with temperature given in Celsius [28]:

$$pK_a = 0.09018 + \left(\frac{2729.92}{273.2 + T}\right) \tag{3.4}$$

And finally, the unionized fraction of NH_3 can be calculated using the following equation developed by Emerson et al. using the pK_a relation described in Eq. (3.4) [28]:

$$\text{Unionized NH}_3 = \frac{100}{\left(1 + 10^{(pK_a - pH)}\right)} \tag{3.5}$$

Table 3.2 expresses revised values of different unionized NH_3 percentages according to temperature and pH using Eqs. (3.4) and (3.5). At pH 7.4 and 20 °C, less than 1% of NH_3 is present while at pH 9 and 20 °C, this rises to almost 30%. Dissolved NH_3 is a weak base while the NH_4^+ ion is considered a weak acid, which explains why in basic pH a higher percentage of NH_3 is available. Salinity also affects the degree of ionization of ammonia and can be especially important in marine environments. While the percentage of unionized NH_3 increases with increasing temperature and pH, the opposite occurs with salinity. It also does not affect the percentage drastically. Bower made a comprehensive table of NH_3 percentage from pH from 7.5 to 8.5 and temperatures from 0 to 25 °C in seawater with different salinities [29]. The author analyzed salinity percentages from 18–22% up to 32–40% and found that at 18–22%, the NH_3 percentage is similar to that found in freshwater; however, at 32–40%, common in seawater culture, there is up to a fifth less amount of NH_3 [29]. This dissociation is important because unionized NH_3 has been shown to be toxic to aquatic life as it is lipid soluble, while its ionized counterpart (NH_4^+) is mostly harmless due to its low permeability in plasma membranes [29].

The threshold concentration of NH_4^+ in water is dependent on the water type and application. Furthermore, different countries/regions have different policies, and the unit used can differ. Some of these regulations are summarized as follows:

- Drinking water:
  - 0.5 mg/l as NH_4^+ (EU) [30]
  - 30 mg/l at pH 7 and 20 °C, lifetime threshold (USA) [31]

- Freshwater:
  - 0.2–2.5 mg/l as total ammonia nitrogen (TAN) (England and Wales) [32]
  - Acute concentration of 17 mg/l as TAN and chronic concentration of 1.9 mg/l as TAN (USA) [5]

- Groundwater:
  - 0.28 mg/l as total ammonia (England and Wales) [32]

- Seawater:
  - Long-term concentration of 21 μg/l as N-NH_3 (England and Wales) [32]
  - Acute concentration of 0.035 mg/l as dissolved NH_3 and chronic concentration of 0.233 mg/l as dissolved NH_3 (USA) [33]

Table 3.2 Percentage of unionized ammonia in aqueous solution for pH 6–9 and temperature 0–30 °C.

Temp (°C)	pKa	pH 6	6.2	6.4	6.6	6.8	7	7.2	7.4	7.6	7.8	8	8.2	8.4	8.6	8.8	9
0	10.083	0.008	0.013	0.021	0.033	0.052	0.083	0.131	0.207	0.328	0.519	0.820	1.294	2.035	3.187	4.958	7.637
2	10.010	0.010	0.015	0.025	0.039	0.062	0.098	0.155	0.245	0.388	0.613	0.968	1.525	2.396	3.745	5.809	8.903
4	9.938	0.012	0.018	0.029	0.046	0.073	0.115	0.182	0.289	0.457	0.722	1.139	1.794	2.813	4.387	6.779	10.334
6	9.868	0.014	0.021	0.034	0.054	0.085	0.135	0.214	0.339	0.537	0.848	1.338	2.103	3.293	5.121	7.880	11.939
8	9.798	0.016	0.025	0.040	0.063	0.100	0.159	0.252	0.398	0.629	0.994	1.566	2.460	3.843	5.957	9.124	13.727
10	9.730	0.019	0.030	0.047	0.074	0.117	0.186	0.294	0.466	0.736	1.162	1.829	2.868	4.471	6.906	10.520	15.706
12	9.662	0.022	0.034	0.055	0.087	0.137	0.217	0.344	0.544	0.859	1.355	2.131	3.335	5.185	7.976	12.077	17.878
14	9.595	0.025	0.040	0.064	0.101	0.160	0.253	0.401	0.634	1.000	1.576	2.475	3.867	5.994	9.177	13.804	20.244
16	9.530	0.030	0.047	0.074	0.117	0.186	0.294	0.466	0.736	1.162	1.829	2.868	4.471	6.905	10.519	15.706	22.798
18	9.465	0.034	0.054	0.086	0.136	0.216	0.342	0.540	0.854	1.347	2.117	3.315	5.154	7.929	12.010	17.785	25.531
20	9.401	0.040	0.063	0.100	0.158	0.250	0.396	0.626	0.988	1.557	2.445	3.821	5.923	9.073	13.655	20.041	28.430
22	9.338	0.046	0.073	0.115	0.183	0.289	0.457	0.723	1.141	1.796	2.817	4.392	6.786	10.344	15.459	22.470	31.475
24	9.276	0.053	0.084	0.133	0.211	0.333	0.527	0.833	1.314	2.067	3.236	5.034	7.750	11.751	17.426	25.064	34.645
26	9.214	0.061	0.097	0.153	0.242	0.384	0.607	0.958	1.511	2.373	3.710	5.755	8.823	13.298	19.555	27.811	37.911
28	9.154	0.070	0.111	0.176	0.279	0.441	0.697	1.100	1.733	2.719	4.241	6.560	10.012	14.990	21.843	30.697	41.246
30	9.094	0.080	0.128	0.202	0.320	0.506	0.799	1.261	1.983	3.108	4.837	7.455	11.322	16.830	24.284	33.700	44.617

Source: Adapted from Ref. [28].

3.3.2 Ammonia Detection Techniques

Ammonia in water is present as NH_4^+ ion and dissolved NH_3 and usually the NH_4^+ ion is determined and the NH_3 concentration is calculated by Eq. (3.5). Some methods have been employed for ammonia detection, including spectrophotometry, fluorometry, and electrochemical sensors (Figure 3.6). Spectrophotometric analysis has been widely utilized for the detection of ammonia in freshwater and seawater, being advised by environmental agencies such as the U.S. EPA as the standard method [36–38]. Nesslerization has been considered a classical method to determine ammonia for more than a century, although it has been dropped as a standard method [39]. It consists of a pale-yellow alkaline mercury solution (K_2HgI_4) that becomes deep yellow in the presence of ammonia. The problem arises because of the use of mercury, which is difficult to dispose of and highly toxic [39]. A standard method is the titrimetric method, where a standard solution (titrant) is prepared and added to the analyte until a change in color occurs. To determine NH_3-N, a sulfuric acid (H_2SO_4) titrant solution is used until a pale lavender color is observed [39]. This method is usually employed for samples with concentration >5 mg NH_3-N/l as the error escalates in lower concentrations and volatile amines (hydrazine and amines) can influence the result [39]. Another method is the phenate or indophenol, where the reaction between ammonia, phenol, and hypochlorite, catalyzed by sodium nitroprusside, produces an intense blue compound, the indophenol [40]. It can be also automated and produce a linear response from 0.02 to 2.0 mg NH_3-N/l [39]. These colorimetric methods require pretreatment of the samples for an accurate measurement, which ranges from adding dechlorinating agents to distilling the water [39]. Simpler colorimetric methods that require no user calibration can also be employed to detect dissolved NH_3. For example, Figure 3.6a provides an example of a disposable paper-based sensor that uses different ammonia-sensitive dyes [40]. The change in dye color can then be correlated to the presence of dissolved NH_3.

Fluorometry is a popular technique for detecting ammonia. Its low limit of detection (LOD), up to the nanomolar range, makes this technique suitable for water bodies with low concentration of ammonia, such as seawater [38]. O-phthalaldehyde (OPA) is the most used chemical for fluorometric detection of ammonia, being first developed in 1971 for amino acid detection [41]. The reaction has since been improved by some of the reagents and technologies used to minimize interference with other amines [42]. An important modification to the original method was substituting mercaptoethanol with sulfite (Figure 3.6b), which provided a more selective and sensitive method for ammonium over amino acids [43]. Different modifications to the OPA method have since been developed, including addition of methoxy groups to the benzene ring of the OPA molecule and changing the buffer solutions [38].

The selection of the sensitive material is essential for the fabrication of a successful electrochemical sensor as it influences the sensor selectivity, sensitivity, and stability. For ammonia detection, these materials include metal oxides,

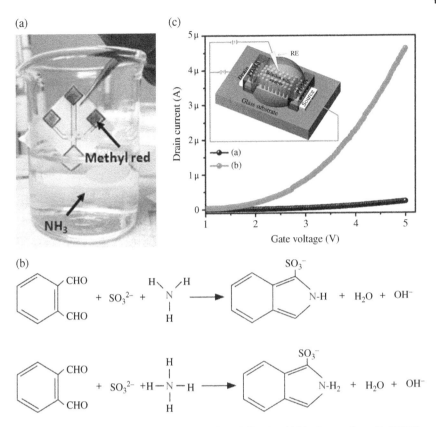

Figure 3.6 (a) Device for colorimetric sensing of dissolved NH_3. *Source:* From Ref. [35]/ Elsevier. (b) Fluorometric detection of ammonia based on OPA-sulfite. *Source:* From Ref. [34]. (c) ZnO-based ion-sensitive field-effect transistor for the determination of dissolved NH_3. *Source:* From Ref. [26]/Royal Society of Chemistry.

carbon nanotubes, metals, and ion-selective membranes, each presenting some advantages and disadvantages [34, 47, 48]. Some examples found in literature include ion-sensitive field-effect transistor (ISFET) ZnO-based ammonia sensors (Figure 3.6c), an amperometric dodecylbenzene sulfonate (DBSA)-doped poly-aniline nanoparticles (nanoPANI)-based sensor, and ionophore-based potentio-metric sensors [16, 26, 45, 47]. The sensitivity range is usually in the micro- or millimolar range for ammonia detection, with commercial sensors usually employing ion-sensitive electrodes (ISEs) [47]. Ammonia electrochemical sensors can also be developed in flexible substrates, and research is ongoing on developing biodegradable disposable ammonia sensors. A summary of the NH_4^+ sensors mentioned here can be seen in Table 3.3.

Table 3.3 A summary of different sensors used for the detection of NH_4^+/dissolved NH_3.

Detection technique	Method	Sensing material	Detection range	LOD	References
Spectrophotometry	Nessler	K_2HgI_4 solution	0.05–1 mg/l/NH_3-N/l	—	[44]
Spectrophotometry	Titrimetric	H_2SO_4	5–100 mg/l/NH_3-N/l	—	[39]
Spectrophotometry	Indophenol	Phenol and hypochlorite	0.02–2.0 mg NH_3-N/l	—	[39]
Spectrophotometry	Fluorometric	O-phthalaldehyde	0.2–60 µM	35 nM	[38]
Spectrophotometry	Colorimetric	Phenolphthalein	17–1700 ppm	17 ppm	[35]
Electrochemistry	Amperometric	DBSA-doped nanoPANI	0.02–10 mM	3.17 µM	[16]
Electrochemistry	ISFET	ZnO nanorods	10 nM to 2.5 mM	70 nM	[26]
Electrochemistry	I–V measurement	ZnO nanopencils	0.05–500 µM	5 nM	[45]
Electrochemistry	Potentiometric	Solid-state nonactin cocktail	7.9 µg–1600 mg/N-NH_4^+/l	7.9 µg/N-NH_4^+/l	[46]
Electrochemistry	Potentiometric	Gas-permeable membrane	0.03–1400 mg/N-NH_3/l	—	[39]

3.4 Nitrate

3.4.1 Nitrate in Aquatic Environments

The World Health Organization has a detailed report on NO_3^- and NO_2^- in drinking water [49]. The NO_3^- ion is the stable form of nitrogen for oxygenated systems and is chemically inert. When reduced by microbes, it produces the NO_2^- ion, a more reactive compound due to its unstable oxidation state, which can then be further used in the nitrogen cycle [49]. This enzymatic process is dependent on pH and temperature. Increased levels of nitrate have been detected due to the expansion of nitrogenous fertilizers, disposal of wastes from animal farm, and changes in land which directly and indirectly impact the quality of water bodies [50]. This is alarming as elevated levels of nitrate can contribute to eutrophication in freshwater and marine ecosystems and nitrite has been found to be toxic to humans [51]. In humans, the toxicity of NO_2^- occurs in the form of methaemoglobinaemia which can cause cyanosis and asphyxia [49, 51]. The reactive NO_2^- ion can interfere with the oxygen transport system, causing irreversible conversion of hemoglobin to methemoglobin in the blood stream. This is particularly problematic for infants and pregnant women [52]. The NO_2^- ion can also form carcinogenic *N*-nitrosamines by reacting with secondary amines and amides in the stomach [52]. As it is considered a hazardous chemical above a certain threshold, many countries have imposed legislations on the use of permitted levels of nitrates and nitrites making them important ions to monitor.

NO_3^- is usually considered toxic to humans once it has been oxidized to NO_2^- [49]. As with ammonium, the nitrate/nitrite environmental threshold changes depending on the country. In some cases, NO_3^-/NO_2^- and NH_4^+ are considered as part of the DIN threshold, but these were not included in this chapter. Following are some of the threshold concentrations for the EU, United States, and England/Wales:

- 50 mg/l for NO_3^- and 0.50 mg/l for NO_2^- for drinking water in the EU. The condition of $[NO_3^-]/50 + [NO_2^-] \leq 1$ in mg/l must be always ensured. For water treatment works, the threshold for NO_2^- is 0.1 mg/l.
- 10 mg/l for NO_3^- and 1 mg/l for NO_2^- for drinking water in the United States [53].
- 37.5 mg/l for NO_3^- for groundwater in England and Wales [32].

3.4.2 Nitrate Detection Techniques

Spectrometric techniques are by far the most employed method for the detection of NO_3^- and NO_2^- ions. It presents the opportunity of low detection limits in the range of 0.01–1 mg/l, for example, by utilizing a continuous-flow spectrometric method that can determine both or each ion separately [49]. Nitrate absorbs UV light at 210 nm making it feasible for UV spectroscopy determination. However, components such as chlorine, nitrite, iron(III), and organic matter absorb in the same region, which can interfere with the experiment [51]. Over the years, modifications to the technique have been made for a more selective and sensitive method. Some of the spectrophotometric methods used are based on the Griess assay, nitrosation, and catalytic processes [52]. The Griess assay involves a diazo-coupling procedure, where a nitrite-acidified aromatic amine suffers diazotization by coupling reaction (Figure 3.7a) [51]. The reaction produces a colored azo chromophore that can be detected at 500–600 nm. The process has been modified in many aspects, including assay conditions, target amines (e.g. sulphanilic acid, nitroaniline, and *p*-aminoacetophenone), coupling agents (phenol, 1-naphthol, 1-naphthol-4-sulfonate, 1-amino naphthalene, and 1,3-diaminobenzene), among others [51]. Although this method detects NO_2^-, enzymatic reduction of NO_2^- to NO_3^- can be employed before the diazotization process to detect nitrates.

The nitrosation reaction is based on the reaction of several indicating species (e.g. barbituric acid, phloroglucinol [1,3,5-trihydroxybenzene]) that react with nitrite to form colored products (e.g. violuric acid) but suffers from interference

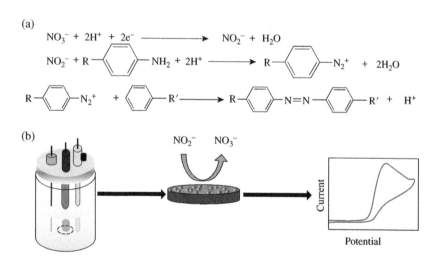

Figure 3.7 (a) The Griess assay method. *Source:* Singh et al. [51]. (b) Electrochemical detection of nitrite. *Source:* Li et al. [54].

from other metal ions [52]. In the catalytic-based spectrophotometric methods, the catalytic effect of nitrite on the oxidation of organic dyes or other indicating species is analyzed. For example, substituted phenothiazine derivatives, triaryl-methane acidic dyes, azo dyes, quinone imide dyes, and brilliant cresyl blue have all been reported as indicating species or organic dyes [52]. However, the determination of nitrite by catalytic-spectrophotometric methods has several drawbacks, including the need for controlled reaction conditions, such as acidity, temperature, and reagent dosage, and interference from ions as Fe^{2+}, Fe^{3+}, Ag^+, SO_3^{2-}, Br^-, and I^- [52]. Recently, electrochemical detection of nitrates has been gaining attention (Figure 3.7b). Carbon-based materials such as graphene oxide, carbon-nanotubes, and carbon black have been employed as the sensitive material, but other metal nanostructures have also been used [24, 54]. Many of the electrochemical sensors developed are based on glassy-carbon electrodes (GCE), which is usually employed in a laboratory setting. To bring these sensors in field, different fabrication techniques, such as screen-printing or cleanroom device fabrication, need to be employed. Other methods utilized for the determination of nitrates and nitrites include the electrochemical, chromatographic, capillary electrophoresis, and electrochemiluminescent methods [52]. Some of the techniques mentioned here are summarized in Table 3.4.

3.5 Urea

3.5.1 Urea in Aquatic Environment

Urea is an important product in the aquatic nitrogen cycle, originating from the metabolic excretion of invertebrates and fish, bacterial decomposition of nitrogenous compounds, and terrestrial run-off from fertilizers [64]. Furthermore, 20–50% of the total nitrogen utilized by phytoplankton is derived from urea, with higher concentrations of dissolved urea in coastal and estuarine environments [64]. Urea is a major element of nitrogen production, with 166 Mt produced in 2014, representing 55% of the total nitrogen output [65]. The average annual growth rate for urea production is around 3.6% and over 80% of the production is used to produce fertilizers [65, 66]. In the environment, urea is usually converted to ammonia by the urease enzyme. When this occurs close to the surface, the ammonia is converted into gas ammonia, escaping to the atmosphere and contributing to air pollution. Although urea is considered nontoxic, ecosystems exposed to indirect, long-term high levels of urea can suffer from

Table 3.4 A summary of different sensors used for the detection of NO_3^-/NO_2^- ions.

Ion	Detection technique	Method	Sensing material	Detection range	LOD	References
NO_2^-	Spectrophotometry	Griess assay	Sulphanilamide +1-naphthol-4-sulfonate	2–40 ng/l	1.4 ng/l	[55]
NO_3^-	Spectrophotometry	Griess assay	Sulphanilamide +1-naphthol-4-sulfonate	1.5–30 ng/l	1.1 ng/l	[55]
NO_2^-	Spectrophotometry	Nitrosation reaction	Barbituric acid	0–3.22 ppm	1.66 µg/l	[56]
NO_2^-	Spectrophotometry	Nitrosation reaction	Phloroglucinol (1,3,5-trihydroxybenzene)	0.03–0.30 µg NO_2^--N/ml	2.9 ng NO_2^--N/ml	[57]
NO_3^-	Spectrophotometry	Nitrosation reaction	Phloroglucinol (1,3,5-trihydroxybenzene)	0.1–1.0 µg NO_3^--N/ml	2.3 ng NO_3^--N/ml	[57]
NO_2^-	Spectrophotometry	Catalytic	Perphenazine	0–4.5 ng/ml	0.07 ng/ml	[58]
NO_2^-	Spectrophotometry	Catalytic	Brilliant cresyl blue	3.7 nM to ~65 µM	1.76 nM	[59]
NO_2^-	Electrochemistry	Amperometric	Au–Fe(III) nanoparticles/GCE	0.2–150 µM	0.1 µM	[60]
NO_2^-	Electrochemistry	Voltammetric	Pt–Fe(III) nanoparticles/GCE	1.1 µM to 11 mM	0.47 µM	[61]
NO_2^-	Electrochemistry	Voltammetric	Reduced graphene oxide/screen-printed electrode	20–1000 µM	0.83	[62]
NO_3^-	Electrochemistry	Voltammetric	Cu nanowires	10–1500 µ	10 µM	[63]

eutrophication, groundwater pollution, and soil acidification [64]. Therefore, monitoring urea can contribute to a safe environment by detecting early-warning signs of nitrogen pollution. Currently, urea does not need to be monitored in water environments.

3.5.2 Urea Detection Techniques

Urea detection is dependent on the type of application, as urea is important in a wide range of fields. The detection is often classified as direct or indirect [64]. Direct detection is defined as those where the urea is detected without prior degradation, whereas the indirect process includes the enzymatic degradation of urea prior to detection. In natural waters, the detection of DON compounds remains a challenge, as most methods rely on the total TDN detection and then estimating the DON concentration by subtracting the DIN concentration [68]. Urea is part of the DON and to selectively determine urea, two main methods are used. The first one is a colorimetric method, where the complexation of urea with diacetyl monoxime results in an imidazole that interacts with thiosemicarbazide to form a red complex [68]. The second method is an enzymatic approach, where the hydrolysis of urea by the urease enzyme releases NH_4^+ which is then determined and correlated to the urea concentration (Figure 3.8). The enzymatic method tends to underestimate urea concentrations because the enzyme is inhibited in certain conditions [68]. The enzymatic approach has since become quite popular, and ISFET biosensors are being developed to increase the sensitivity and stability of the method [11]. Nonenzymatic, catalytic approaches are also being researched for long-term monitoring [69–71]. A summary of the sensors mentioned here can be seen in Table 3.5.

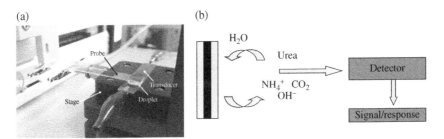

Figure 3.8 (a) Ionovoltaic urea sensors. *Source:* From Ref. [67]/Elsevier. (b) Enzymatic biosensor using urease. *Source:* From Ref. [11]/with permission of Elsevier.

Table 3.5 A summary of different sensors used for the detection of urea.

Detection technique	Method	Sensing material	Detection range	LOD	References
Spectrophotometry	Colorimetric	Imidazolone + thiosemicarbizide	0–5 μM	0.14 μM	[72]
Electrochemistry	Amperometric	Urease-ZnO nanorods	1–20 mM	0.13 mM	[73]
Electrochemistry	Potentiometric	Urease-Poly(vinyl ferrocenium) matrix	50 μM to 100 mM	5 μM	[74]
Electrochemistry	Voltammetric	$NiCo_2O_4$ nanoneedles	0.01–5 mM	1 μM	[71]

3.6 Conclusion and Future Perspectives

In this chapter, the major processes in the nitrogen cycle and the toxicity of its dissolved inorganic and organic components were discussed. We identified three significant nitrogenous compounds: ammonia as dissolved NH_3 gas and NH_4^+ ion, the stable NO_3^- and its reactive counterpart NO_2^-, and urea, a main component of the DON. The most used techniques for the detection of DIN and DON were then briefly explained and each nitrogenous component was then introduced in its own section. The chapter covered the ammonia dissociation in water and some of the detection techniques, including spectrophotometry, fluorometry, and electrochemical sensors. The toxicity of nitrite was then explained as well as some of the common detection methods. Finally, sources of urea and detection techniques for dissolved urea in natural waters were elucidated.

Acknowledgment

This work was supported by the European Commission through the AQUASENSE (H2020-MSCA-ITN-2018-813680) project.

References

1 Stein, L.Y. and Klotz, M.G. (2016). The nitrogen cycle. *Curr. Biol.* 26 (3): R94–R98. https://doi.org/10.1016/j.cub.2015.12.021.

2 Timmer, B., Olthuis, W., and van den Berg, A. (2005). Ammonia sensors and their applications—a review. *Sensors Actuators B Chem.* 107 (2): 666–677.

3 Holmer, M., Black, K., Duarte, C.M. et al. (2007). *Aquaculture in the Ecosystem.* Springer Science & Business Media.

4 Li, T., Panther, J., Qiu, Y. et al. (2017). Gas-permeable membrane-based conductivity probe capable of in situ real-time monitoring of ammonia in aquatic environments. *Environ. Sci. Technol.* 51 (22): 13265–13273.

5 U.S. Environmental Protection Agency (2013). *Aquatic Life Ambient Water Quality Criteria for Ammonia – Freshwater.* USEPA.

6 World Health Organization (1996). *Guidelines for Drinking-Water Quality: Health Criteria and Other Supporting Information.* Geneva: World Health Organization.

7 Muñoz-Huerta, R.F., Guevara-Gonzalez, R.G., Contreras-Medina, L.M. et al. (2013). A review of methods for sensing the nitrogen status in plants: advantages, disadvantages and recent advances. *Sensors* 13 (8): 10823–10843.

8 Kafle, B.P. (ed.) (2020). Theory and instrumentation of absorption spectroscopy: UV–VIS spectrophotometry and colorimetry. In: *Chemical Analysis and Material*

Characterization by Spectrophotometry, 17–38. Elsevier https://doi.org/10.1016/B978-0-12-814866-2.00002-6.

9 Lakowicz, J.R. (ed.) (2006). Instrumentation for fluorescence spectroscopy. In: *Principles of Fluorescence Spectroscopy*, 27–61. Boston, MA: Springer https://doi.org/10.1007/978-0-387-46312-4_2.

10 Amornthammarong, N., Zhang, J.-Z., and Ortner, P.B. (2011). An autonomous batch analyzer for the determination of trace ammonium in natural waters using fluorometric detection. *Anal. Methods* 3 (7): 1501–1506. https://doi.org/10.1039/C1AY05095H.

11 Pundir, C.S., Jakhar, S., and Narwal, V. (2019). Determination of urea with special emphasis on biosensors: a review. *Biosens. Bioelectron.* 123: 36–50.

12 Blanco-López, M.C., Lobo-Castañón, M.J., Miranda-Ordieres, A.J., and Tuñón-Blanco, P. (2004). Electrochemical sensors based on molecularly imprinted polymers. *TrAC Trends Anal. Chem.* 23 (1): 36–48. https://doi.org/10.1016/S0165-9936(04)00102-5.

13 Manjakkal, L., Djurdjic, E., Cvejin, K. et al. (2015). Electrochemical impedance spectroscopic analysis of RuO_2 based thick film pH sensors. *Electrochim. Acta* 168: 246–255.

14 Mei, B.-A., Munteshari, O., Lau, J. et al. (2018). Physical interpretations of Nyquist plots for EDLC electrodes and devices. *J. Phys. Chem. C* 122 (1): 194–206.

15 Manjakkal, L., Dervin, S., and Dahiya, R. (2020). Flexible potentiometric pH sensors for wearable systems. *RSC Adv.* 10 (15): 8594–8617. https://doi.org/10.1039/D0RA00016G.

16 Crowley, K., O'Malley, E., Morrin, A. et al. (2008). An aqueous ammonia sensor based on an inkjet-printed polyaniline nanoparticle-modified electrode. *Analyst* 133 (3): 391–399. https://doi.org/10.1039/B716154A.

17 Franco, F.F., Manjakkal, L., Shakthivel, D., and Dahiya, R. (2019). ZnO based screen printed aqueous ammonia sensor for water quality monitoring. *Proceedings of the 2019 IEEE Sensors*, Montreal, QC (27–30 October 2019), 1–4. IEEE. https://doi.org/10.1109/SENSORS43011.2019.8956763.

18 Zen, J.-M., Yang, C.-C., and Kumar, A.S. (2002). Voltammetric behavior and trace determination of Pb^{2+} at a mercury-free screen-printed silver electrode. *Anal. Chim. Acta* 464 (2): 229–235. https://doi.org/10.1016/S0003-2670(02)00472-5.

19 Schwarz, J., Trommer, K., Gerlach, F., and Mertig, M. (2018). All-solid-state screen-printed sensors for potentiometric calcium(II) determinations in environmental samples. *Am. J. Anal. Chem.* 09 (03): https://doi.org/10.4236/ajac.2018.93010.

20 Smith, S., Madzivhandila, P., Ntuli, L. et al. (2019). Printed paper–based electrochemical sensors for low-cost point-of-need applications. *Electrocatalysis* 10 (4): https://doi.org/10.1007/s12678-019-0512-8.

21 Ma, X., Gao, F., Liu, G. et al. (2019). Sensitive determination of nitrite by using an electrode modified with hierarchical three-dimensional tungsten disulfide and reduced graphene oxide aerogel. *Microchim. Acta* 186 (5): https://doi.org/10.1007/s00604-019-3379-8.

22 Farahani, H., Wagiran, R., and Hamidon, M.N. (2014). Humidity sensors principle, mechanism, and fabrication technologies: a comprehensive review. *Sensors* 14 (5): https://doi.org/10.3390/s140507881.

23 Manjakkal, L., Szwagierczak, D., and Dahiya, R. (2020). Metal oxides based electrochemical pH sensors: current progress and future perspectives. *Prog. Mater. Sci.* 109: 100635. https://doi.org/10.1016/j.pmatsci.2019.100635.

24 Yilong, Z., Dean, Z., and Daoliang, L. (2015). Electrochemical and other methods for detection and determination of dissolved nitrite: a review. *Int. J. Electrochem. Sci.* 10: 1144–1168.

25 Singh, K.V., Whited, A.M., Ragineni, Y. et al. (2010). 3D nanogap interdigitated electrode array biosensors. *Anal. Bioanal. Chem.* 397 (4): 1493–1502. https://doi.org/10.1007/s00216-010-3682-z.

26 Ahmad, R., Tripathy, N., Khan, M.Y. et al. (2016). Ammonium ion detection in solution using vertically grown ZnO nanorod based field-effect transistor. *RSC Adv.* 6 (60): 54836–54840.

27 Körner, S., Das, S.K., Veenstra, S., and Vermaat, J.E. (2001). The effect of pH variation at the ammonium/ammonia equilibrium in wastewater and its toxicity to *Lemna gibba*. *Aquat. Bot.* 71 (1): 71–78.

28 Emerson, K., Russo, R.C., Lund, R.E., and Thurston, R.V. (1975). Aqueous ammonia equilibrium calculations: effect of pH and temperature. *J. Fish. Board Can.* 32 (12): 2379–2383.

29 Bower, C.E. and Bidwell, J.P. (1978). Ionization of ammonia in seawater: effects of temperature, pH, and salinity. *J. Fish. Board Can.* 35 (7): 1012–1016.

30 European Commission (1998). Council directive 98/83/EC of 3 November 1998 on the quality of water intended for human consumption. *Off. J. Eur. Communities* 330: 32–54.

31 U.S. Environmental Protection Agency (2018). Edition of the drinking water standards and health advisories tables. 2018 Edition of the Drinking Water Standards and Health Advisories Tables (EPA 822-F-18-001) (accessed 3 April 2023).

32 T. W. F. Directive (2015). The water framework directive (Standards and Classification) directions (England and Wales). The Water Framework Directive (Standards and Classification) Directions (England and Wales) 2015 (legislation. gov.uk) (accessed 3 April 2023).

33 U.S. Environmental Protection Agency (1989). *Ambient Water Quality Criteria for Ammonia (Saltwater)*. Washington, DC: U.S. Environmental Protection Agency.

34 Li, D., Xu, X., Li, Z. et al. (2020). Detection methods of ammonia nitrogen in water: a review. *TrAC Trends Anal. Chem.* 115890.

35 Chen, Y., Zilberman, Y., Mostafalu, P., and Sonkusale, S.R. (2015). Paper based platform for colorimetric sensing of dissolved NH_3 and CO_2. *Biosens. Bioelectron.* 67: 477–484. https://doi.org/10.1016/j.bios.2014.09.010.

36 Ma, J., Adornato, L., Byrne, R.H., and Yuan, D. (2014). Determination of nanomolar levels of nutrients in seawater. *TrAC Trends Anal. Chem.* 60: 1–15.

37 Xia, L.I.U., Qing, X.U., Yafei, G.U.O. et al. (2016). Ammonia nitrogen speciation analysis in aquatic environments. *Proceedings of the International Conference on Biological Engineering and Pharmacy 2016 (BEP 2016)* (December 2016). Atlantis Press.

38 Zhu, Y., Chen, J., Yuan, D. et al. (2019). Development of analytical methods for ammonium determination in seawater over the last two decades. *TrAC Trends Anal. Chem.* 119: 115627. https://doi.org/10.1016/j.trac.2019.115627.

39 Rice, E.W., Baird, R.B., Eaton, A.D., and Clesceri, L.S. (2012). *Standard Methods for the Examination of Water and Wastewater*, vol. 541. Washington, DC: American Public Health Association.

40 O'Dell, J.W. (1993). Determination of ammonia nitrogen by semi-automated colorimetry. EPA Method 350.1: Nitrogen, Ammonia (Colorimetric, Automated Phenate) (accessed 3 April 2023)

41 Roth, M. (1971). Fluorescence reaction for amino acids. *Anal. Chem.* 43 (7): 880–882. https://doi.org/10.1021/ac60302a020.

42 Hu, H., Liang, Y., Li, S. et al. (2014). A modified *o*-phthalaldehyde fluorometric analytical method for ultratrace ammonium in natural waters using EDTA-NaOH as buffer. *J. Anal. Methods Chem.* 2014: 728068. https://doi.org/10.1155/2014/728068.

43 Genfa, Z. and Dasgupta, P.K. (1989). Fluorometric measurement of aqueous ammonium ion in a flow injection system. *Anal. Chem.* 61 (5): 408–412.

44 U.S. Environmental Protection Agency (1983). *Methods for Chemical Analysis of Water and Wastes*, EPA/600/4-79/020. Cincinnati: U.S. Environmental Protection Agency.

45 Dar, G.N., Umar, A., Zaidi, S.A. et al. (2012). Ultra-high sensitive ammonia chemical sensor based on ZnO nanopencils. *Talanta* 89: 155–161.

46 Huang, Y., Wang, T., Xu, Z. et al. (2019). Real-time in situ monitoring of nitrogen dynamics in wastewater treatment processes using wireless, solid-state, and ion-selective membrane sensors. *Environ. Sci. Technol.* 53 (6): 3140–3148. https://doi.org/10.1021/acs.est.8b05928.

47 Cuartero, M., Colozza, N., Fernández-Pérez, B.M., and Crespo, G.A. (2020). Why ammonium detection is particularly challenging but insightful with ionophore-based potentiometric sensors – an overview of the progress in the last 20 years. *Analyst* 145 (9): 3188–3210. https://doi.org/10.1039/D0AN00327A.

48 Lin, K., Zhu, Y., Zhang, Y., and Lin, H. (2019). Determination of ammonia nitrogen in natural waters: recent advances and applications. *Trends Environ. Anal. Chem.* 24: e00073.

49 World Health Organization (2011). *Nitrite in Drinking-Water. Background Document for Development of WHO Guidelines for Drinking-Water Quality.* Geneva: World Health Organization (WHO/SDE/WSH/07.01/16/Rev/1).

50 Mahmud, M.A.P., Ejeian, F., Azadi, S. et al. (2020). Recent progress in sensing nitrate, nitrite, phosphate, and ammonium in aquatic environment. *Chemosphere* 259: 127492. https://doi.org/10.1016/j.chemosphere.2020.127492.

51 Singh, P., Singh, M.K., Beg, Y.R., and Nishad, G.R. (2019). A review on spectroscopic methods for determination of nitrite and nitrate in environmental samples. *Talanta* 191: 364–381.

52 Wang, Q.-H., Yu, L.-J., Liu, Y. et al. (2017). Methods for the detection and determination of nitrite and nitrate: a review. *Talanta* 165: 709–720.

53 Agardy, F.J. and Sullivan, P.J. (2003). Appendix a: summary tables of drinking water standards and health advisories. In: *Drinking Water Regulation and Health* (ed. F.W. Pontius), 583–619. Hoboken, NJ: Wiley https://doi.org/10.1002/0471721999.app1.

54 Li, D., Wang, T., Li, Z. et al. (2020). Application of graphene-based materials for detection of nitrate and nitrite in water – a review. *Sensors* 20 (1): https://doi.org/10.3390/s20010054.

55 Wang, G.F., Horita, K., and Satake, M. (1998). Simultaneous spectrophotometric determination of nitrate and nitrite in water and some vegetable samples by column preconcentration. *Microchem. J.* 58 (2): https://doi.org/10.1006/mchj.1997.1544.

56 Aydin, A., Ercan, Ö., and Taşcioğlu, S. (2005). A novel method for the spectrophotometric determination of nitrite in water. *Talanta* 66 (5): https://doi.org/10.1016/j.talanta.2005.01.024.

57 Burakham, R., Oshima, M., Grudpan, K., and Motomizu, S. (2004). Simple flow-injection system for the simultaneous determination of nitrite and nitrate in water samples. *Talanta* 64 (5): 1259–1265. https://doi.org/10.1016/j.talanta.2004.03.059.

58 Mubarak, A.T., Mohamed, A.A., Fawy, K.F., and Al-Shihry, A.S. (2007). A novel kinetic determination of nitrite based on the perphenazine-bromate redox reaction. *Microchim. Acta* 157 (1–2): https://doi.org/10.1007/s00604-006-0661-3.

59 Ensafi, A.A. and Amini, M. (2010). A highly selective optical sensor for catalytic determination of ultra-trace amounts of nitrite in water and foods based on brilliant cresyl blue as a sensing reagent. *Sensors Actuators B Chem.* 147 (1): https://doi.org/10.1016/j.snb.2010.03.014.

60 Liu, T.S., Kang, T.F., Lu, L.P. et al. (2009). Au–Fe(III) nanoparticle modified glassy carbon electrode for electrochemical nitrite sensor. *J. Electroanal. Chem.* 632 (1–2): https://doi.org/10.1016/j.jelechem.2009.04.023.

61 Wang, S., Yin, Y., and Lin, X. (2004). Cooperative effect of Pt nanoparticles and Fe(III) in the electrocatalytic oxidation of nitrite. *Electrochem. Commun.* 6 (3): https://doi.org/10.1016/j.elecom.2003.12.008.

62 Gholizadeh, A., Voiry, D., Weisel, C. et al. (2017). Toward point-of-care management of chronic respiratory conditions: electrochemical sensing of nitrite content in exhaled breath condensate using reduced graphene oxide. *Microsyst. Nanoeng.* 3: https://doi.org/10.1038/micronano.2017.22.

63 Patella, B., Russo, R.R., O'Riordan, A. et al. (2021). Copper nanowire array as highly selective electrochemical sensor of nitrate ions in water. *Talanta* 221: https://doi.org/10.1016/j.talanta.2020.121643.

64 Francis, P.S., Lewis, S.W., and Lim, K.F. (2002). Analytical methodology for the determination of urea: current practice and future trends. *TrAC Trends Anal. Chem.* 21 (5): 389–400. https://doi.org/10.1016/S0165-9936(02)00507-1.

65 Heffer, P. and Prud'homme, M. (2015). Fertilizer outlook 2015–2019. *Proceedings of the 83rd IFA Annual Conference*, Istanbul (Turkey), vol. 415.

66 Antonetti, E., Iaquaniello, G., Salladini, A. et al. (2017). Waste-to-chemicals for a circular economy: the case of urea production (waste-to-urea). *ChemSusChem* 10 (5): 912–920. https://doi.org/10.1002/cssc.201601555.

67 Yang, Y., Yoon, S.G., Shin, C.H. et al. (2019). Ionovoltaic urea sensor. *Nano Energy* 57: 195–201.

68 Berman, T. and Bronk, D.A. (2003). Dissolved organic nitrogen: a dynamic participant in aquatic ecosystems. *Aquat. Microb. Ecol.* 31 (3): 279–305.

69 Tong, Y., Chen, P., Zhang, M. et al. (2018). Oxygen vacancies confined in nickel molybdenum oxide porous nanosheets for promoted electrocatalytic urea oxidation. *ACS Catal.* 8 (1): 1–7.

70 Zhu, D., Guo, C., Liu, J. et al. (2017). Two-dimensional metal–organic frameworks with high oxidation states for efficient electrocatalytic urea oxidation. *Chem. Commun.* 53 (79): 10906–10909.

71 Amin, S., Tahira, A., Solangi, A. et al. (2019). A practical non-enzymatic urea sensor based on $NiCo_2O_4$ nanoneedles. *RSC Adv.* 9 (25): https://doi.org/10.1039/c9ra00909d.

72 Goeyens, L., Kindermans, N., Abu Yusuf, M., and Elskens, M. (1998). A room temperature procedure for the manual determination of urea in seawater. *Estuar. Coast. Shelf Sci.* 47 (4): https://doi.org/10.1006/ecss.1998.0357.

73 Palomera, N., Balaguera, M., Arya, S.K. et al. (2011). Zinc oxide nanorods modified indium tin oxide surface for amperometric urea biosensor. *J. Nanosci. Nanotechnol.* 11 (8): 6683–6689. https://doi.org/10.1166/jnn.2011.4248.

74 Kuralay, F., Özyörük, H., and Yildiz, A. (2005). Potentiometric enzyme electrode for urea determination using immobilized urease in poly(vinylferrocenium) film. *Sensors Actuators B Chem.* 109 (2): https://doi.org/10.1016/j.snb.2004.12.043.

4

Monitoring of Pesticides Presence in Aqueous Environment

Yuqing Yang, Pierre Lovera, and Alan O'Riordan

Nanotechnology Group, Tyndall National Institute, University College Cork, Cork, Ireland

4.1 Introduction: Background on Pesticides

Pesticides are routinely administrated in fields to protect crops and plant products from damage caused by insects, fungi, weeds, and other pests. They are required for the production of sufficient volumes of food to meet consumer demands at a reasonable price. The global use of pesticides was measured to be 4.11 million tons in 2017 [1]. In the EU, the sale of pesticides was estimated to be 360 000 tons per year from 2011 to 2018. As the world population is set to increase to 9.8 billion by 2050, the food production as well as the use of pesticides are bound to increase. Pesticides could go through the underground water from vegetable and fruit surface with the rain, or through the soil, then pollute water sources that human and other animals have access to, and so it is essential to monitor pesticides in water environments.

4.1.1 Types and Properties

There are five main groups of pesticides based on their chemical compositions: organochlorines, organophosphorus, carbamates, pyrethroids, and neonicotinoids. Figure 4.1 shows the chemical structures and examples of each type of pesticides. Among the five groups, organochlorides have long-term residual effect in the environment [2], organophosphates are biodegradable [3], carbamates can be easily degraded under natural environment with minimum environmental

Sensing Technologies for Real Time Monitoring of Water Quality, First Edition.
Libu Manjakkal, Leandro Lorenzelli, and Magnus Willander.
© 2023 The Institute of Electrical and Electronics Engineers, Inc.
Published 2023 by John Wiley & Sons, Inc.

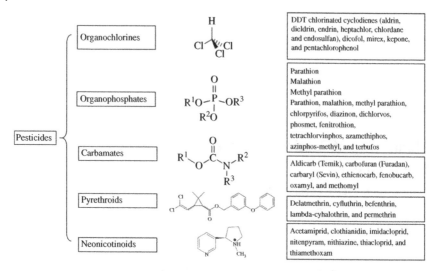

Organochlorines	H, Cl, Cl, Cl	DDT chlorinated cyclodienes (aldrin, dieldrin, endrin, heptachlor, chlordane and endosulfan), dicofol, mirex, kepone, and pentachlorophenol
Organophosphates	$R^1O\!-\!\overset{\displaystyle O}{\underset{\displaystyle R^2O}{P}}\!-\!OR^3$	Parathion, Malathion, Methyl parathion, Parathion, malathion, methyl parathion, chlorpyrifos, diazinon, dichlorvos, phosmet, fenitrothion, tetrachlorvinphos, azamethiphos, azinphos-methyl, and terbufos
Carbamates	$R^1\!-\!O\!-\!\overset{\displaystyle O}{C}\!-\!\underset{\displaystyle R^3}{N}\!-\!R^2$	Aldicarb (Temik), carbofuran (Furadan), carbaryl (Sevin), ethienocarb, fenobucarb, oxamyl, and methomyl
Pyrethroids		Delatmethrin, cyfluthrin, befenthrin, lambda-cyhalothrin, and permethrin
Neonicotinoids		Acetamiprid, clothianidin, imidacloprid, nitenpyram, nithiazine, thiacloprid, and thiamethoxam

Figure 4.1 Five categories of pesticides: organochlorines, organophosphates, carbamates, pyrethroids, and neonicotinoids.

pollution [4], pyrethroids are nonpersistent, and get broken easily on exposure to light [5, 6]. Neonicotinoids have been developed three decades ago because pests developed resistance against the other types of pesticides. Most neonicotinoids can dissolve in water and break down slowly in the environment [7, 8].

4.1.2 Risks

Pesticides can be found in liquid state that is sprayed directly on crops by tractor or planes as well as in directly pre-coated seeds. Because of their long half-life and high water solubility, organochlorides and neonicotinoids can enter the water bodies such as river or lakes, mainly through runoff and drainage in agricultural areas. Unfortunately, pesticides lead to the degradation of soil ecosystem, affecting the bio-environment by killing beneficial bacteria and microorganisms.

Pesticides also do harms to human and fauna health. Organochlorine (OC) pesticides were reported to increase the possibilities of hormone-related cancers including prostate, breast, stomach, and lung cancer [9]. OC pesticides may affect the thyroid system by gender-specific mechanisms [10]. It was reported as a potential risk factor for gallstone disease in humans [11]. Organophosphate neurotoxins are related to organophosphate-induced delayed neurotoxicity (OPIDN) in humans and susceptible species. Carbamate insecticides were also found to affect human melatonin receptors and inhibiting acetylcholinesterase [12]. Exposure to high-concentration pyrethrum may cause symptoms such as asthmatic headache, breathing, sneezing, nausea, nasal stuffiness, and loss of coordination. There is also evidence that neonicotinoids are contributing to the downfall of bee populations and that of other pollinators [13, 14].

4.1.3 Regulation and Legislation

Countries have established strict legislations to control the sale and use of pesticides. In the United States, the Federal Insecticide, Fungicide, and Rodenticide Act (FIFRA) governs the registration, manufacture, distribution, sale, and use of pesticides [15]. Similarly, in the European Union (EU), the rules for the use of pesticides are set in the Sustainable Use of Pesticides Directive (Directive 2009/128/EC). It identifies specific measures that member states are required to implement such as training, education, and information exchange, as well as controls on application equipment, storage, supply, and use of pesticides. It also promotes an Integrated Pest Management (IPM) aiming at lowering pesticide input, giving wherever possible preference to nonchemical methods such as organic farming. Other European legislations related to pesticides include:

- Regulation (EC) No.1107/2009 (concerning the placing of plant protection products on the market)
- Regulation (EC) No.1185/2009 (concerning statistics on plant protection products)
- Regulation (EC) No.396/2005 (on maximum residue levels of pesticides in or on food and feed)
- Regulation (EC) No.1272/2008 (on product classification, labeling, and packaging)
- Directive 2000/60/EC (establishing a framework for water policy)
- Directive 98/83/EC (on quality of water intended for human consumption)
- Directive 2006/118/EC (on the protection of groundwater)
- Directive 2004/35/CE (on environmental liability)

To ensure safety of environment and consumption, Maximum Residue Levels (MRL) have been fixed by the European Commission. These correspond to the highest level of pesticide residue that is legally tolerated, in or on food or feed, when pesticides are applied correctly. For foodstuff, these MRL are set out in Regulation No. 396/2005 and can be found in the MRL database on the Commission website. For water, the Water Directive Framework (WFD) is the overarching legislation. It is supported by more targeted legislation such as the Drinking Water Directive that states the limit for a single pesticide in drinking water should not exceed $0.1 \, \mu g/l$ and for the sum of all pesticides $0.5 \, \mu g/l$. These MRL are regularly reviewed and adjusted in light of new knowledge acquired. For instance, in April of 2018, member states of the EU agreed upon a total ban on neonicotinoid insecticide use, except within closed greenhouses [16].

4.1.4 Occurrence of Pesticide Exceedance

Despite the regulations and guidelines on how to administer these pesticides, trace amounts finding their way in water bodies have been reported in wide-scale surveys. The Irish EPA detected OC herbicides 2-methyl-4-chlorophenoxyacetic acid (MCPA) in 55% of rivers monitored in 2013–2018 (1292 samples) and 2,4-dichlorophenoxyacetic acid (2,4-D) in 29% of river monitored (1158 samples) [17]. In 2018, the drinking water standard for individual pesticides (0.1 µg/l) exceeded in Ireland in 42 public drinking water supplies, and MCPA was the chemical responsible for 75% of all failures detected [10].

4.2 Current Pesticides Detection Methods

Chromatography is a technique used to separate, identify, and quantify chemical components in mixed phases based on the difference in molecular affinity, molecular sizes, ionic strength, etc. Current pesticides residues detection methods include gas chromatography (GC), liquid chromatography (LG) [11], high-pressure liquid chromatography (HPLC), and mass chromatography (MC) [18]. Chromatography coupled with mass spectroscopy based technologies are nowadays the gold standard and main analytical method used by regulation agency to guarantee high selectivity and detectability [19]. They are able to analyze varieties of pesticides in a single sample, with repeatable and highly accurate results. Chromatography methods are highly sensitive but sample extractions are needed to minimize matrix effects (matrix effects are caused by co-eluting matrix components, leading to reduced or increased sensitivity of the analysis) [20]. Extra extraction steps are needed before the chromatography analysis [16].

Besides, chromatography-based technique requires bulky instrumentation that is not easily deployed for *in situ* analysis and the samples thus have to be sent to dedicated laboratories. It also suffers from long analysis time, high reagent consumption with associated high costs, and require complicated sample preparations or pre-concentration steps. Therefore, other detection technologies that have portable size and lower running costs are being investigated. Recently, devices based on electrochemical or optical (fluorescence and Raman spectroscopy) enabling capabilities have shown tremendous potential for *in situ* pesticide monitoring. In the following part, the advantages and disadvantages of electrochemical and optical spectrum detection method are discussed, along with the examples of pesticides detection in the summary tables. Different detection methods are shown in Figure 4.2.

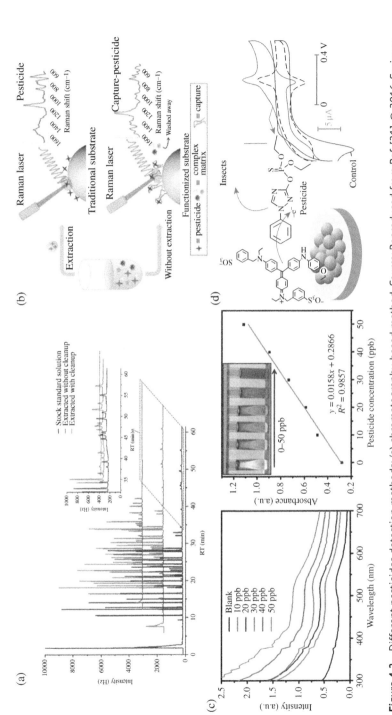

Figure 4.2 Different pesticides detection methods: (a) chromatography-based method. *Source:* Reproduced from Ref. [21], © 2016, Springer Nature/Public Domain CC BY 4.0; (b) SERS-based methods. *Source:* Reproduced from Ref. [22], © 2016, Elsevier. (c) Fluorescence spectrum-based method. *Source:* Reproduced from Ref. [23], © 2020, Springer Nature/Public Domain CC BY 4.0. (d) Electrochemical-based method [24].

4.2.1 Detection of Pesticides Based on Electrochemical Methods

Electrochemical methods have been widely used to test physical parameters such as pH and conductivity, chemical contaminants such as heavy metals or nitrates, and pathogens. They are particularly well suited to *in situ* monitoring pesticides owing to their sensitivity and low cost, as well as the small footprint and connectivity of the readout electronics. They have the potential to overcome the shortcomings of the chromatography-based methods.

4.2.1.1 Brief Overview of Electrochemical Methods

Electrochemical reactions are typically undertaken in a three electrodes system, composed of a working electrode, a counter electrode, and a reference electrode. The working electrode is where the electrochemical reaction to be monitored occurs. The counter electrode (CE) acts as a supply of electrons (source/sink of electrons) to the working electrode for the redox reaction. The reference electrode provides a stable potential for the electrochemical system [25]. The working electrode can be tailored to improve the sensitivity or selectivity of the sensor. Materials that are typically used as working electrodes are inert conductors, such as gold, carbon, silver, and platinum. Other materials can also be used according to the need for different functions (e.g. selectivity, wider potential range, and catalytic behavior) [26, 27]. For example, graphene or carbon-based electrode, carbon nanotube-based electrode, boron-doped diamond electrode (BDD), MoS_2-based or other 2-D nanomaterials, and mixed metal oxide electrode (MMO) [28–32].

There are four main types of electrochemical techniques: voltammetric, potentiometric, coulometric, and impedimetric techniques. Of those, voltammetric techniques have been the most widely used for the detection of pesticides. These techniques monitor reduction and/or oxidation processes of targeted chemical compounds after the application of a certain potential. Different sub-techniques (cyclic voltammetry, square wave voltammetry, differential pulse voltammetry [DPV], etc.) can be applied to optimize signal to noise ratio. Impedance spectroscopy is another kind of electrochemical technique that is also being used for sensing applications and typically monitor the change of impedance upon binding of a target analyte.

4.2.1.2 Detection of Pesticides by Electrochemistry

Monitoring different pesticide concentrations using direct electrochemical reactions at bare electrodes can be readily achieved in samples spiked with the known target. Sometimes, however, in real samples, other unknown compounds can generate electrochemical signals that overlap with the one from the target analyte. To circumvent this, functionalization of the working electrode should be undertaken to improve selectivity and sensitivity. The functional layer depends on the electrode material and the target analyte. Electrode modification based on amide link [33, 34] (e.g. aptamer-modified electrode), enzyme-modified electrode [35], or electrode modification based on Molecular Imprinted Polymer (MIP) has been reported [36,37].

For detection of pesticides by electrochemistry, Geto et al. determined bentazone in 10 seconds by screen-printed carbon electrode. They demonstrated a sensitivity of 0.0987 μA/μM and a limit of detection (LOD) of 0.034 μM. Those sensors have proved to be reusable and stable with a drop of only 2% after 15 times reuse [38]. Akyuz et al. detected pesticides fenitrothion based on the hybrid of manganese phthalocyanine and polyaniline, reaching an LOD of 0.049 μmol/dm^3 [39]. Akyuz et al. developed indium tin oxide coated glass/4-azido polyaniline/Terminal alkynyl substituted manganese phthalocyanine (ITO/PANI-N-3/TA-MnPc) electrode for electrochemical sensing fenitrothion [40]. Huixiang et al. developed three-dimensional graphene–gold nanoparticles (Au NPs) composites for ultrasensitive and selective detection of organophosphorus; this sensor was able to detect diethyl cyanophosphonate with LOD 3.45×10^{-12} M [41].

Fares et al. developed the glyphosate sensor based on an electrodeposited molecularly imprinted chitosan [42]. Qader et al. developed electrochemical molecularly imprinted polymer sensors to determine disulfoton (DSN) in both spiked synthetic human plasma and human urine samples. The LOD was 0.183 μM, while the LOD was 1.64 μM with cyclic voltammetry for the bare glassy carbon electrode [43]. The electrochemical behavior of carbendazim was studied by Luo et al. on hybrid nanomaterial graphene oxide–multi-walled nanotubes/glassy carbon GO-MWNTs/GC-modified electrode with cyclic voltammetry and DPV, with LOD 5 nM reported [44]. Gu et al. reported on a novel sensor based on glass carbon electrode (GCE)-modified by copper oxide @ mesoporous carbon (CuOx@mC) composite to detect glyphosate with LOD 7.69×10^{-16} M [45]. Copper ions can coordinate with glyphosate with the DPV method used to collect the electrochemical response signal, and the glyphosate content is measured as the amount of current reduction. Calfuman et al. used glassy carbon/1-butyl-3-methylimidazolium bis(trifluoromethylsulfonyl) imide/tetrarutenated metalloporphyrin (Ni(II) and Zn(II)) (GC/BMIMNTF2/MTRP)-modified electrodes to detect atrazine in neutral media [46]. Durovic et al. used chronopotentiometry on thin film mercury and glassy carbon electrode to determine imidacloprid, the LOD were 0.17 mg/l and 0.93 mg/l for thin film mercury and glassy carbon electrode, respectively [13]. Further examples of electrochemical based detection of pesticide can be found in Table 4.1

4.2.2 Detection of Pesticides Based on Optical Methods

Aside from electrochemical-based techniques, detection methods based on optical spectrum have recently attracted attention owing to the increasing availability of miniature and cost-effective photonic components, the reliability, and high sensitivity of the sensing results.

Table 4.1 Examples of pesticides detection by electrochemical technique.

Pesticides	Electrochemical technique	Electrode	Sample type	Limit of detection (LOD)	References
Dinoterb	Stripping voltammetric	Boron-doped diamond electrode	River water and soil matrices	0.0022 µg/ml (9.16×10^{-9} mol/l)	[47]
Diazinon	Differential pulse voltammetry (DPV) and electrochemical impedance spectroscopy (EIS)	VS_2 QDs doped GNP/MWCNTs	Various real samples	11 fmol/l (DPV) and 2 fmol/l (EIS)	[48]
Methyl parathion	Voltammetric	Biochar	Pre-concentrated sample	/	[49]
Glyphosate	Electrochemical Impedance Spectroscopy (EIS)	Chitosan MIP on gold microelectrode	Spiked river waters	0.001 pg/ml (S/N = 3)	[42]

Compound (structure)	Method	Electrode	Sample	Sensitivity/LOD	Ref.
Thiram	Cyclic Voltammetry	Glassy carbon electrode	River waters	0.039 µA/µmol/l	[50]
Glyphosate	Cyclic Voltammetry; Voltammetry	Organosmectite-modified carbon paste electrode	Soils	0.9 µM/l	[51]
Nitenpyram	Cyclic voltammetry (CV) and differential pulse voltammetry (DPV)	Hydroxylated multiwall carbon nanotubes/single-wall carbon nanohorns (HCNT/CNH)	Corn sample	4.0 nM/l	[52]
Malathion	Cyclic voltammetry (CV), differential pulse voltammetry (DPV), and electrochemical impedance spectroscopy (EIS)	Screen-printed gold electrodes (Au-SPE)	Olive oil	0.06 pg/ml	[36]

(Continued)

Table 4.1 (Continued)

Pesticides	Electrochemical technique	Electrode	Sample type	Limit of detection (LOD)	References
Aldicarb	Cyclic voltammetry (CV); differential pulse voltammetry (DPV)	Polyaniline nanofibers	Pesticides sample	500 pM/l	[53]
Pymetrozine	Cyclic voltammetry (CV)	L-Lysine and beta-cyclodextrin film on glassy carbon electrode	Real sample	1.3×10^{-8} mol/l (S/N = 3)	[54]
2,4-D	Differential pulse voltammetry (DPV)	MIP and dsDNA on pencil graphite electrode [55]	Spiked environmental water and soil samples	4.0 fM	[55]

4.2.2.1 Detection of Pesticides Based on Fluorescence

The general principles of fluorescence phenomena are illustrated in the Jablonsky diagram in Figure 4.3. The material absorbs light and electrons move from the ground state to the excited state. Following this, vibrational relaxation or internal conversion occur in a time frame of 10^{-9} to 10^{-15} seconds. Finally, fluorescence emission occurs after 10^{-8} seconds of the excitation, as the excited electron returns to its ground state. The emitted phonon shows the characteristic peaks for the analytes in the fluorescence spectrum.

As pesticides are generally not fluorescent, their presence is detected through their interaction with fluorescent probes. There are a few frequently used fluorescent probes for pesticides detection, including enzyme employed AChE [56], dyes [57], or other non-emissive chemical compounds combining with the pesticides to form the emissive molecular to have the fluorescent effect [58, 59]. The concentration of the target pesticides can be identified and quantified based on the spectral shift or change in intensity of the fluorescence peak.

Fluorescence-based method has been used for screening and quantification of different pesticides in water, food, and pharmaceuticals. For example, Paliwal et al. developed a new fluorescence sensor for the direct detection of p-nitrophenol and p-nitrophenol-substituent organophosphorus neurotoxins [60]. In this research, Coumarin1 was used as the fluorescent compound and competitive inhibitor. Organophosphorus hydrolase can catalyze the hydrolysis of the P–S, P–O, P–CN, and P–F bonds of OP-neurotoxins. The fluorescence intensity of coumarin1 varies with different concentrations of p-nitrophenol and paraoxon in the range of 7.0×10^{-7} to 1.7×10^{-4} M. Similarly, Ashrafi et al. used carbon dots as the identifier, and achieved LOD of 0.25, 0.5, and 2 ng/ml for diazinon, amicarbazone, and glyphosate, respectively [61]. Yang et al. used molecularly imprinted polymers and fluorescence measurement to study the sensing principles and signal transduction mechanisms, so as to improve the sensitivity of the sensors [62]. Chu et al. developed a rapid, sensitive, instrument-free, and visual quantitative platform for detection of pesticide by fluorescence device. The fluorescence

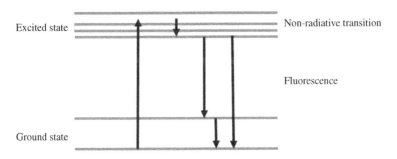

Figure 4.3 Fluorescence band diagram.

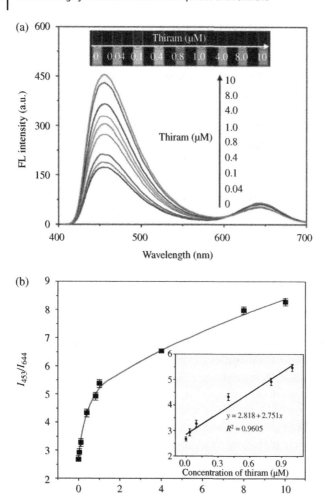

Figure 4.4 (a) Fluorescence spectra (λex = 350 nm) of rQDs@SiO$_2$@CDs with addition of different concentrations. The inset photo shows the corresponding fluorescence photographs. (b) Fluorescence intensity ratio (I_{453}/I_{644}) versus the concentration of thiram (μM). The inset was the linear plot of the fluorescence intensity ratio (I_{453}/I_{644}) versus the concentration of thiram from 0 to 1 μM. *Source:* Republished with permission from Ref. [63], American Chemical Society.

spectrum of red-emitting CdTe quantum dots@SiO$_2$@carbon dots (rQDs@SiO$_2$@ CDs) with addition of different concentrations of thiram is shown in Figure 4.5. Sensitive LOD of 59 nM was reported. The authors integrated the platform onto a smartphone using 3D printing, which allowed remote detection and could be used in various environments [63].

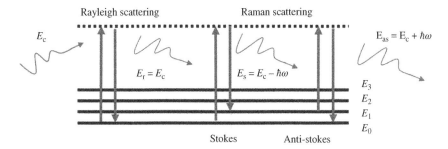

Figure 4.5   Raman scattering band diagram.

Fluorescence-based technologies are convenient in operation and have the potential to be integrated into portable devices such as smartphones. However, they rely on fluorescent probes that can be complex or expensive to synthesize. Moreover, they have lower sensitivities than chromatography-based technology. In order to achieve high sensitivity and fast detection, other emerging optical technologies such as surface enhanced Raman spectroscopy (SERS) sensing devices are currently being investigated.

4.2.3   Detection of Pesticides Based on Raman Spectroscopy

4.2.3.1   Introduction to SERS

Raman spectroscopy was first discovered by Sir Chandrasekhara Venkata Raman in 1928 [64]. The Raman process is an inelastic scattering and probe molecular vibrations of sample under investigation, see Figure 4.5 [65, 66]. Raman spectroscopy is often referred to as fingerprinting method [67]. Unfortunately, Raman spectroscopy suffers from low signal/noise ratio as typically only 1 in 10^8 incident photons is Raman scattered [68].

Fortunately, Fleischman et al. discovered in 1974 the Surface Enhanced Raman Spectroscopy (SERS) effect that can overcome the limits of conventional Raman spectroscopy [69]. SERS relies on plasmon phenomena that can be found in noble metallic nanostructures, such as silver, gold, copper, etc. Some new plasmonic materials combining metallic and semiconducting or metal oxide material are also being investigated nowadays [70–73]. Briefly speaking, when the distance between the metallic nanocluster is between 0 and 100 nm, the free electrons move to the metal surface to oscillate, strong electromagnetic field generates in between the gap. See Figure 4.6. It is commonly accepted that the SERS enhancement is a combination of chemical enhancement [69, 74–76] and the electromagnetic enhancement [77, 78].

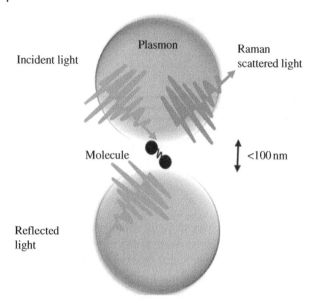

Figure 4.6 SERS phenomenon between two metallic nanoparticles.

The principles of the chemical enhancement are not completely understood [79–82]. Most common explanation is that the ejection of hot electrons from the metallic nanostructure to the targeted molecule results in an increase of the molecular polarizability [83]. The chemical enhancement therefore depends on the orientation and the band structure of the molecule at the surface of the substrate. Regarding the electromagnetic enhancement [84], the theory of localized surface plasmon resonance (LSPR) is introduced [85]. The excited LSPR induces dipole moments in the molecule on the nanostructures, and sequentially produces the surface Raman enhancement.

SERS substrates are usually characterized by a Figure of Merit called the enhancement factor (EF). The EF depends on the roughness and homogeneity of the plasmonic surface, the molecule adsorption states, the morphology, the laser intensity and wavelength, etc. It is defined as

$$ EF = \frac{I_{SERS}/\left(\mu_M \mu_S A_M\right)}{I_{RS}/\left(C_{RS} H_{eff}\right)} $$

where I_{RS} is the intensity of the normal Raman signal, I_{SERS} is the intensity of Surface Enhanced Raman signal, μ_S (m^{-2}) is the surface density of molecules adsorbed to NP, μ_M (m^{-2}) is the surface density of NPs contributing to enhancement, H_{eff} (m) is the effective height of the scattering volume, A_M (m²) is the surface area of metallic NPs, and C_{RS} (M) is the concentration of the solution used for non-SERS measurements.

4.2.3.2 Fabrication of SERS Substrates
There are several methods to fabricate the SERS substrates, including chemical and physical processes. Chemical methods encompass chemical synthesis [86], colloidal method [87], electrodeposition [88–91], etc. Physical method relate to electron-beam (E-Beam) lithography [92–94], inkjet printing [95–104], template method [105], etc. Both methods have advantages and disadvantages. For chemical methods, synthesis of metallic nanostructures can be readily achieved, but spreading the nanocluster homogeneously on the substrate is difficult and it is also hard to control the distance between adjacent nanoclusters. For the physical method, it is easy to get periodic substrate with plasmonic nanostructures, but the process is complex, time-consuming, and sometimes expensive. For inkjet printing, the choices for the substrate and ink are limited.

Among these different fabrication methods, electrochemical-based ones are more cost effective compared to E-Beam lithography, and offer a wider choice of substrates and plasmonic material compared to the inkjet method. The morphology and the shape of the plasmon structure can be controlled through control of the deposition time, deposition voltage, and concentration of the precursor solvent. Figure 4.7 demonstrates the electrochemical-based fabrication process of SERS substrates.

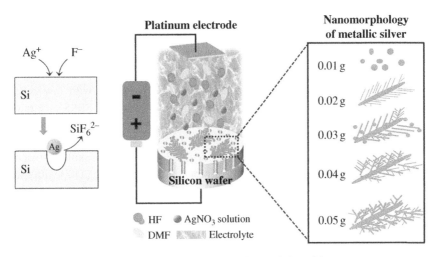

Figure 4.7 SERS substrate fabrication by electrochemical deposition.
Source: Republished with permission of American Chemical Society, from Ref. [89], 1 Mar 2020 permission conveyed through Copyright Clearance Center, Inc.

For electrochemical method in SERS substrate fabrication, Wang et al. used electrochemically roughened nano-Au film as the SERS substrate, which had high reproducibility and good stability for the detection of organic pollutants [106]. Li et al. electrodeposited a high density of ZnO-nanorods onto polystyrene microspheres, then Ag-nanoparticles (Ag-NPs) were formed onto the surface of each ZnO-nanorod through electrochemical deposition, resulting in ordered hierarchical Ag/ZnO hybrid arrays [107]. Wu et al. used ammonia reduction nitridation and electrochemical deposition method to deposit Au nanoparticles on the TiN surface to form the Au/TiN composite as effective SERS substrate [108]. Lu et al. reported the electrochemical deposition method for fabricating super hydrophobic flexible SERS substrates based on 3D Ag nanodendrites (Ag NDs)/carbon fiber cloth. This micro/nanostructures substrate could easily obtain ultra-wet surfaces [109]. Zavatski et al. deposited copper on porous aluminum oxide using the electrochemical method and protected the metal nanostructure surface from oxidation by covering the surface with polyethylene glycol and silver [110]. Lu et al. synthesized hierarchical flower-like gold microstructures on a flexible PET film substrate through the electrochemical deposition method [111].

SERS signals will be affected by the background noise from the environment. Besides, some pesticides do not have strong interaction with the metallic SERS substrate surface. As in electrochemistry, functionalization can overcome these shortcomings in some SERS detection [112].

4.2.3.3 Detection of Pesticide by SERS

The pesticides used in agriculture contaminate the water bodies as well as the food products produced. Detection of pesticides in food products is more difficult than in water bodies because of the sophisticated sample preparation process for food products. SERS has been widely used in agriculture and food industry for pesticides monitoring. A variety of SERS substrates have been used for the detection of pesticides in fruit surface, soil, and food matrix, with low LOD reported, see Table 4.2. The detection steps typically involve the SERS substrate preparation, sample preparation (for soil, fruit juice, food matrix, etc.), detection, and data analysis.

For pesticides detection on fruit surfaces, see Table 4.2. Parnsubsakul et al. used the 2D Au@Ag nanodot array to detect thiram and thiabendazole (TBZ) with the LOD 0.0011 ppm and 0.051 ppm, respectively [113]. Xie et al. used filter paper and Au NPs to detect methyl parathion with LOD $0.011\,\mu g/cm^2$ [114]. Chen et al. synthesized flower-like two-dimensional molybdenum sulfide and Ag (MoS_2@Ag) for thiram and methyl parathion detection and LOD is 6.4×10^{-7} mg/ml and 9.8×10^{-7} mg/ml, respectively. The substrate was sequentially used for recyclable detection of pesticide residues on several kinds of fruits and vegetables [115]. Teixeira et al. developed Au NPs on common office paper using a wax printer.

Table 4.2 Examples of pesticides detection on fruit surface by SERS.

	Fruit surface		
Pesticides	Substrate	LOD	References
Thiram and thiabendazole	2D Au@Ag nanodot array	0.0011 and 0.051 ppm	[113]
Methyl parathion	Filter paper and gold nanoparticles (Au NPs)	0.011 µg/cm^2	[114]
Thiram and methyl parathion	Flower-like two-dimensional molybdenum sulfide and Ag (MoS$_2$@Ag)	6.4×10^{-7} and 9.8×10^{-7} mg/ml	[115]
Thiabendazole	Gold nanoparticles on common office paper were prepared using a wax printer	—	[116]
Parathion-methyl, triazophos, and phosmet	Snowflake-like gold nanoparticles	0.026 ng/cm^2, 0.031 ng/cm^2, and 0.032 ng/cm^2	[117]
Thiabendazole	Ag@SiO$_2$ nanocubes on Fe-TiO$_2$ nanosheets-modified paper	19 µg/l	[118]
Phosmet and carbaryl	Nanostructures with spikes	—	[119]

These substrates were used for the detection of TBZ on varieties of mango samples [116]. Huang et al. synthesized snowflake-like Au NPs for the detection of parathion-methyl, triazophos, and phosmet with LOD of 0.5 0.026 ng/cm^2, 0.031 ng/cm^2, and 0.032 ng/cm^2, respectively [117]. Sitjar et al. used nanostructures with spikes for phosmet and carbaryl monitoring. They synthesized Ag NPs in varying degrees of agglomeration that were then deposited onto a transparent adhesive tape as a flexible substrate for SERS applications. An EF of 1.7×10^7 and a detection of the pesticides down to 10^{-7} M were reported [119].

For pesticides detection in food matrix, see Table 4.3. Chen et al. used cellulose nanofibers decorated with silver nanoparticles to detect flusilazole in Oolong tea with an LOD of 0.5 mg/kg [120]. Quan et al. used 3-D tilted ZnO micro rods with an Ag hierarchical structure to detect thiram and methamidophos compounds in various juices, with a LOD for thiram residue of 10^{-11} M [121]. Muhammad et al. used label-free SiO$_2$@Au core/shell nanoparticle SERS substrates to quantitatively detect fipronil in chicken eggs membrane and analyze the characteristic Raman band by density functional theory (DFT) calculations. Their work may provide a practical solution to quick inspection of fipronil contamination in

Table 4.3 Examples of pesticides detection in food matrix and soil by SERS.

		Food matrix		
Pesticides	**Matrix**	**Substrate**	**LOD**	**References**
Flusilazole	Oolong tea	Cellulose nanofibers decorated with silver nanoparticles (AgNPs)	0.5 mg/kg	[120]
Thiram and methamidophos compounds	Various juices	3-D tilted ZnO micron rods with an Ag hierarchical structure	10^{-11} M	[121]
Fipronil	Chicken eggs	SiO_2@Au core/shell nanoparticles	—	[122]
Imazalil	Agricultural products	Silver-coated gold nanocubes (Au@Ag NCs)	2 mg/l	[123]
Micro-extraction of CV; thiram	Fish skin; fruit surfaces	Ag NPs/polydimethylsiloxane probes	10^{-7} M and 10^{-5} M	[123]

	Soil		
Pesticides	**Substrate**	**LOD**	**References**
Hexachlorobenzene residue	Magnetic force assisting self-assembled nanoparticles in arrayed holes	—	[19]
Organophosphorus pesticides	Gold grating surface with the metal–organic framework (MOF-5)	10^{-12} M	[124]

chicken eggs or other foods [122]. Zhao et al. used Ag NPs/polydimethylsiloxane probes for the detection of micro-extraction of crystal violet and thiram in the fish skin and fruit surfaces with LOD of 1×10^{-7} M and 1×10^{-5} M, respectively [123]. Wang et al. used the silver-coated gold-nanoparticle (Au@Ag NP) plasmonic array to sense the TBZ residues in fruit juices [15].

For detection of pesticides residual in soil, see Table 4.3. The sample is typically prepared by adding 1 g of natural soil to the mixed water and ethanol (1 : 1) solvent. The mixture is sonicated for one hour, then allowed to stay still for one hour so that the material settles at the bottom [19, 124]. Gong et al. used magnetic force assisting self-assembled nanoparticles in arrayed hole to detect hexachlorobenzene residue in soil [19]. Guselnikova et al. used gold grating surface with metal–organic framework (MOF-5) for organophosphorus pesticides detection [124].

For detecting pesticides in the aqua environment using SERS, Hussain et al. used cysteamine-modified silver-coated gold nanoparticles (Au@Ag-CysNPs) detection of oxamyl and thiacloprid in liquid samples of milk. The LOD for oxamyl and thiacloprid were 0.031 ppm and 0.023 ppm, respectively [125]. Fu et al. used chitosan nanofibrils anchored with Au NPs grew into a three-dimensional architecture. The EF approached 10^7 and spot-to-spot reproducibility could be as low as 5.66% for detecting 4-mercaptobenzoic acid. Due to its hierarchical porosity and mechanical flexibility, straightforward trace detections of melamine in milk and pesticides on the fruit surface were realized using this SERS substrate [126]. Liu et al. combined a coffee-ring structure with the waveguide effect of optical fibers. A detection limit lower than 10^{-8} M in aqueous solvent for thiram and methyl parathion was achieved. LOD of 10^{-7} M for thiram in lake water and in orange juice and 10^{-6} M for melamine and tetracycline in liquid milk were achieved [127]. Zhang et al. used commercially available filter paper functionalized with silver nanoparticles to detect melamine in diluted milk, with an instrument LOD down to 1 ppm This SERS substrate could also be applied successfully for the determination of pesticide (thiram) and dye (malachite green) in real environment [128].

Overall, SERS is fast, cheap, easy to operate, and has high sensitivity. Another attractive feature of SERS is that portable Raman device is now commercially available and affordable [19]. Thus, SERS can be widely used on the field for detection of the pesticides residues in food or in water bodies.

4.2.3.4 Challenges and Future Perspectives

As discussed above, electrochemistry or optical spectrum-based techniques are good candidates in fabricating the cheap, ultrafast, ultrasensitive pesticides monitoring sensors that can be used in varieties of detection environments. But special efforts are needed to improve the selectivity, the signal to noise ratio, and the durability of the sensor. The technical challenges faced in portable electrochemical device and portable Raman device in signal transformation from the lab-used ones to the portable ones are also the bottleneck for the in-site sensing activities. The surface modification and functionalization of the work electrode in electrochemistry and the SERS substrates in surface enhanced Raman are important steps to improve the selectivity and the signal to noise ratio in signal readout.

4.3 Conclusion

Pesticides are harmful to both human and environment, and as a result need to be applied with precaution. Because of their toxicities, it is important to use sensitive, fast, and reliable devices to monitoring the pesticides in nature environment (e.g. river water, sea, and soil), food matrix, fruit/vegetable surface, etc. In this chapter, different technologies for pesticides detection have been discussed, including chromatography-based technologies, fluorescence-based technology, electrochemical technologies, and SERS. Among those technologies, electrochemical technologies and SERS are very sensitive, fast, user friendly, and can be integrated in portable device. They have the potential to be used *in situ* and replace chromatography-based technologies that are mostly used in environment agency nowadays.

Acknowledgments

Authors gratefully acknowledge the financial support and funding from the European Union's Horizon 2020 research and innovation program under the Marie Skłodowska-Curie grant agreement No: H2020-MSCA-ITN-2018-813 680.

References

1 Ritchie, H., Roser, M. and Rosado, P. (2022). Pesticides. OurWorldInData.org. https://ourworldindata.org/pesticides (accessed 12 April 2023).
2 Jayaraj, R., Megha, P., and Sreedev, P. (2016). Organochlorine pesticides, their toxic effects on living organisms and their fate in the environment. *Interdiscip. Toxicol.* 9 (3–4): 90–100.
3 Singh, B.K. and Walker, A. (2006). Microbial degradation of organophosphorus compounds. *FEMS Microbiol. Rev.* 30 (3): 428–471.
4 Mustapha, M.U., Halimoon, N., Johari, W.L.W., and Shukor, M.Y.A. (2019). An overview on biodegradation of carbamate pesticides by soil bacteria. *Pertanika J. Sci. Technol.* 27 (2): 547–563.

5 Cycoń, M. and Piotrowska-Seget, Z. (2016). Pyrethroid-degrading microorganisms and their potential for the bioremediation of contaminated soils: a review. *Front. Microbiol.* 7: 1463–1463.

6 Miyamoto, J. (1976). Degradation, metabolism and toxicity of synthetic pyrethroids. *Environ. Health Perspect.* 14: 15–28.

7 Acero, J.L., Real, F.J., Benitez, F.J., and Matamoros, E. (2019). Degradation of neonicotinoids by UV irradiation: kinetics and effect of real water constituents. *Sep. Purif. Technol.* 211: 218–226.

8 Hussain, S., Hartley, C.J., Shettigar, M., and Pandey, G. (2016). Bacterial biodegradation of neonicotinoid pesticides in soil and water systems. *FEMS Microbiol. Lett.* 363 (23): fnw252.

9 Wolff, M.S., Toniolo, P.G., Lee, E.W. et al. (1993). Blood levels of organochlorine residues and risk of breast cancer. *J. Natl. Cancer Inst.* 85 (8): 648–652.

10 Health Service Executive, National Drinking Water Group (2018). Pesticides in drinking water. https://www.hse.ie/eng/health/hl/water/drinkingwater/faq-pesticides.pdf (accessed 12 April 2023).

11 Velkoska-Markovska, L. and Petanovska-Ilievska, B. (2020). Rapid resolution liquid chromatography method for determination of malathion in pesticide formulation. *Acta Chromatogr.* 32 (4): 256–259.

12 Popovska-Gorevski, M., Dubocovich, M.L., and Rajnarayanan, R.V. (2017). Carbamate insecticides target human melatonin receptors. *Chem. Res. Toxicol.* 30 (2): 574–582.

13 Durovic, A., Stojanovic, Z., Kravic, S. et al. (2016). Development and validation of chronopotentiometric method for imidacloprid determination in pesticide formulations and river water samples. *Int. J. Anal. Chem.* 2016: 5138491.

14 Lundin, O., Rundlof, M., Smith, H.G. et al. (2015). Neonicotinoid insecticides and their impacts on bees: a systematic review of research approaches and identification of knowledge gaps. *PLoS One* 10 (8): 20.

15 Wang, K., Sun, D., Pu, H. et al. (2019). Stable, flexible, and high-performance SERS chip enabled by a ternary film-packaged plasmonic nanoparticle array. *ACS Appl. Mater. Interfaces* 11 (32): 29177–29186.

16 Moldoveanu, S. and David, V. (ed.) (2015). Other sample preparation techniques not involving chemical modifications of the analyte. In: *Modern Sample Preparation for Chromatography*, 287–305. Amsterdam: Elsevier.

17 EPA (2019). *Water Quality in Ireland 2013-2018*. Wexford Ireland: Environmental Protection Agency.

18 Han, L.J. and Sapozhnikova, Y. (2020). Semi-automated high-throughput method for residual analysis of 302 pesticides and environmental contaminants in catfish by fast low-pressure GC-MS/MS and UHPLC-MS/MS. *Food Chem.* 319: 126592.

19 Gong, T.X., Huang, Y., Wei, Z. et al. (2020). Magnetic assembled 3D SERS substrate for sensitive detection of pesticide residue in soil. *Nanotechnology* 31 (20): 205501.

20 Panuwet, P., Hunter Jr, R.E., D'Souza, P.E. et al. (2016). Biological matrix effects in quantitative tandem mass spectrometry-based analytical methods: advancing biomonitoring. *Crit. Rev. Anal. Chem.* 46 (2): 93–105.

21 Kong, W.-J., Liu, Q.-T., Kong, D.-D. et al. (2016). Trace analysis of multi-class pesticide residues in Chinese medicinal health wines using gas chromatography with electron capture detection. *Sci. Rep.* 6 (1): 21558.

22 Pang, S., Yang, T., and He, L. (2016). Review of surface enhanced Raman spectroscopic (SERS) detection of synthetic chemical pesticides. *TrAC Trends Anal. Chem.* 85: 73–82.

23 Mane, P.C., Shinde, M.D., Varma, S. et al. (2020). Highly sensitive label-free bio-interfacial colorimetric sensor based on silk fibroin-gold nanocomposite for facile detection of chlorpyrifos pesticide. *Sci. Rep.* 10 (1): 4198.

24 Shalini Devi, K.S., Anusha, N., Raja, S., and Senthil Kumar, A. (2018). A new strategy for direct electrochemical sensing of a organophosphorus pesticide, triazophos, using a coomassie brilliant-blue dye surface-confined carbon-black-nanoparticle-modified electrode. *ACS Appl. Nano Mater.* 1 (8): 4110–4119.

25 Bard, A.J., Faulkner, L.R., Leddy, J., and Zoski, C.G. (1980). *Electrochemical Methods: Fundamentals and Applications*, vol. 2. New York: Wiley.

26 Jafari, M., Tashkhourian, J., and Absalan, G. (2020). Electrochemical sensor for enantioselective recognition of naproxen using l-cysteine/reduced graphene oxide modified glassy carbon electrode. *Anal. Bioanal. Chem. Res.* 7 (1): 1–15.

27 Roushani, M., Mohammadi, F., and Valipour, A. (2020). Electroanalytical sensing of Asulam based on nanocomposite modified glassy carbon electrode. *J. Nanostruct.* 10 (1): 128–139.

28 Kurbanoglu, S., Ozkan, S.A., and Merkoci, A. (2017). Nanomaterials-based enzyme electrochemical biosensors operating through inhibition for biosensing applications. *Biosens. Bioelectron.* 89: 886–898.

29 Shestakova, M. and Sillanpaa, M. (2017). Electrode materials used for electrochemical oxidation of organic compounds in wastewater. *Rev. Environ. Sci. Biotechnol.* 16 (2): 223–238.

30 Wong, A., Silva T.A. Caetano, F.R. et al. (2017). An overview of pesticide monitoring at environmental samples using carbon nanotubes-based electrochemical sensors. *C. J. Carbon Res.* 3 (1): 8.

31 Zhao, F.N., Wu, J., Ying, Y. et al. (2018). Carbon nanomaterial-enabled pesticide biosensors: design strategy, biosensing mechanism, and practical application. *TrAC Trends Anal. Chem.* 106: 62–83.

32 Su, S., Chen, S.M., and Fan, C.H. (2018). Recent advances in two-dimensional nanomaterials-based electrochemical sensors for environmental analysis. *Green Energy Environ.* 3 (2): 97–106.

33 Fu, J.Y., Yao, Y. An, X. et al. (2020). Voltammetric determination of organophosphorus pesticides using a hairpin aptamer immobilized in a graphene oxide-chitosan composite. *Microchim. Acta* 187 (1): 36.

34 Zeng, Y.B., Yu, D., Yu, Y. et al. (2012). Differential pulse voltammetric determination of methyl parathion based on multiwalled carbon nanotubes-poly(acrylamide) nanocomposite film modified electrode. *J. Hazard. Mater.* 217: 315–322.

35 Hou, L., Zhang, X., Kong, M. et al. (2020). A competitive immunoassay for electrochemical impedimetric determination of chlorpyrifos using a nanogold-modified glassy carbon electrode based on enzymatic biocatalytic precipitation. *Microchim. Acta* 187 (4): 204.

36 Aghoutane, Y., Diouf, A., Osterlund, L. et al. (2020). Development of a molecularly imprinted polymer electrochemical sensor and its application for sensitive detection and determination of malathion in olive fruits and oils. *Bioelectrochemistry* 132: 107404.

37 Xu, L.P., Li, J., Zhang, J. et al. (2020). A disposable molecularly imprinted electrochemical sensor for the ultra-trace detection of the organophosphorus insecticide phosalone employing monodisperse Pt-doped UiO-66 for signal amplification. *Analyst* 145 (9): 3245–3256.

38 Geto, A., Noori, J.S., Mortensen, J. et al. (2019). Electrochemical determination of bentazone using simple screen-printed carbon electrodes. *Environ. Int.* 129: 400–407.

39 Akyüz, D. and Koca, A. (2019). An electrochemical sensor for the detection of pesticides based on the hybrid of manganese phthalocyanine and polyaniline. *Sensors Actuators B Chem.* 283: 848–856.

40 Akyuz, D. and Koca, A. (2018). Construction of modified electrodes with click electrochemistry based on the hybrid of 4-azido aniline and manganese phthalocyanine and electrochemical pesticide sensor applications. *J. Electrochem. Soc.* 165 (11): B508–B514.

41 Huixiang, W., Danqun, H., Yanan, Z. et al. (2017). A non-enzymatic electro-chemical sensor for organophosphorus nerve agents mimics and pesticides detection. *Sensors Actuators B Chem.* 252: 1118–1124.

42 Zouaoui, F., Bourouina-Bacha, S., Bourouina, M. et al. (2020). Electrochemical impedance spectroscopy determination of glyphosate using a molecularly imprinted chitosan. *Sensors Actuators B Chem.* 309: 127753.

43 Qader, B., Baron, M., Hussain, I. et al. (2019). Electrochemical determination of disulfoton using a molecularly imprinted poly-phenol polymer. *Electrochim. Acta* 295: 333–339.

44 Luo, S.X., Wu, Y.H., and Gou, H. (2013). A voltammetric sensor based on GO-MWNTs hybrid nanomaterial-modified electrode for determination of carbendazim in soil and water samples. *Ionics* 19 (4): 673–680.

45 Gu, C., Wang, Q., Zhang, L. et al. (2020). Ultrasensitive non-enzymatic pesticide electrochemical sensor based on HKUST-1-derived copper oxide @ mesoporous carbon composite. *Sensors Actuators B Chem.* 305: 127478.

46 Calfumán, K., Honores, J., Isaacs, M. et al. (2019). Quick and easy modification of glassy carbon electrodes with ionic liquid and tetraruthenated porphyrins for the electrochemical determination of atrazine in water. *Electroanalysis* 31 (4): 671–677.

47 Pınar, P.T., Allahverdiyeva, S., Yardim, Y. et al. (2020). Voltammetric sensing of dinitrophenolic herbicide dinoterb on cathodically pretreated boron-doped diamond electrode in the presence of cationic surfactant. *Microchem. J.* 155: 104772.

48 Khosropour, H., Rezaei, B., Rezaei, P. et al. (2020). Ultrasensitive voltammetric and impedimetric aptasensor for diazinon pesticide detection by VS_2 quantum dots-graphene nanoplatelets/carboxylated multiwalled carbon nanotubes as a new group nanocomposite for signal enrichment. *Anal. Chim. Acta* 1111: 92–102.

49 Kalinke, C., de Oliveira, P.R., Mangrich, A.S. et al. (2020). Chemically-activated biochar from *Ricinus communis* L. cake and their potential applications for the voltammetric assessment of some relevant environmental pollutants. *J. Braz. Chem. Soc.* 31 (5): 941–952.

50 Alves Sa da Silva, V., da Silva Santos, A., Ferreira, T.L. et al. (2020). Electrochemical evaluation of pollutants in the environment: interaction between the metal ions Zn(II) and Cu(II) with the fungicide thiram in billings dam. *Electroanalysis* 32 (7): 1582–1589.

51 Mbokana, J.G.Y., Dedzo, G.K., and Ngameni, E. (2020). Grafting of organophilic silane in the interlayer space of acid-treated smectite: application to the direct electrochemical detection of glyphosate. *Appl. Clay Sci.* 188: 105513.

52 Wang, H.W., Pan, L., Liu, Y. et al. (2020). Electrochemical sensing of nitenpyram based on the binary nanohybrid of hydroxylated multiwall carbon nanotubes/single-wall carbon nanohorns. *J. Electroanal. Chem.* 862: 113955.

53 Saxena, S., Lakshmi, G.B.V.S., Chauhan, D. et al. (2020). Molecularly imprinted polymer-based novel electrochemical sensor for the selective detection of aldicarb. *Phys. Status Solidi A Appl. Mater. Sci.* 217 (6): 1900599.

54 Gao, Y.D. (2020). A voltammetry sensor platform for environmental pollution detection based on simultaneous electropolymerisation of L-lysine and beta-cyclodextrin. *Int. J. Environ. Anal. Chem.* 102 (6): 1467.

55 Azadmehr, F. and Zarei, K. (2019). An imprinted polymeric matrix containing DNA for electrochemical sensing of 2,4-dichlorophenoxyacetic acid. *Microchim. Acta* 186 (12): 814.

56 Caballero-Diaz, E., Benitez-Martinez, S., and Valcarcel, M. (2017). Rapid and simple nanosensor by combination of graphene quantum dots and enzymatic inhibition mechanisms. *Sensors Actuators B Chem.* 240: 90–99.

57 Zhang, J., Zhou, W.D., Zhai, L.J. et al. (2020). A stable dual-emitting dye@LMOF luminescence probe for the rapid and visible detection of organophosphorous pesticides in aqueous media. *CrystEngComm* 22 (6): 1050–1056.

58 Wang, B.H., Lian, X., and Yan, B. (2020). Recyclable Eu^{3+} functionalized Hf-MOF fluorescent probe for urinary metabolites of some organophosphorus pesticides. *Talanta* 214: 120856.

59 Xu, H.M., Xiao, K., Zhang, Q. et al. (2020). Rapid and visual detection of bipyridylium herbicides based on polyelectrolyte-induced nanoassemblies of pyrenyl probes. *ACS Sustain. Chem. Eng.* 8 (17): 6861–6867.

60 Paliwal, S., Wales M., Good, T. et al. (2007). Fluorescence-based sensing of *p*-nitrophenol and *p*-nitrophenyl substituent organophosphates. *Anal. Chim. Acta* 596 (1): 9–15.

61 Ashrafi Tafreshi, F., Fatahi, Z., Ghasemi, S.F. et al. (2020). Ultrasensitive fluorescent detection of pesticides in real sample by using green carbon dots. *PLoS One* 15 (3): e0230646.

62 Yang, Q., Li, J., Wang, X. et al. (2018). Strategies of molecular imprinting-based fluorescence sensors for chemical and biological analysis. *Biosens. Bioelectron.* 112: 54–71.

63 Chu, S., Wang, H., Ling, X. et al. (2020). A portable smartphone platform using a ratiometric fluorescent paper strip for visual quantitative sensing. *ACS Appl. Mater. Interfaces* 12 (11): 12962–12971.

64 Raman, C.V. and Krishnan, K.S. (1928). A new type of secondary radiation *Nature* 121 (3048): 501–502.

65 Kalantar-zadeh, K. and Fry, B. (2007). *Nanotechnology-Enabled Sensors*. Springer Science & Business Media.

66 Settle, F.A. (1997). *Handbook of Instrumental Techniques for Analytical Chemistry*. Prentice Hall PTR.

67 Laserna, J. (1996). *Modern Techniques in Raman Spectroscopy*. Wiley.

68 Gardiner, D.J. (1989). *Practical Raman Spectroscopy*. Spinger-Verlag.

69 Fleischmann, M., Hendra, P.J., and McQuillan, A.J. (1974). Raman spectra of pyridine adsorbed at a silver electrode. *Chem. Phys. Lett.* 26 (2): 163–166.

70 Kannan, P.K., Shankar, P., Blackman, C., and Chung, C.H. (2019). Recent advances in 2D inorganic nanomaterials for SERS sensing. *Adv. Mater.* 31 (34): 1803432.

71 Kumar, N.V.S. and Singh, H. (2020). An APT charge based descriptor for atomic level description of chemical Raman enhancement by adsorption of 4-Mercaptopyridine on semiconducting nanoaclusters: a theoretical study. *Vib. Spectrosc.* 107: 103019.

72 Rakkesh, R.A., Durgalakshmi, D., Karthe, P., and Balakumar, S. (2020). Anisotropic growth and strain-induced tunable optical properties of Ag-ZnO hierarchical nanostructures by a microwave synthesis method. *Mater. Chem. Phys.* 244: 122720.

73 Wang, W.E., Sang, Q.Q., Yang, M. et al. (2020). Detection of several quinolone antibiotic residues in water based on Ag-TiO$_2$ SERS strategy. *Sci. Total Environ.* 702: 134956.

74 Moskovits, M. (1985). Surface-enhanced spectroscopy. *Rev. Mod. Phys.* 57 (3): 783–826.

75 Otto, A., Mrozek, I., Grabhorn, H., and Akemann, W. (1992). Surface-enhanced Raman scattering. *J. Phys. Condens. Matter* 4 (5): 1143.

76 Stiles, P.L., Dieringer, J.A., Shah, N.C., and Duyne, R.P.V. (2008). Surface-enhanced Raman spectroscopy. *Annu. Rev. Anal. Chem.* 1 (1): 601–626.

77 Gersten, J.I. (1980). The effect of surface roughness on surface enhanced Raman scattering. *J. Chem. Phys.* 72 (10): 5779–5780.

78 Gersten, J. and Nitzan, A. (1980). Electromagnetic theory of enhanced Raman scattering by molecules adsorbed on rough surfaces. *J. Chem. Phys.* 73 (7): 3023–3037.

79 Le Ru, E.C., Blackie, E., Meyer, M., and Etchegoin, P.G. (2007). Surface enhanced Raman scattering enhancement factors: a comprehensive study. *J. Phys. Chem. C* 111 (37): 13794–13803.

80 Sun, M., Wan, S., Liu, Y. et al. (2008). Chemical mechanism of surface-enhanced resonance Raman scattering via charge transfer in pyridine–Ag_2 complex. *J. Raman Spectrosc.* 39 (3): 402–408.

81 Kambhampati, P., Child, C., Foster, M.C., and Campion, A. (1998). On the chemical mechanism of surface enhanced Raman scattering: experiment and theory. *J. Chem. Phys.* 108 (12): 5013–5026.

82 Ikeda, K., Suzuki, S., and Uosaki, K. (2013). Enhancement of SERS background through charge transfer resonances on single crystal gold surfaces of various orientations. *J. Am. Chem. Soc.* 135 (46): 17387–17392.

83 Lombardi, J.R., Birke, R.L., Sanchez, L.A. et al. (1984). The effect of molecular structure on voltage induced shifts of charge transfer excitation in surface enhanced Raman scattering. *Chem. Phys. Lett.* 104 (2): 240–247.

84 Campion, A., Ivanecky, J. III, Child, C., and Foster, M. (1995). On the mechanism of chemical enhancement in surface-enhanced Raman scattering. *J. Am. Chem. Soc.* 117 (47): 11807–11808.

85 Hutter, E. and Fendler, J.H. (2004). Exploitation of localized surface Plasmon resonance. *Adv. Mater.* 16 (19): 1685–1706.

86 Banholzer, M.J., Millstone, J.E., Qin, L., and Mirkin, C.A. (2008). Rationally designed nanostructures for surface-enhanced Raman spectroscopy. *Chem. Soc. Rev.* 37 (5): 885–897.

87 Bell, S.E. and McCourt, M.R. (2009). SERS enhancement by aggregated Au colloids: effect of particle size. *Phys. Chem. Chem. Phys.* 11 (34): 7455–7462.

88 Choi, S., Jeong, H., Choi, K.H. et al. (2014). Electrodeposition of triangular Pd rod nanostructures and their electrocatalytic and SERS activities. *ACS Appl. Mater. Interfaces* 6 (4): 3002–3007.

89 Ge, D., Wei, J., Ding, J. et al. (2020). Silver nano-dendrite-plated porous silicon substrates formed by single-step electrochemical synthesis for surface-enhanced Raman scattering. *ACS Appl. Nano Mater.* 3 (3): 3011–3018.

90 Jeong, H. and Kim, J. (2015). Electrodeposition of nanoflake Pd structures: structure-dependent wettability and SERS activity. *ACS Appl. Mater. Interfaces* 7 (13): 7129–7135.

91 Raveendran, J., Stamplecoskie, K.G., and Docoslis, A. (2020). Tunable fractal nanostructures for surface-enhanced Raman scattering via templated electrodeposition of silver on low-energy surfaces. *ACS Appl. Nano Mater.* 3 (3): 2665–2679.

92 Kahl, M., Voges, E., Kostrewa, S. et al. (1998). Periodically structured metallic substrates for SERS. *Sensors Actuators B Chem.* 51 (1–3): 285–291.

93 Yu, Q., Guan, P., Qin, D. et al. (2008). Inverted size-dependence of surface-enhanced Raman scattering on gold nanohole and nanodisk arrays. *Nano Lett.* 8 (7): 1923–1928.

94 Wu, T. and Lin, Y.W. (2018). Surface-enhanced Raman scattering active gold nanoparticle/nanohole arrays fabricated through electron beam lithography. *Appl. Surf. Sci.* 435: 1143–1149.

95 Chisanga, M., Linton, D., Muhamadali, H. et al. (2020). Rapid differentiation of *Campylobacter jejuni* cell wall mutants using Raman spectroscopy, SERS and mass spectrometry combined with chemometrics. *Analyst* 145 (4): 1236–1249.

96 D'Apuzzo, F., Sengupta, R.N., Overbay, M. et al. (2020). A generalizable single-chip calibration method for highly quantitative SERS via inkjet dispense. *Anal. Chem.* 92 (1): 1372–1378.

97 Joshi, P. and Santhanam, V. (2018). Inkjet-based fabrication process with control over the morphology of SERS-active silver nanostructures. *Ind. Eng. Chem. Res.* 57 (15): 5250–5258.

98 Kumar, A. and Santhanam, V. (2019). Paper swab based SERS detection of non-permitted colourants from dals and vegetables using a portable spectrometer. *Anal. Chim. Acta* 1090: 106–113.

99 Lan, L.L., Hou, X., Gao, Y. et al. (2020). Inkjet-printed paper-based semiconducting substrates for surface-enhanced Raman spectroscopy. *Nanotechnology* 31 (5): 055502.

100 Li, L. and Xiao, G.N. (2019). Research progress of preparing surface-enhanced Raman scattering active substrates by printing technologies. *Spectrosc. Spectr. Anal.* 39 (11): 3326–3332.

101 Kumar, S., Namura, K., Kumani, D. et al. (2020). Highly reproducible, large scale inkjet-printed Ag nanoparticles-ink SERS substrate. *Results in Materials* 8: 100139.

102 Micciche, C., Arrabito, G., Amato, F. et al. (2018). Inkjet printing Ag nanoparticles for SERS hot spots. *Anal. Methods* 10 (26): 3215–3223.

103 Oravec, M., Sasinkova, V., Tomanova, K. et al. (2018). *In-situ* surface-enhanced Raman scattering and FT-Raman spectroscopy of black prints. *Vib. Spectrosc.* 94: 16–21.

104 Weng, G.J., Yang, Y., Zhao, J. et al. (2018). Preparation and SERS performance of Au NP/paper strips based on inkjet printing and seed mediated growth: the effect of silver ions. *Solid State Commun.* 272: 67–73.

105 Lee, S.J., Morrill, A.R., and Moskovits, M. (2006). Hot spots in silver nanowire bundles for surface-enhanced Raman spectroscopy. *J. Am. Chem. Soc.* 128 (7): 2200–2201.

106 Wang, J.C., Qiu, C., Mu, X. et al. (2020). Ultrasensitive SERS detection of rhodamine 6G and *p*-nitrophenol based on electrochemically roughened nano-Au film. *Talanta* 210: 120631.

107 Li, Z.B., Zhang, L.J., He, X., and Chen, B.S. (2020). Urchin-like ZnO-nanorod arrays templated growth of ordered hierarchical Ag/ZnO hybrid arrays for surface-enhanced Raman scattering. *Nanotechnology* 31 (16): 165301.

108 Wu, Z.G., Liu, Y.M., Wu, M.M. et al. (2020). Preparation and surface enhanced Raman spectroscopy of Au/TiN composite films. *Spectrosc. Spectr. Anal.* 40 (2): 420–426.

109 Lu, S.C., You, T., Yang, N. et al. (2020). Flexible SERS substrate based on Ag nanodendrite-coated carbon fiber cloth: simultaneous detection for multiple pesticides in liquid droplet. *Anal. Bioanal. Chem.* 412 (5): 1159–1167.

110 Zavatski, S., Redko, S., and Bandarenka, H. (2019). Shelf life improvement of SERS-active substrates based on copper and porous aluminum oxide. *Int. J. Nanosci.* 18 (3–4): 1940074.

111 Lu, S.C., You, T., Gao, Y. et al. (2019). Rapid fabrication of three-dimensional flower-like gold microstructures on flexible substrate for SERS applications. *Spectrochim. Acta A Mol. Biomol. Spectrosc.* 212: 371–379.

112 Klutse, C.K., Mayer, A., Wittkamper, J., and Cullum, B.M. (2012). Applications of self-assembled monolayers in surface-enhanced Raman scattering. *J. Nanotechnol.* 2012: 319038.

113 Parnsubsakul, A., Ngoensawat, U., Wutikhun, T. et al. (2020). Silver nanoparticle/bacterial nanocellulose paper composites for paste-and-read SERS detection of pesticides on fruit surfaces. *Carbohydr. Polym.* 235: 115956.

114 Xie, J., Li, L., Khan, I.M. et al. (2020). Flexible paper-based SERS substrate strategy for rapid detection of methyl parathion on the surface of fruit. *Spectrochim. Acta A Mol. Biomol. Spectrosc.* 231: 118104.

115 Chen, Y., Liu, H., Tian, Y. et al. (2020). In situ recyclable surface-enhanced Raman scattering-based detection of multicomponent pesticide residues on fruits and vegetables by the flower-like MoS2@Ag hybrid substrate. *ACS Appl. Mater. Interfaces* 12 (12): 14386–14399.

116 Teixeira, C.A. and Poppi, R.J. (2020). Paper-based SERS substrate and one-class classifier to monitor thiabendazole residual levels in extracts of mango peels. *Spectrochim. Acta A Mol. Biomol. Spectrosc.* 229: 117913.

117 Huang, D.D., Zhao, J.C., Wang, M.L., and Zhu, S.H. (2020). Snowflake-like gold nanoparticles as SERS substrates for the sensitive detection of organophosphorus pesticide residues. *Food Control* 108: 106835.

118 Mekonnen, M.L., Chen, C.H., Osada, M. et al. (2020). Dielectric nanosheet modified plasmonic-paper as highly sensitive and stable SERS substrate and its application for pesticides detection. *Spectrochim. Acta A Mol. Biomol. Spectrosc.* 225: 117484.

119 Sitjar, J., Liao, J., Lee, H. et al. (2019). Ag nanostructures with spikes on adhesive tape as a flexible sers-active substrate for *in situ* trace detection of pesticides on fruit skin. *Nano* 9 (12): 1750.

120 Chen, X., Lin, H., Xu, T. et al. (2020). Cellulose nanofibers coated with silver nanoparticles as a flexible nanocomposite for measurement of flusilazole residues in Oolong tea by surface-enhanced Raman spectroscopy. *Food Chem.* 315: 126276.

121 Quan, Y.N., Yao, J.C., Yang, S. et al. (2020). Detect, remove and re-use: sensing and degradation pesticides via 3D tilted ZMRs/Ag arrays. *J. Hazard. Mater.* 391: 122222.

122 Muhammad, M., Yao, G.H., Zhong, J. et al. (2020). A facile and label-free SERS approach for inspection of fipronil in chicken eggs using SiO_2@Au core/shell nanoparticles. *Talanta* 207: 120324.

123 Zhao, H., Hasi, W.L.J., Li, N. et al. (2019). *In situ* analysis of pesticide residues on the surface of agricultural products via surface-enhanced Raman spectroscopy using a flexible Au@Ag-PDMS substrate. *New J. Chem.* 43 (33): 13075–13082.

124 Guselnikova, O., Postnikov, P., Elashnikov, R. et al. (2019). Metal-organic framework (MOF-5) coated SERS active gold gratings: a platform for the selective detection of organic contaminants in soil. *Anal. Chim. Acta* 1068: 70–79.

125 Hussain, A., Pu, H.B., and Sun, D.W. (2020). Cysteamine modified core-shell nanoparticles for rapid assessment of oxamyl and thiacloprid pesticides in milk using SERS. *J. Food Meas. Charact.* 14 (4): 2021–2029.

126 Fu, F.Y., Yang, B.B., Hu, X.M. et al. (2020). Biomimetic synthesis of 3D Au-decorated chitosan nanocomposite for sensitive and reliable SERS detection. *Chem. Eng. J.* 392: 123693.

127 Liu, Y., Zhou, F., Wang, H.C. et al. (2019). Micro-coffee-ring-patterned fiber SERS probes and their *in situ* detection application in complex liquid environments. *Sensors Actuators B Chem.* 299: 126990.

128 Zhang, C.M., You, T.T., Yang, N. et al. (2019). Hydrophobic paper-based SERS platform for direct-droplet quantitative determination of melamine. *Food Chem.* 287: 363–368.

5

Waterborne Bacteria Detection Based on Electrochemical Transducer

Nasrin Razmi, Magnus Willander, and Omer Nur

Department of Sciences and Technology, Physics and Electronics, Linköping University, Norrköping, Sweden

5.1 Introduction

Water and food are important resources and vital components of life, and they are highly susceptible to contamination by pathogenic bacteria [1, 2]. Pathogenic microorganisms are the major global public health and ecosystems problem as they are the key concern for environmental biology, food industry, water supplies, and hospitals as they cause a wide variety of infectious diseases even at infinitesimal amounts and leading to hospitalization and millions of deaths each year [3–5]. Contamination of many sources of water with *Shigella, Campylobacter, Staphylococcus, Salmonella, Bacillus, Vibrio, Escherichia coli*, etc., leads to gastroenteritis, typhoid fever, cholera, and serious diarrheal illnesses [6]. Based on the 2016 report of World Health Organization (WHO), 829 000 annual deaths were due to unsafe water, sanitation, and hygiene [7]. The reports of WHO highlights the urgent requirement of simple, rapid, accurate, easy-to-use, sensitive, and selective system for pathogenic microorganisms detection in food safety and water monitoring. The conventional standard microbiological methods involve preenrichment, selective enrichment, biochemical screening, serological and toxin confirmation lengthening the experimental time from two to six days for the confirmation and results [8]. Routine pathogenic bacteria detection methods are plate culture, enzyme-linked immunosorbent assay (ELISA), and polymerase chain reaction (PCR) which require skilled personnel and are time-consuming [9]. Other less common methods for pathogenic bacteria detection are ATP

Sensing Technologies for Real Time Monitoring of Water Quality, First Edition.
Libu Manjakkal, Leandro Lorenzelli, and Magnus Willander.
© 2023 The Institute of Electrical and Electronics Engineers, Inc.
Published 2023 by John Wiley & Sons, Inc.

bioluminescence, solid-phase cytometry, and flow cytometry. Although PCR and ELISA approaches are highly sensitive, selective, and accepted methods, they have disadvantages and limitations such as the need of specific instrumentation, trained technicians, high amount of biological waste production, incorrect sampling period, etc. [8, 10, 11]. To overcome these drawbacks, rapid, sensitive, reliable, simple, and selective methods which provide accurate detection and real-time monitoring of pathogenic microorganisms are required. Biosensors with the improvement of nanotechnology have provided an easy-to-use, cost-effective, and simple platform. This chapter aims to overview the most recent advances in electrochemical biosensors for some of the most important waterborne pathogens.

5.2 Typical Waterborne Pathogens

The waterborne diseases caused by waterborne bacteria are the global public health concern for their morbidity, mortality, and high treatment cost. Table 5.1 summarizes the most important waterborne bacteria and their associated diseases.

5.3 Traditional Diagnostic Tools

Food safety control and water quality monitoring are the primary factors for the prevention of infections and diseases associated with pathogenic microorganisms. A variety of methods have been studied and established for pathogen detection. These methods can be categorized into three main parts, including culture-based testing, molecular methods, and immunological methods [1]. Culture-based method, also known as the golden standard, is the first developed method for pathogen detection and this method has been widely employed for bacteria detection due to its cost-effectiveness and simplicity [31]. Enrichment, plating, incubation of plates, counting colonies, biochemical screening, and serological confirmation are the main steps in culture-based testing to provide accurate, sensitive, and inexpensive detection strategy, making this method labor-intensive and time-consuming.

Moreover, culture-based method is not suitable to detect viable but noncultur-able bacteria such as *Helicobacter pylori* and *E. coli*. Due to these limitations, the application of culture-based testing for on-site, rapid, and multiplexed bacteria detection is not possible, and therefore molecular and immunological methods are more promising [1]. With the advancements in biotechnology, nucleic acid-based molecular methods have offered rapid detection techniques compared to culture-based methods. Hybridization and amplification are the main strategies

Table 5.1 Microbial bacteria associated with waterborne diseases.

Pathogen	Disease	References
Helicobacter Pylori	Peptic ulcers, gastric cancer	[12]
Shigella spp.	Pneumonia, diarrhea, acute gastroenteritis, fever, abdominal pain, shigellosis	[13, 14]
Clostridium difficile	Sepsis, antibiotic-related diarrhea, toxic megacolon, ileus, pseudomembranous colitis	[15, 16]
Escherichia coli 0157	Abdominal cramps, kidney failure, acute gastritis, gastric cancer, hemorrhagic colitis, hemolytic uremic syndrome	[17–20]
Klebsiella spp.	Urinary tract infections, pneumonia, nosocomial infections, meningitis, sepsis	[21]
Vibrio spp.	Acute gastroenteritis diarrhea, vomiting, headache	[22]
Salmonella enterica	Diarrhea, salmonellosis	[23–25]
Legionella pneumophila	Legionnaires' disease	[26]
Salmonella typhi	Typhoid fever	[27]
Yersinia enterocolitica	Yersiniosis, abdominal pain, fever, diarrhea	[28]
Pseudomonas aeruginosa	Otitis, pneumonia, keratitis, endocarditis, septicaemia	[29]
Nontuberculous mycobacteria	Skin infection, pulmonary disease	[25, 30]

in nucleic acid-based molecular methods. Oligonucleotide microarray, DNA sequencing and matching, *in situ* fluorescence hybridization (FISH), and PCR are the molecular techniques used for pathogen detection. Despite offering high specificity and selectivity, all abovementioned molecular-based techniques have some major drawbacks such as high cost, false-positive results, pre-enrichment steps, and the need of trained technician [31]. Immunological techniques based on antigen and antibody binding are other established methods used for foodborne and waterborne pathogen detection. Enzyme-linked fluorescent assay (ELFA), bioluminescent enzyme immunoassay (BEIA), enzyme immunoassay (EIA), and ELISA are different formats of the immunological method developed for pathogen detection. Competitive ELISAs, sandwich ELISAs, and direct ELISAs are different types of commercially available ELISA which is the most prevalent method for sensitive and selective pathogen detection. Although these techniques are reliable and error-proof for detection of the pathogens, they are expensive and time-consuming. Blocking, washing, incubation, and substrate development are the typical steps of ELISA which are problematic and requiring a long time to complete [10, 32]. Thus, the approachable rapid detection strategies are of great importance for foodborne and waterborne bacteria detection.

5.4 Biosensors for Bacteria Detection in Water

During the last decades, biosensor technology has gained a lot of attention for simple, selective, sensitive, and rapid bacteria detection. The concept of a biosensor, namely "enzyme electrode" was first demonstrated by Clark and Lyon [33]. International Union of Pure and Applied Chemistry (IUPAC) defined a biosensor as a tool which employs a biorecognition part or receptor in direct spatial contact with a transduction part to produce thermal, optical, or electrical signal [34]. A biosensor for bacteria detection can be described as an analytical device consisting of a biorecognition element such as antibodies, antigens, nucleic acids, enzymes, whole-cell and tissues, and a transducer converting a biochemical reaction with the bacteria as a target analyte into a measurable signal. In a label-free biosensor, a measurable signal can be generated directly from the biorecognition event, but in a labeled biosensor, a label as a secondary molecule is employed to generate a measurable signal [35]. Simplicity, compatibility, specificity, robustness, reproducibility, sensitivity, no need of skilled personnel, the ability of automation, providing a real-time response, and accuracy are the main and important features of the desired biosensor for bacteria detection [36]. The appropriate choice of a transducer, bioreceptor, and their immobilization strategies are the key elements of a selective biosensor for bacteria detection. Generally, transducers can be categorized into four groups, including thermal, piezoelectric, optical, and electrochemical [3, 37]. In the last decades, different kinds of nanomaterials and bioreceptors have been employed for the development of biosensors for foodborne and waterborne pathogenic bacteria detection to overcome the drawbacks of conventional detection methods. Biosensors, especially electrochemical transducers, have gained selective, sensitive, and real-time detection of bacteria in food safety and environmental monitoring and are expected as next-generation tools for bacteria detection.

5.4.1 Common Bioreceptors for Electrochemical Sensing of Foodborne and Waterborne Pathogenic Bacteria

The recognition element or bioreceptor such as aptamers, enzymes, antibodies, etc., is the crucial part of a biosensor and its interaction with the analyte results in a biorecognition event. Selectivity and specificity of the bioreceptors toward the target molecule and their capability of being easily functionalized are the main key advantages of them. However, for better application, batch-to-batch variation, stability, and the cost of them need to be considered and improved [38]. Biocatalytic and biocomplexing are the typical classes of biorecognition elements for electrochemical biosensors. The generated response when using a biocatalytic biorecognition element is based on the biorecognition reaction while the interaction between analytes and organized molecular assemblies or macromolecules generates the biosensor response in the case of biocomplexing [39]. In this section, the most common bioreceptors used in electrochemical biosensors for pathogen detection are introduced.

5.4.1.1 Antibodies

Antibodies also named as immunoglobulins which can recognize antigens are one of the most preferred bioreceptors for waterborne pathogen detection using an electrochemical transducer. Monoclonal and polyclonal antibodies are the different types of antibodies that widely have been used in electrochemical biosensors named immunosensors [38]. Recognition sites of antibodies selectively can bind to specific regions of antigens, namely epitopes. Polyclonal antibodies are capable of binding to multiple epitopes in an antigen, while monoclonal antibodies only bind to a single epitope [35]. Comparing to polyclonal antibodies, although monoclonal antibodies have higher selectivity as they only bind to a single epitope, they require a long producing time, and their cost is high [39]. Due to the binding affinity and selectivity of the antibodies toward the target molecules, biosensors based on antibodies are highly proposed and even commercialized for foodborne and waterborne bacteria detection.

5.4.1.2 Enzymes

Enzymes are bioreceptors which are specific to target molecules, and by decreasing the activation energy, they increase the rate of the reaction. An appropriate substrate with a catalytic enzyme generates electrons which pass on to the transducer [40]. Enzymes mostly are used as labels conjugated with aptamers or antibodies to increase their sensitivity, rather than a recognition element [41].

5.4.1.3 DNA and Aptamers

DNA biorecognition elements match with their target DNA and by triggering, the transduction part of a biosensor can be detected offering high specificity, cost-effective, simple, and rapid measurements. However, in the case of the DNA damage, specificity and sensitivity would decrease. Aptamers are single-stranded RNA, DNA, or peptide molecules which can bind to the target molecules with high selectivity and affinity and they are commonly used as a bioreceptor for bacteria detection using an electrochemical transducer. They are selected from a library of random sequences through a systematic revolution of ligands by exponential enrichment strategy (SELEX) [42]. Selectivity and specificity toward the target molecule, ease of functionalization, no batch to batch variability, low cost, and stability toward the temperature are the main advantages of aptamers over antibodies [43]. Looking at recent publications, a variety of studies have been reported using aptamers as a recognition element toward the bacteria detection.

5.4.1.4 Phages
Bacteriophages (phages) are viruses that specifically recognize and capture the bacteria through tail-spike proteins and replicate in bacteria [44]. Considering the advantages of specificity and host selectivity, different kinds of electrochemical-based biosensors using phages as biorecognition element for bacteria detection have been reported.

5.4.1.5 Cell and Molecularly Imprinted Polymers
Recently, efforts have been focused to make engineered molecular bioreceptors. Cell- and molecularly imprinted polymers are the major common strategies in materials-based biorecognition elements. Biomimetic recognition elements are artificially synthetic biorecognition elements to mimic enzymes, nucleic acids, and antibodies as a recognition element [10]. Molecularly imprinted polymer (MIP) is a synthetic polymer that has been created by the molecular imprinting technique around a molecule as a synthetic recognition site. MIPs overcome the challenges of biorecognition elements due to their reusability, stability, the possibility to be fabricated for any target molecule and selectivity [38]. Cell-based biosensors are basically engineered living cells for detection of targets with high specificity and sensitivity in a fast, inexpensive, and noninvasive manner. Cells which could be sensitive to an analyte maintain in their natural relevant environment conditions to stay alive for long period and survive [45, 46]. In general, a cell-based biosensor for bacteria detection uses living cells as recognition elements which is combined with a transducer via the interaction between stimulus and cells. Due to the advantages of cell-based biosensors, they have been utilized in medical diagnosis and environmental monitoring [47, 48]. Optimization of a cell-based biosensor and the improvement of its selectivity still remain challenging.

5.4.2 Nanomaterials for Electrochemical Sensing of Waterborne Pathogenic Bacteria

The recent progress in nanotechnology has led to the development of different nanomaterial-based biosensors for rapid, inexpensive, and accurate bacteria detection. Compared to the bulk counterparts of the nanomaterials, nanoparticles and nanomaterials possess enhanced surface reactivity, and their integration into the biosensors have offered improvement in the selectivity and sensitivity, multiple detection, rapidness, and miniaturization for bacteria detection in environmental monitoring [3, 32]. According to the International Organization for Standardization (ISO), a nanomaterial is made of many nanoparticles, which are in the range of 1–100 nm [49]. The unique physical, chemical, mechanical, magnetic, catalytic, and optical properties of nanomaterials with high surface-to-volume ratio, reactivity, and high penetrability allowed the use of a variety of advanced nanomaterials to improve the specificity and sensitivity of the biosensors for microbial detection [32, 50]. Moreover, these technological advancements have allowed for tuning the nanomaterial properties by changing the morphology, shape, size, composition and surface area, surface charge, modifying different adsorption capacities with proper functionalization with the utilization of surface chemistry by reducing sensing time and multiplexing capability [3, 51]. In majority of the fabricated biosensors, nanomaterials have the role of transducers, and in some biosensors, functionalization of nanomaterials and nanoparticles give the role of both recognition element and transducer [38]. The selection of the nanomaterials for the fabrication of a biosensor depends on the properties of the nanomaterials and their application. Up to date, different kinds of nanomaterials have been used in the design of microbial biosensors and in this section, various types of nanomaterials are discussed.

5.4.2.1 Metal and Metal Oxide Nanoparticles

Different kinds of metal and metal oxide nanoparticles such as gold nanoparticles, silver nanoparticles, iron oxide nanoparticles, etc., have been used for designing of the electrochemical biosensors for bacteria detection. Among them, gold nanomaterials also known as gold sol (colloid in which solid particles are dispersed in continuous liquid phase) are one of the widely used nanomaterials for bacterial detection owing to their distinct physiochemical, optical, and electronic features, ease of synthesis and modification and manipulation while synthesizing, high surface-to-volume ratio, biocompatibility, functionalization with various biological recognition elements, low toxicity, and the possibility of controlling the physicochemical properties [43, 52]. Various biosensors have been successfully designed for the bacteria detection employing different kinds of gold nanomaterials including gold nanoparticles, gold nanorods, gold nanoclusters, gold composites, gold alloy nanoparticles, and gold nanomaterials of different composition and dimension. Silver (Ag) nanoparticles and nanostructures are other kinds of noble nanomaterials which can be prepared through different physical, biological, and chemical procedures and utilized for bacteria detection [3]. Advantages of inexpensive synthesis, efficient functionalization due to their large surface-to-volume ratio, and physicochemical stability of magnetic nanoparticles (MNPs) also have attracted a lot of consideration for their application in bacteria detection [53, 54]. Metal oxides are another promising materials that aroused much attention as immobilizing matrix with desired orientation and high biological activity for biosensor development. Metal oxides are electroactive materials which provide many advantages such as enhanced electrochemical reversible redox behavior, large surface-to-volume ratio, wide potential window, surface charge properties, etc., leading to enhanced electron transfer and improved biosensing features for bacteria detection [55].

5.4.2.2 Conducting Polymeric Nanoparticles

The property of electrical conductance of intrinsically conductive polymers (CPs) or synthetic metals like poly(3,4-ethylenedioxythiophene) (PEDOT), polypyrrole, polyaniline, etc., have made them one of the most widely studied electronic materials which have received significant attention in the fabrication of biosensors [56, 57]. Various kinds of CPs such as nanofibers, nanowires, nanotubes, and nanoparticles have been synthesized and employed in the fabrication of biosensors [38, 58]. To improve the electrical properties of the electrochemical biosensors, different nanocomposites including CNs with graphene, carbon nanotubes (CNTs), and metal nanoparticles have been designed. Different recognition elements such as antibodies, enzymes, etc., have been used to improve the magnetic, optical, and electronic properties of conducting polymers for the design of inexpensive, simple, sensitive, selective, and easy CP-based electrochemical biosensors providing a direct electrical readout platform [38].

5.4.2.3 Carbon Nanomaterials

The mechanical and electrical properties, surface area, the possibility for real-time applications, being inexpensive and stable with time, various shapes, and characteristics of carbon-based nanomaterials have made carbon nanomaterials one of the widely used nanomaterials for biosensor fabrication. As an important kind of carbon nanomaterials, graphene has received great attention due to its interesting electrical, magnetic, optical, and mechanical properties. Graphene quantum dots (GQDs), reduced graphene oxide (rGO), and graphene oxide are the different forms of graphene which have been incorporated mostly in the fabrication of electrochemical biosensors [59]. The interesting catalytic, mechanical, and electrical properties of the allotropes of carbon-like CNTs composing of either multiple walls or single wall and fullerenes have made them as one of the increasingly used nanomaterials to improve the performance of different electrochemical biosensors [38, 60].

5.4.2.4 Silica Nanoparticles

Taking the advantages of non-toxicity, biocompatibility, and high chemical and physical stability of silica nanoparticles (SiNPs) against various conditions, different sensing platforms in conjugation with various recognition elements have been proposed [61, 62]. SiNPs with a size range of 5–1000 nm have been used in electrochemical biosensors for microbial detection [43].

5.5 Various Electrochemical Biosensors Available for Pathogenic Bacteria Detection in Water

Among various forms of biosensors, electrochemical transducer due to its selectivity, high sensitivity, measurability in complex and turbid samples, simple structure and miniaturization, rapid response, and low cost is the most prominent type of biosensors allowing the detection of microorganisms in real time [63]. Electrochemical biosensors are based on measuring the interaction between the biorecognition element and samples, which generates detectable parameters including impedance, conductance, current, and potential difference. Accordingly, electrochemical biosensors are divided into four categories of impedimetric, conductometric, amperometric, and potentiometric [10]. Table 5.2 summarizes the different waterborne bacteria detection based on the electrochemical transducer in different samples.

5.5.1 Amperometric Detection

Amperometric detection methodology is one of the common studied methods for the detection of microorganisms after the first simple amperometric detection of glucose by glucose enzyme electrode developed by Clark and Lyons [33]. The principle of the amperometric biosensor for pathogen detection is based on the current measurement as a result of electrochemical reactions on the electrode surface under a given potential for further analysis [116, 117]. Low fabrication cost, selectivity, simplicity, high sensitivity, integration with other biosensing technologies, and portability are the advantages of amperometric detection [116]. Generation of false signal and signal reduction from fouling agents and interferents are the major drawbacks of the amperometric detection. During the detection at the constant potential, the presence of the charging current results in the minimized background signal and accordingly affects the detection limit [116, 118]. Various amperometric methods for the detection of pathogens have been reported. Alexandre et al. proposed an amperometric biosensor based on self-assembled monolayers technique and labeled with peroxidase enzyme for sensitive and rapid detection of *Salmonella typhimurium* in milk samples. For the proper orientation of polyclonal antibodies, protein A has been used. With a detection time of 125 minutes, low limit of detection of 10 CFU/ml was achieved. The fabricated sensor does not require any pre-enrichment steps which is an

Table 5.2 Summary of the reported electrochemical-based biosensors for waterborne bacteria detection in different samples.

Transducer	Target	Sample	Material	Linear range	LOD	References
Amperometric	E. coli	Apple juice	Au NPs	10 to 10^9 CFU/ml	10 CFU/ml	[64]
Amperometric	E. coli O157:H7	Standard sample	3-Aminipropyl triethoxysilane (APTES)	1 fM to 10 μM	0.8 fM	[65]
Amperometric	E. coli O157:H7	Standard sample	Au@Pt/SiO$_2$ NPs and Fe$_3$O$_4$@ SiO$_2$ NPs	3.5×10^2 to 3.5×10^8 CFU/ml	1.83×10^2 CFU/ml	[66]
Amperometric	E. coli O157:H7	Milk	Au@Pt/SiO$_2$ NPs and Fe$_3$O$_4$@ SiO$_2$ NPs	–	3.5×10^2 CFU/ml	[66]
Amperometric	E. coli O157:H7	Standard sample	Silica coated Fe$_3$O$_4$ magnetic nanoparticles and Au@Pt nanoparticles	4×10^3 to 4×10^8 CFU/ml	4.5×10^2 CFU/ml	[67]
Amperometric	E. coli O157:H7	Standard sample	Carboxylated graphene nanoflakes (Cx-Gnfs)	10^{-6} to 10^{-17} M	10^{-17} M	[68]
Amperometric	E. coli O157:H7	Culture	Core–shell magnetic beads and Au NPs	10^2 to 10^6 CFU/ml	52 CFU/ml	[69]
Amperometric	E. coli O157:H7	Food	Core–shell magnetic beads and Au NPs	10^3 to 10^6 CFU/ml	190 CFU/ml	[69]
Amperometric	E. coli O157:H7	PBS buffer	PPy/AuNP/MWCNT/Chi bionanocomposite	3×10 to 3×10^7 CFU/ml	30 CFU/ml	[57]
Amperometric	Salmonella	PBS	Chitosan (Chit)/gold nanoparticles (GNPs)	1.0×10^1 to 5.0×10^4 CFU/ml	5 CFU/ml	[70]
Amperometric	S. aureus	Culture	SWCNT	10^2 to 10^5 CFU/ml	10^2 CFU/ml	[71]
Amperometric	S. aureus	Culture	SWCNT	10^2 to 10^5 CFU/ml	10^2 CFU/ml	[72]
Amperometric	Vibrio cholerae	Standard sample	–	1 to 10^4 CFU/ml	10 CFU/ml	[73]

Method	Target	Sample	Material	Linear range	LOD	Ref.
Amperometric	Clostridium tetani	Standard sample	—	0.01 IU/ml and 0.25 IU/ml	0.011 IU/ml	[74]
Amperometric	Listeria monocytogenes	Standard sample	Silica MBs	—	$0.13\,\mu l^{-1}$	[75]
Amperometric	E. coli	Standard sample	Silica MBs	—	$0.05\,\mu l^{-1}$	[75]
Amperometric	Salmonella typhimurium	Milk	—	—	10 CFU/ml	[76]
Amperometric	Salmonella pullorum	Standard sample	(rGO/AuNPs)	10^2 to 10^6 CFU/ml	89 CFU/ml	[77]
Amperometric	E. coli O157:H7	Standard sample	Nickel oxide	10^1 to 10^7 cells/ml	1 cell/ml	[78]
Conductometric	E. coli	Spiked tap water samples	Magnetic beads	2.5×10^3 to 2.5×10^8 CFU/ml	2.3×10^4 CFU/ml	[79]
Conductometric	Staphylococcus aureus	Spiked tap water samples	Magnetic beads	4.1×10^3 to 4.1×10^8 CFU/ml	4.0×10^3 CFU/ml	[79]
Impedimetric	E. coli O157:H7	Standard sample	Gold nanofilm	50–500 CFU/ml	50 CFU/ml	[80]
Impedimetric	E. coli O157:H7	Standard sample	Streptavidin modified magnetic nanoparticles, Gold NPs	10 to 10^5 CFU/ml	12 CFU/ml	[81]
Impedimetric	E. coli O157:H7	Spiked milk	Streptavidin modified magnetic nanoparticles, Gold NPs	1.2×10^1 to 1.2×10^5 CFU/ml	—	[81]
Impedimetric	E. coli O157:H7	Standard sample	Gold thin film	—	13 CFU/ml	[55]
Impedimetric	E. coli O157:H7	Spiked milk	Streptavidin modified MNPs, Au NPs	10 to 10^5 CFU/ml	10 CFU/ml	[82]
Impedimetric	E. coli O157:H7	Standard sample	Graphene oxide Chitosan hybrid nanocomposite	1×10^{-14} to 1×10^{-8} M	3.584×10^{-15} M	[83]
Impedimetric	E. coli O157:H7	Standard sample	Magnetic nanobeads	10^4–10^7 CFU/ml	$10^{4.45}$ CFU/ml	[84]

(Continued)

Table 5.2 (Continued)

Transducer	Target	Sample	Material	Linear range	LOD	References
Impedimetric	E. coli O157:H7	Ground beef sample	Magnetic nanobeads	10^5 to 10^6 CFU/ml	—	[84]
Impedimetric	E. coli	Milk samples	—	7.2×10^0 to 7.2×10^8 cells/ml	7 cells/ml	[18]
Impedimetric	E. coli O157:H7	Ground beef	Streptavidin coated magnetic beads (MBs)	10^2 to 10^6 CFU/ml	2.05×10^3 CFU/ml	[85]
Impedimetric	S. typhimurium	Chicken rinse water	Streptavidin coated magnetic beads (MBs)	10^2 to 10^6 CFU/ml	1.04×10^3 CFU/ml	[85]
Impedimetric	E. coli	Drinking water	Gold print	10 to 10^8 CFU/ml	3×10 CFU/ml	[86]
Impedimetric	E. coli O157:H7	Cultured Bacteria	Au NPs	300 to 10^5 CFU/ml	100 CFU/ml	[87]
Impedimetric	E. coli O157:H7	Spiked potable water sample	3-Aminipropyl trimethoxysilane (APTES) and GA	0.5 to 25 pg/10 ml	0.1 pg/10 ml	[88]
Impedimetric	E. coli	Standard sample	$Cu_3(BTC)_2$/PANI	2 to 2×10^8 CFU/ml	2 CFU/ml	[89]
Impedimetric	E. coli O157:H7	Standard sample	3-Dithiobis-(sulfosuccinimidyl-propionate) (DTSP)	10^2 to 10^7 CFU/ml	10^2 CFU/ml	[90]
Impedimetric	E. coli O157:H7	Standard sample	—	10 to 10^6 CFU/ml	1 CFU/ml	[91]
Impedimetric	E. coli O157:H7	Standard sample	Graphene wrapped copper(II) assisted cysteine hierarchical structure	10 to 10^8 CFU/ml	3.8 CFU/ml	[92]
Impedimetric	E. coli	Standard sample	11-Mercaptoundecanoic acid (MUA) and dithiothreitol (DTT)	1×10^2 to 1×10^5 cells/ml	75 cells/ml	[9]
Impedimetric	Shigella dysenteriae	Standard sample	AuNPs	10^1 to 10^6 CFU/ml	10^0 CFU/ml	[93]

Method	Target	Sample	Material	Linear range	LOD	Ref.
Impedimetric	*S. typhimurium*	Standard sample	MWCNT	—	6.7×10^1 CFU/ml	[60]
Impedimetric	*Clostridium perfringens*	Standard sample	CeO_2/chitosan	1.0×10^{-14} to 1.0×10^{-7} mol/l	7.06×10^{-15} mol/l	[94]
Impedimetric	*E. coli* O157:H7	Standard sample	Polypyrrole (PPy)	10^3 to 10^8 CFU/ml	10^3 CFU/ml	[95]
Impedimetric	*E. coli* O157:H7	Artificially contaminated real samples	Polypyrrole (PPy)	10^3 to 10^5 CFU/ml	—	[95]
Impedimetric	*S. typhimurium*	Spiked water and egg samples	Au NPs	10 to 10^5 CFU/ml	10 CFU/ml	[96]
Impedimetric	*Salmonella typhimurium*	Standard sample	Nanoporous gold (NPG)	6.5×10^2 to 6.5×10^8 CFU/ml	1 CFU/ml	[97]
Impedimetric	*S. aureus*	Spiked pork sausage	—	1.0×10^9 to 1.2 CFU/ml	1.26 CFU/ml	[98]
Impedimetric	*Salmonella typhimurium*	Standard sample	Poly[pyrrole-co-3-carboxyl-pyrrole] copolymer	10^2 to 10^8 CFU/ml	3 CFU/ml	[99]
Impedimetric	*Salmonella typhimurium*	Spiked apple juice sample	Poly[pyrrole-co-3-carboxyl-pyrrole] copolymer	10^2 CFN/ml-(NA)	1.4×10^2 CFU/ml	[99]
Impedimetric	*Salmonella typhimurium*	Spiked apple juice sample	Poly[pyrrole-co-3-carboxyl-pyrrole] copolymer	10^6 CPF/ml-(NA)	4.1×10^6 CFU/ml	[99]
Impedimetric	*Salmonella typhimurium*	Standard sample	Diazonium salt	1×10^1 to 1×10^8 CFU/ml	6 CFU/ml	[100]
Impedimetric	*Salmonella*	Standard sample	Graphene oxide (rGO) and carboxy-modified multi-walled carbon nanotubes (MWCNTs)	75 to 7.5×10^5 CFU/ml	25 CFU/ml	[101]
Impedimetric	*Salmonella*	RTE Turkey food matrix	—	—	300 Cells/ml	[102]

(Continued)

Table 5.2 (Continued)

Transducer	Target	Sample	Material	Linear range	LOD	References
Impedimetric	E. coli, P. aeruginosa, S. aureus, and S. epidermidis	Standard sample	Synthetic antimicrobial peptides (sAMPs)	—	10^2 CFU/ml	[103]
Impedimetric	Vibrio cholerae O1	Standard sample	CeO_2 nanowires	1.0×10^2 to 1.0×10^4 CFU/ml	1.0×10^2 CFU/ml	[104]
Impedimetric	Staphylococcus aureus	NaCl	—	3.6×10^7 to 9.3×10^7 CFU/ml	3.6×10^6 CFU/ml	[105]
Impedimetric	Salmonella typhimurium	Culture sample	—	1.41×10^1 to 1.41×10^5 CFU/50 μl	1.14×10^1 CFU/50 μl	[106]
Impedimetric	E. coli K12	Frozen chicken meat	—	10^3 to 10^5 CFU/ml	10^3 CFU/ml	[107]
Impedimetric	Salmonella typhimurium	Water and juice sample	Graphene-graphene oxide (G-GO)	10^1 to 10^6 CFU/ml	10 CFU/ml	[108]
Potentiometric	E. coli	Orange juice	Poly(vinyl alcohol)/poly(acrylic acid) (PVA/PAA) hydrogel NFs	10^2 to 10^6 CFU/ml	10^2 CFU/ml	[109]
Potentiometric	Salmonella typhimurium	Standard sample	PEDOT:PSS	1 to 1.28×10^5 cells/ml	5 cells/ml	[110]
Potentiometric	Salmonella typhimurium	PB and apple juice	(AuNPs-PIM)	13 to 1.3×10^6 cells/ml	6 cells/ml	[111]
Potentiometric	E. coli	Standard sample	Poly acrylic acid/polyvinyl alcohol (PAA/PVA) hydrogel nanofibers	—	20 CFU/ml	[112]
Potentiometric	E. coli	Standard sample	Carbon quantum dots	2.9 to 2.9×10^6 CFU/ml	0.66 CFU/ml	[113]
Potentiometric	Listeria monocytogenes	Standard sample	MBs	1.0×10^2 to 1.0×10^6 CFU/ml	10 CFU/ml	[114]
Potentiometric	E. coli O157:H7	Standard sample	ZnO nanorod arrays	10 to 10^5 CFU/ml	1.0×10^2 CFU/ml	[115]

important aspect for food quality monitoring [76]. Zhu et al. reported an nonenzymatic sandwich-based immunoassay for *E. coli* O157:H7 (Figure 5.1). To reduce the detection time and increase the sensitivity of the fabricated biosensor, silica-coated Fe_3O_4 magnetic nanoparticles modified with anti-*E. coli* O157:H7 monoclonal antibody (Ab_1) act as a capture probe. Then, rGO-NR-Au@Pt composite with biocompatibility and high surface area was fabricated to act as carriers of detection antibodies (Ab_2). The reduction of hydrogen peroxide was catalyzed by Au@Pt for *E. coli* O157:H7 detection with the thionine (TH) as an electron mediator for current signal amplification. Under the optimized conditions, the fabricated immunoassay showed the LOD of 4.5×10^2 CFU/ml. The performance of the fabricated assay in milk and pork samples was investigated, confirming its application in food safety [67]. Very recently, another amperometric biosensor using NanoZyme-mediated sensing platform and gold nanoparticles reported by Das et al. for the rapid detection of *E. coli* in fruit juice. Low limit of detection of 10 cells/ml was obtained for the *E. coli* detection in apple juice. The proposed rapid and simple assay is capable of being applied in packaged food for monitoring bacterial contamination [64].

5.5.2 Impedimetric Detection

Recently, electrochemical impedance biosensors by employing different types of molecules such as aptamers, antibodies, imprinted polymers, etc., have been widely exploited for the bacteria detection. Sluyters, for the first time, presented the electrochemical impedance spectrum and then in 1975, the electrochemical impedance spectroscopy was defined by Lorenz and Schulze [116, 119]. The basic principle of the impedimetric biosensor is applying a small signal potential in a wide frequency range and then the generated signal is measured as a function of frequencies [63, 118]. Directly monitoring of binding reactions in EIS overcomes the drawback of interference problems. Accuracy of the detection based on EIS depends on the operating process and the instrumental technical precision. Slow response is another disadvantage of the EIS technique. Several studies based on the EIS technique have been reported for various bacteria detection. Chiriacò et al. developed a portable impedimetric biosensor for *Listeria monocytogenes* detection in milk samples based on microfluidic device. Specific antibodies of *L. monocytogenes* was immobilized on the microelectrodes. A low limit of detection of 5.5 CFU/ml in a range of 2.2×10^3 CFU/ml to 1×10^2 was obtained without the requirement of complex instrumentation, PCR amplification, fluorescent tagging of reagents, and skilled operators. The proposed portable and robust platform is highly promising as a point-of-care device for food safety and other areas [120]. Very recently, Cimafonte et al. proposed screen print-based impedimetric immunosensor for *E. coli* detection in drinking water (Figure 5.2). By using the photochemical immobilization

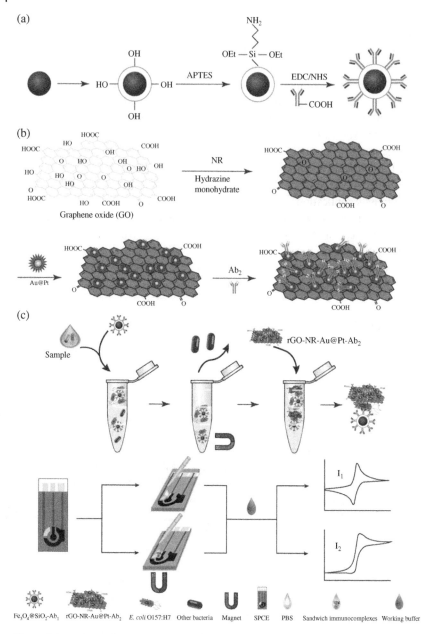

Figure 5.1 Schematic illustration of (a) Fe$_3$O$_4$@SiO$_2$-Ab$_1$ preparation, (b) rGO-NR-Au@ Pt-Ab$_2$ preparation, and (c) detection strategy. *Source:* From Ref. [67].

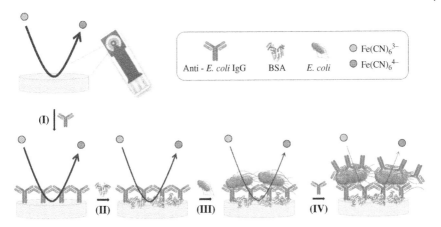

Figure 5.2 Representation of functionalization and detection process of impedimetric biosensor. The black line shows the redox reaction intensity, which is inhibited as the surface covering grows. The reduction of its thickness is related to a reduction of the available "effective" area for the electrolyte current, which is measured as an increase of the charge transfer resistance. *Source:* From Ref. [86]/MDPI/CC BY 4.0.

technique, antibodies covalently were immobilized on the gold electrode surface. Charge transfer resistance change was observed in the redox probe $Fe(CN)_6^{3-}/Fe(CN)_6^{4-}$ using the EIS. Low limit of detection of 3×10^1 CFU/ml was obtained by the fabricated biosensor. Detection time of six hours is needed for the proposed biosensor, which makes the strategy applicable for out-of-lab use [86].

5.5.3 Conductometric Detection

The conductometric biosensor is an analytical device in which electrical conductance is generated from the specific biological recognition event. By employing a low-amplitude alternating electrical potential or voltage, the conductivity of the sample solution is monitored. The consumption or production of charged species changes the conductivity, which is the principle of this technique [39, 116]. The major advantages of conductometric biosensors are the use of inexpensive thin film in their production, no requirement of a reference electrode which enables easy miniaturization, and easily integrated structure [116]. Conductometric biosensors are rarely applied for pathogen detection due to their disadvantages of low specificity and low selectivity. These disadvantages are due to the significant background conductivity while analyzing liquid samples and low signal to noise ratio [63]. Zhang et al. proposed a conductometric assay based on magnetic analyte separation via aptamer for viable *E. coli* and *Staphylococcus aureus* determination (Figure 5.3). By employing aptamer-functionalized magnetic beads, the bacteria

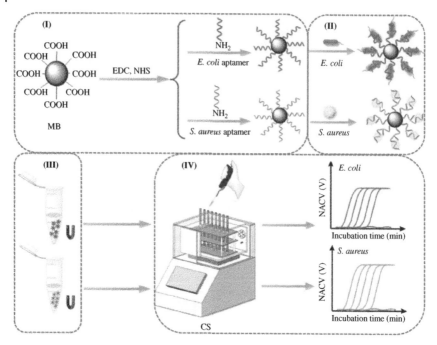

Figure 5.3 Illustration of the process of the MAS@CS method, (I) the aptamer-functionalized magnetic beads preparation, (II) selective capture, (III) bacterial cells separation, and (IV) viable bacteria determination via CS. *Source:* Republished with permission of Springer Nature, from [79], 2019 permission conveyed through Copyright Clearance Center, Inc.

were separated and captured. Then the measurement of their growth kinetics through time-dependent conductivity changes of culture media was recorded. The linear ranges were 4.1×10^3 to 4.1×10^8 CFU/ml and 2.5×10^3 to 2.5×10^8 CFU/ml for *S. aureus* and *E. coli*, respectively. The lower limits of detection of 4.0×10^3 CFU/ml and 2.3×10^4 CFU/ml with recoveries of 92.5–105.0% and 87.0–108.7% in spiked tap water samples for *S. aureus* and *E. coli* was achieved, respectively. The relative standard deviation of the results was 10.0% which was lower than of the 13.9% for the plate counting method. Although the performance of the reported biosensor was facile and accurate, more improvements are required in the sensitivity of the assay by reducing the analyte volume and increasing the incubation time [79].

5.5.4 Potentiometric Detection

Potentiometric methods or controlled-current methods measure the electrical potential in response to an applied current (low amplitude) [39]. The changes in ionic concentration between reference electrode and an ion-selective electrode under the conditions of no current flow is measured by the three-electrode potentiometric system. Potentiometric technique offers the detection of minute concentration changes and direct current changes without the requirement of redox mediators; however, the slow dynamic response is the drawback of this technique [39, 118]. Shaibani et al. reported a portable nanofiber-light addressable potentiometric platform for the detection of *E. coli* in orange juice by employing electrospun pH-sensitive polyvinyl alcohol/polyacrylic acid (PVA/PAA) hydrogel NFs with the detection time of less than one hour (Figure 5.4). The low limit of detection of 10^2 CFU/ml was achieved by the fabricated sensor. The performance of the sensor in orange juice was investigated, confirming its application as a monitoring device in juice safety control [109]. Silva et al. developed a label-free disposable paper-based potentiometric immunosensor for the detection of *S. typhimurium* by employing the blocking surface principle. As a reservoir of the internal solution, a filter paper pad was used. The paper-strip electrode was immobilized with two different immunosensing interfaces. The first interface was based on the direct conjugation of the antibody to the polymer membrane, and the other one was reported to the polyamidoamine dendrimer intermediate layer with an ethylenediamine core of fourth generation. The inexpensive and simple proposed platform consisting of a paper-strip electrode with a carboxylated PVC membrane joined with a filter paper pad could detect the *Salmonella* with a limit of detection of 5 cells/ml in phosphate buffer. The biosensor was applied for *Salmonella* detection in apple juice samples, and 54.00% of average recovery value and 15.00% of

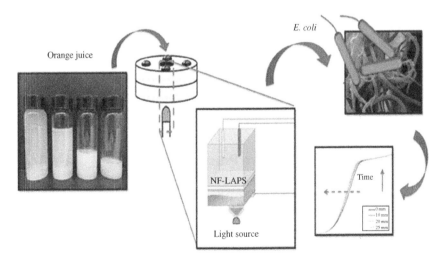

Figure 5.4 Illustration of the potentiometric sensor by employing electrospun pH-sensitive polyvinyl alcohol/polyacrylic acid hydrogel NFs for detection of *E. coli*. *Source:* From Ref. [109]/ P. M. Shaibani et al., (2018), American Chemical Society.

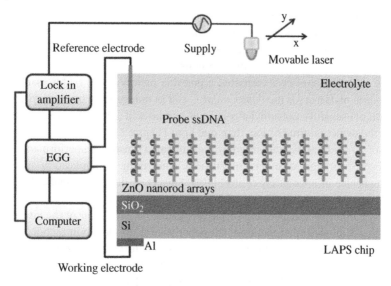

Figure 5.5 Illustration of the measurement setup of the light-addressable potentiometric sensor for detection of *E. coli* O157:H7 DNA by employing ZnO nanorod arrays. *Source:* From Ref. [115]/MDPI/CC BY 4.0.

the coefficient of variation was obtained which was close to the obtained value of phosphate buffer [110]. In another study, Tian et al. proposed a label-free light-addressable potentiometric detection of *E. coli* O157:H7 DNA by employing ZnO nanorod arrays (Figure 5.5). The specific probe single-stranded DNA of the target bacteria was immobilized on the LAPS surface and then by hybridization of the probe DNA with target *E. coli* ssDNA molecule, the surface charge changes induced and was detected by employing LAPS measurements. 1.0×10^2 CFU/ml of *E. coli* O157:H7 cells were detected by the fabricated platform confirming its application in the quality monitoring of water and food [115].

5.6 Conclusion and Future Prospective

Waterborne pathogens are a major global health problem which are responsible for specific health problems and economic losses. The traditional methods of culture plating, PCR, and ELISA require specialist personnel, expensive instrumentation, and are laborious and time-consuming. Thus, methods for rapid, inexpensive, and accurate detection of bacteria are demanded. Electrochemical biosensors, as a powerful analytical devices, provide portable, cost-effective instrumentation, miniaturized, sensitive, rapid on-site, real-time, and multiple analysis for bacteria detection. In this chapter, recent advances in detection techniques for waterborne bacteria were provided. Although various electrochemical biosensors by using different recognition elements along with nanomaterials have been reported in the last five years, including immunosensors, aptasensors, etc., in the field of electrochemical lab-on-a-chip and point-of-care devices for bacteria detection is still an open area for more advancements and progress. More research and improvement in nanotechnology will enable the development of highly sensitive, selective, and inexpensive, and portable nanosensors for waterborne bacteria detection. Looking at the proposed biosensors, although some of the electrochemical biosensors show LODs as low as a single CFU, there are some limitations and weaknesses which should be considered for future studies. Most of the reported biosensors are not able to differentiate between dead and living cells which is one of the important criteria in food safety and water quality monitoring. The accuracy and reliability of the fabricated biosensors is another important criteria for commercialization which require further studies in different complex samples. The selectivity of the most of overviewed biosensors are not studied or only is assessed by testing a limited number of nontarget pathogens. Moreover, the low limit of quantification in the majority of studies have not been reported. Enrichment steps and pretreatments still in most of the reported publications are a limitation. Looking at Table 5.2, the majority of the studies are conducted for *E. coli* O157:H7 and *salmonella*, while there are only a few studies for other pathogenic bacteria detection. Multiplex sensing is another area which needs more improvements. Overall, for water quality monitoring, improvements are required for fabrication of a portable point of care devices with high sensitivity and capability of multiplex sensing and simultaneous detection of multiple pathogens. Incorporation of smartphones for interpreting the measurements from a single system like microfluidic chips will offer rapid, easy, cost affordable analysis and monitoring of bacterial contamination in water and food safety.

Acknowledgments

Authors gratefully acknowledge the financial support and funding from the European Union's Horizon 2020 research and innovation program under the Marie Skłodowska-Curie grant agreement No: H2020-MSCA-ITN-2018-813680.

References

1 Bhardwaj, N., Bhardwaj, S.K., Bhatt, D. et al. (2019). Optical detection of waterborne pathogens using nanomaterials. *TrAC Trends Anal. Chem.* 113: 280–300.
2 Sun, J., Ji, J., Sun, Y. et al. (2015). DNA biosensor-based on fluorescence detection of *E. coli* O157:H7 by Au@Ag nanorods. *Biosens. Bioelectron.* 70: 239–245.

3 Sai-Anand, G., Sivanesan, A., Benzigar, M.R. et al. (2019). Recent progress on the sensing of pathogenic bacteria using advanced nanostructures. *Bull. Chem. Soc. Jpn.* 92 (1): 216–244.

4 Yang, Z., Sun, X.-C., Wang, T. et al. (2015). A giant magnetoimpedance-based biosensor for sensitive detection of *Escherichia coli* O157:H7. *Biomed. Microdevices* 17 (1): 5.

5 Taniguchi, M., Saito, H., and Mitsubayashi, K. (2017). Repetitive Immunosensor with a fiber-optic device and antibody-coated magnetic beads for semi-continuous monitoring of *Escherichia coli* O157:H7. *Sensors* 17 (9): 2145.

6 Banerjee, T., Sulthana, S., Shelby, T. et al. (2016). Multiparametric magneto-fluorescent nanosensors for the ultrasensitive detection of *Escherichia coli* O157:H7. *ACS Infect. Dis.* 2 (10): 667–673.

7 World Health Organization. Drinking water. https://www.who.int/NEWS-ROOM/FACT-SHEETS/DETAIL/FOOD-SAFETY (accessed 2020).

8 Xu, M., Wang, R., and Li, Y. (2017). Electrochemical biosensors for rapid detection of *Escherichia coli* O157:H7. *Talanta* 162: 511–522. https://doi.org/10.1016/j.talanta.2016.10.050.

9 Yang, H., Zhou, H., Hao, H. et al. (2016). Detection of *Escherichia coli* with a label-free impedimetric biosensor based on lectin functionalized mixed self-assembled monolayer. *Sensors Actuators B Chem.* 229: 297–304.

10 Yahaya, M.L., Noordin, R., and Razak, K.A. (2019). Advanced nanoparticle-based biosensors for diagnosing foodborne pathogens. In: *Advanced Biosensors for Health Care Applications* (ed. Inamuddin, R. Khan, A. Mohammad, and A.M. Asiri), 1–43. Elsevier.

11 Guo, Y., Wang, Y., Liu, S. et al. (2015). Electrochemical immunosensor assay (EIA) for sensitive detection of *E. coli* O157:H7 with signal amplification on a SG-PEDOT-AuNPs electrode interface. *Analyst* 140 (2): 551–559. https://doi.org/10.1039/c4an01463d.

12 Twing, K.I., Kirchman, D.L., and Campbell, B.J. (2011). Temporal study of *Helicobacter pylori* presence in coastal freshwater, estuary and marine waters. *Water Res.* 45 (4): 1897–1905. https://doi.org/10.1016/j.watres.2010.12.013.

13 Malham, S.K., Rajko-Nenow, P., Howlett, E. et al. (2014). The interaction of human microbial pathogens, particulate material and nutrients in estuarine environments and their impacts on recreational and shellfish waters. *Environ. Sci. Processes Impacts* 16 (9): 2145–2155.

14 Basri, Z.D.M., Yunus, R.N.R., and Hadi, S. (2018). Detection of pathogenic bacteria in water bodies. *Politeknik Kolej Komuniti J. Eng. Technol.* 3 (1): 98–103.

15 Romano, V., Pasquale, V., Krovacek, K. et al. (2012). Toxigenic *Clostridium difficile* PCR ribotypes from wastewater treatment plants in southern Switzerland. *Appl. Environ. Microbiol.* 78 (18): 6643–6646. https://doi.org/10.1128/AEM.01379-12.

16 Janezic, S., Potocnik, M., Zidaric, V., and Rupnik, M. (2016). Highly divergent *Clostridium difficile* strains isolated from the environment. *PLoS One* 11 (11): e0167101. https://doi.org/10.1371/journal.pone.0167101.

17 Balakrishnan, B., Barizuddin, S., Wuliji, T., and El-Dweik, M. (2016). A rapid and highly specific immunofluorescence method to detect *Escherichia coli* O157:H7 in infected meat samples. *Int. J. Food Microbiol.* 231: 54–62. https://doi. org/10.1016/j.ijfoodmicro.2016.05.017.

18 Liu, J.-T., Settu, K., Tsai, J.-Z., and Chen, C.-J. (2015). Impedance sensor for rapid enumeration of *E. coli* in milk samples. *Electrochim. Acta* 182: 89–95.

19 Kim, J., Kim, M., Kim, S., and Ryu, S. (2017). Sensitive detection of viable *Escherichia coli* O157:H7 from foods using a luciferase-reporter phage phiV10lux. *Int. J. Food Microbiol.* 254: 11–17.

20 Ma, A., Glassman, H., and Chui, L. (2020). Characterization of *Escherichia coli* possessing the locus of heat resistance isolated from human cases of acute gastroenteritis. *Food Microbiol.* 88: 103400. https://doi.org/10.1016/j.fm.2019.103400.

21 Niu, L., Zhao, F., Chen, J. et al. (2018). Isothermal amplification and rapid detection of *Klebsiella pneumoniae* based on the multiple cross displacement amplification (MCDA) and gold nanoparticle lateral flow biosensor (LFB). *PLoS One* 13 (10): e0204332. https://doi.org/10.1371/journal.pone.0204332.

22 Wu, W., Jing, Z., Yu, X. et al. (2019). Recent advances in screening aquatic products for *Vibrio* spp. *TrAC Trends Anal. Chem.* 111: 239–251.

23 Shachar, Y., Wu, W., Chen, T. et al. (2016). Method and apparatus for forming of an automated sampling device for the detection of *Salmonella enterica* utilizing an electrochemical aptamer biosensor. US Patents, filed 10 April 2009.

24 Wang, B., Park, B., Xu, B., and Kwon, Y. (2017). Label-free biosensing of *Salmonella enterica* serovars at single-cell level. *J. Nanobiotechnol.* 15 (1): 40.

25 Deshmukh, R.A., Joshi, K., Bhand, S., and Roy, U. (2016). Recent developments in detection and enumeration of waterborne bacteria: a retrospective minireview. *Microbiologyopen* 5 (6): 901–922. https://doi.org/10.1002/mbo3.383.

26 Lei, K.F. and Leung, P.H. (2012). Microelectrode array biosensor for the detection of *Legionella pneumophila*. *Microelectron. Eng.* 91: 174–177.

27 Singh, A., Sinsinbar, G., Choudhary, M. et al. (2013). Graphene oxide-chitosan nanocomposite based electrochemical DNA biosensor for detection of typhoid. *Sensors Actuators B Chem.* 185: 675–684.

28 Sobhan, A., Lee, J., Park, M.-K., and Oh, J.-H. (2019). Rapid detection of *Yersinia enterocolitica* using a single–walled carbon nanotube-based biosensor for Kimchi product. *LWT* 108: 48–54.

29 Zhou, H., Duan, S., Huang, J., and He, F. (2020). An ultrasensitive electrochemical biosensor for *Pseudomonas aeruginosa* assay based on a rolling circle amplification-assisted multipedal DNA walker. *Chem. Commun. (Camb.)* 56 (46): 6273–6276. https://doi.org/10.1039/d0cc01619e.

30 Chuensirikulchai, K., Laopajon, W., Phunpae, P. et al. (2019). Sandwich antibody-based biosensor system for identification of *Mycobacterium tuberculosis* complex and *Nontuberculous mycobacteria*. *J. Immunoassay Immunochem.* 40 (6): 590–604. https://doi.org/10.1080/15321819.2019.1659814.

31 Kumar, S., Nehra, M., Mehta, J. et al. (2019). Point-of-care strategies for detection of waterborne pathogens. *Sensors (Basel)* 19 (20): 4476. https://doi.org/10.3390/s19204476.

32 Pandey, V.K. and Mishra, P.K. (2020). Nanoconjugates for detection of waterborne bacterial pathogens. In: *Waterborne Pathogens*, 363–384. Elsevier. https://doi.org/10.1016/B978-0-12-818783-8.00018-9.

33 Clark, L.C. Jr. and Lyons, C. (1962). Electrode systems for continuous monitoring in cardiovascular surgery. *Ann. N. Y. Acad. Sci.* 102 (1): 29–45.

34 Koyun, A., Ahlatcolu, E., Koca, Y., and Kara, S. (2012). Biosensors and their principles. In: *A Roadmap of Biomedical Engineers and Milestones* (ed. S. Kara), 117–142. IntechOpen.

35 Riu, J. and Giussani, B. (2020). Electrochemical biosensors for the detection of pathogenic bacteria in food. *TrAC Trends Anal. Chem.* 126: 115863.

36 Mobed, A., Baradaran, B., de la Guardia, M. et al. (2019). Advances in detection of fastidious bacteria: from microscopic observation to molecular biosensors. *TrAC Trends Anal. Chem.* 113: 157–171.

37 Zhang, Y., Tan, W., Zhang, Y. et al. (2019). Ultrasensitive and selective detection of *Staphylococcus aureus* using a novel IgY-based colorimetric platform. *Biosens. Bioelectron.* 142: 111570.

38 Ragavan, K. and Neethirajan, S. (2019). Nanoparticles as biosensors for food quality and safety assessment. In: *Nanomaterials for Food Applications* (ed. L.R. Amparo, M.J. Rovira, M.M. Sanz, and L.G. Gómez-Mascaraque), 147–202. Elsevier.

39 Cesewski, E. and Johnson, B.N. (2020). Electrochemical biosensors for pathogen detection. *Biosens. Bioelectron.* 159: 112214. https://doi.org/10.1016/j.bios.2020.112214.

40 Chauhan, N., Tiwari, S., Narayan, T., and Jain, U. (2019). Bienzymatic assembly formed@Pt nano sensing framework detecting acetylcholine in aqueous phase. *Appl. Surf. Sci.* 474: 154–160.

41 Velusamy, V., Arshak, K., Korostynska, O. et al. (2010). An overview of foodborne pathogen detection: in the perspective of biosensors. *Biotechnol. Adv.* 28 (2): 232–254. https://doi.org/10.1016/j.biotechadv.2009.12.004.

42 Majdinasab, M., Hayat, A., and Marty, J.L. (2018). Aptamer-based assays and aptasensors for detection of pathogenic bacteria in food samples. *TrAC Trends Anal. Chem.* 107: 60–77.

43 Sharifi, S., Vahed, S.Z., Ahmadian, E. et al. (2020). Detection of pathogenic bacteria via nanomaterials-modified aptasensors. *Biosens. Bioelectron.* 150: 111933. https://doi.org/10.1016/j.bios.2019.111933.

44 Wang, D., Hinkley, T., Chen, J. et al. (2019). Phage based electrochemical detection of *Escherichia coli* in drinking water using affinity reporter probes. *Analyst* 144 (4): 1345–1352.

45 Liu, Q. and Wang, P. (2009). *Cell-Based Biosensors: Principles and Applications.* Artech House.

46 Bousse, L. (1996). Whole cell biosensors. *Sensors Actuators B Chem.* 34 (1–3): 270–275.

47 Liu, Q., Wu, C., Cai, H. et al. (2014). Cell-based biosensors and their application in biomedicine. *Chem. Rev.* 114 (12): 6423–6461.

48 Rogers, K.R. (1995). Biosensors for environmental applications. *Biosens. Bioelectron.* 10 (6–7): 533–541.

49 Gill, A.A.S., Singh, S., Thapliyal, N., and Karpoormath, R. (2019). Nanomaterial-based optical and electrochemical techniques for detection of methicillin-resistant *Staphylococcus aureus*: a review. *Mikrochim. Acta* 186 (2): 114. https://doi.org/10.1007/s00604-018-3186-7.

50 Vikesland, P.J. and Wigginton, K.R. (2010). Nanomaterial enabled biosensors for pathogen monitoring – a review. *Environ. Sci. Technol.* 44 (10): 3656–3669. https://doi.org/10.1021/es903704z.

51 Chen, J., Andler, S.M., Goddard, J.M. et al. (2017). Integrating recognition elements with nanomaterials for bacteria sensing. *Chem. Soc. Rev.* 46 (5): 1272–1283. https://doi.org/10.1039/c6cs00313c.

52 Elahi, N., Kamali, M., and Baghersad, M.H. (2018). Recent biomedical applications of gold nanoparticles: a review. *Talanta* 184: 537–556. https://doi.org/10.1016/j.talanta.2018.02.088.

53 Pastucha, M., Farka, Z., Lacina, K. et al. (2019). Magnetic nanoparticles for smart electrochemical immunoassays: a review on recent developments. *Mikrochim. Acta* 186 (5): 312. https://doi.org/10.1007/s00604-019-3410-0.

54 Zhu, F., Zhao, G., and Dou, W. (2018). Electrochemical sandwich immunoassay for *Escherichia coli* O157:H7 based on the use of magnetic nanoparticles and graphene functionalized with electrocatalytically active Au@Pt core/shell nanoparticles. *Microchim. Acta* 185 (10): 455.

55 Abdullah, A., Jasim, I., Alalem, M. et al. (2017). MEMS based impedance biosensor for rapid detection of low concentrations of foodborne pathogens. *Proceedings of the 2017 IEEE 30th international conference on micro electro mechanical systems (MEMS)*, Las Vegas, NV (22–26 January 2017), 381–385. IEEE.

56 El-Said, W.A., Abdelshakour, M., Choi, J.H., and Choi, J.W. (2020). Application of conducting polymer nanostructures to electrochemical biosensors. *Molecules* 25 (2): 307. https://doi.org/10.3390/molecules25020307.

57 Güner, A., Çevik, E., Şenel, M., and Alpsoy, L. (2017). An electrochemical immunosensor for sensitive detection of *Escherichia coli* O157:H7 by using chitosan, MWCNT, polypyrrole with gold nanoparticles hybrid sensing platform. *Food Chem.* 229: 358–365.

58 Han, J., Wang, M., Hu, Y. et al. (2017). Conducting polymer-noble metal nanoparticle hybrids: synthesis mechanism application. *Progress Polym. Sci.* 70: 52–91.

59 Morales-Narvaez, E., Baptista-Pires, L., Zamora-Galvez, A., and Merkoci, A. (2017). Graphene-based biosensors: going simple. *Adv. Mater.* 29 (7): 1604905. https://doi.org/10.1002/adma.201604905.

60 Hasan, M.R., Pulingam, T., Appaturi, J.N. et al. (2018). Carbon nanotube-based aptasensor for sensitive electrochemical detection of whole-cell Salmonella. *Anal. Biochem.* 554: 34–43. https://doi.org/10.1016/j.ab.2018.06.001.

61 Luo, Y., Dou, W., and Zhao, G. (2017). Rapid electrochemical quantification of *Salmonella pullorum* and *Salmonella gallinarum* based on glucose oxidase and antibody-modified silica nanoparticles. *Anal. Bioanal. Chem.* 409 (17): 4139–4147. https://doi.org/10.1007/s00216-017-0361-3.

62 Huang, F., Guo, R., Xue, L. et al. (2020). An acid-responsive microfluidic salmonella biosensor using curcumin as signal reporter and ZnO-capped mesoporous silica nanoparticles for signal amplification. *Sensors Actuators B Chem.* 312: 127958.

63 Chen, Y., Wang, Z., Liu, Y. et al. (2018). Recent advances in rapid pathogen detection method based on biosensors. *Eur. J. Clin. Microbiol. Infect. Dis.* 37 (6): 1021–1037. https://doi.org/10.1007/s10096-018-3230-x.

64 Das, R., Chaterjee, B., Kapil, A., and Sharma, T.K. (2020). Aptamer-NanoZyme mediated sensing platform for the rapid detection of *Escherichia coli* in fruit juice. *Sensing Bio Sensing Res.* 27: 100313.

65 Rajapaksha, R., Hashim, U., Uda, M.A. et al. (2017). Target ssDNA detection of *E. coli* O157:H7 through electrical based DNA biosensor. *Microsyst. Technol.* 23 (12): 5771–5780.

66 Ye, L., Zhao, G., and Dou, W. (2018). An electrochemical immunoassay for *Escherichia coli* O157:H7 using double functionalized Au@Pt/SiO$_2$ nanocomposites and immune magnetic nanoparticles. *Talanta* 182: 354–362.

67 Zhu, F., Zhao, G., and Dou, W. (2018). A non-enzymatic electrochemical immunoassay for quantitative detection of *Escherichia coli* O157:H7 using Au@Pt and graphene. *Anal. Biochem.* 559: 34–43.

68 Jaiswal, N., Pandey, C.M., Soni, A. et al. (2018). Electrochemical genosensor based on carboxylated graphene for detection of water-borne pathogen. *Sensors Actuators B Chem.* 275: 312–321.

69 Xu, M., Wang, R., and Li, Y. (2016). An electrochemical biosensor for rapid detection of *E. coli* O157:H7 with highly efficient bi-functional glucose oxidase-polydopamine nanocomposites and Prussian blue modified screen-printed interdigitated electrodes. *Analyst* 141 (18): 5441–5449.

70 Lu, D., Pang, G., and Xie, J. (2017). A new phosphothreonine lyase electrochemical immunosensor for detecting *Salmonella* based on horseradish peroxidase/GNPs-thionine/chitosan. *Biomed. Microdevices* 19 (1): 12. https://doi.org/10.1007/s10544-017-0149-4.

71 Yamada, K., Choi, W., Lee, I. et al. (2016). Rapid detection of multiple foodborne pathogens using a nanoparticle-functionalized multi-junction biosensor. *Biosen. Bioelectron.* 77: 137–143.

72 Lee, I. and Jun, S. (2016). Simultaneous detection of *E. coli* K12 and *S. aureus* using a continuous flow multijunction biosensor. *J. Food Sci.* 81 (6): N1530–N1536.

73 Yu, C.Y., Ang, G.Y., Chan, K.G. et al. (2015). Enzymatic electrochemical detection of epidemic-causing *Vibrio cholerae* with a disposable oligonucleotide-modified screen-printed bisensor coupled to a dry-reagent-based nucleic acid amplification assay. *Biosens. Bioelectron.* 70: 282–288. https://doi.org/10.1016/j.bios.2015.03.048.

74 Patris, S., Vandeput, M., Kenfack, G.M. et al. (2016). An experimental design approach to optimize an amperometric immunoassay on a screen printed electrode for *Clostridium tetani* antibody determination. *Biosens. Bioelectron.* 77: 457–463. https://doi.org/10.1016/j.bios.2015.09.064.

75 Liebana, S., Brandao, D., Cortes, P. et al. (2016). Electrochemical genosensing of *Salmonella*, *Listeria* and *Escherichia coli* on silica magnetic particles. *Anal. Chim. Acta* 904: 1–9. https://doi.org/10.1016/j.aca.2015.09.044.

76 Alexandre, D., Melo, A.M.A., Furtado, R.F. et al. (2016). Amperometric biosensor for *Salmonella typhimurium* detection in milk. *Embrapa Agroindústria Tropical-Artigo em anais de congresso (ALICE)*, 2016: In: CONGRESSO BRASILEIRO DE CIÊNCIA E TECNOLOGIA DE ALIMENTOS, 25, 2016, FAURGS-Gramado/RS (24–27 October 2016),

77 Fei, J., Dou, W., and Zhao, G. (2016). Amperometric immunoassay for the detection of *Salmonella pullorum* using a screen-printed carbon electrode modified with gold nanoparticle-coated reduced graphene oxide and immunomagnetic beads. *Microchim. Acta* 183 (2): 757–764.

78 Dhull, N., Kaur, G., Jain, P.U. et al. (2019). Label-free amperometric biosensor for *Escherichia coli* O157:H7 detection. *Appl. Surf. Sci.* 495: 143548.

79 Zhang, X., Wang, X., Yang, Q. et al. (2020). Conductometric sensor for viable *Escherichia coli* and *Staphylococcus aureus* based on magnetic analyte separation via aptamer. *Microchimica Acta* 187 (1): 43.

80 Yang, Z., Liu, Y., Lei, C. et al. (2016). Ultrasensitive detection and quantification of *E. coli* O157:H7 using a giant magnetoimpedance sensor in an open-surface microfluidic cavity covered with an antibody-modified gold surface. *Microchim. Acta* 183 (6): 1831–1837.

81 Yao, L., Wang, L., Huang, F. et al. (2018). A microfluidic impedance biosensor based on immunomagnetic separation and urease catalysis for continuous-flow detection of *E. coli* O157:H7. *Sensors Actuators B Chem.* 259: 1013–1021.

82 Wang, L., Huang, F., Cai, G. et al. (2017). An electrochemical aptasensor using coaxial capillary with magnetic nanoparticle, urease catalysis and PCB electrode for rapid and sensitive detection of *Escherichia coli* O157:H7. *Nanotheranostics* 1 (4): 403.

83 Xu, S., Zhang, Y., Dong, K. et al. (2017). Electrochemical DNA biosensor based on graphene oxide-chitosan hybrid nanocomposites for detection of *Escherichia coli* O157:H7. *Int. J. Electrochem. Sci.* 12: 3443–3458.

84 Wang, R., Lum, J., Callaway, Z. et al. (2015). A label-free impedance immunosensor using screen-printed interdigitated electrodes and magnetic nanobeads for the detection of *E. coli* O157:H7. *Biosensors* 5 (4): 791–803.

85 Xu, M., Wang, R., and Li, Y. (2016). Rapid detection of *Escherichia coli* O157:H7 and *Salmonella Typhimurium* in foods using an electrochemical immunosensor based on screen-printed interdigitated microelectrode and immunomagnetic separation. *Talanta* 148: 200–208.

86 Cimafonte, M., Fulgione, A., Gaglione, R. et al. (2020). Screen printed based impedimetric immunosensor for rapid detection of *Escherichia coli* in drinking water. *Sensors (Basel)* 20 (1): 274. https://doi.org/10.3390/s20010274.

87 Wan, J., Ai, J., Zhang, Y. et al. (2016). Signal-off impedimetric immunosensor for the detection of *Escherichia coli* O157:H7. *Sci. Rep.* 6: 19806. https://doi.org/10.1038/srep19806.

88 Deshmukh, R., Prusty, A.K., Roy, U., and Bhand, S. (2020). A capacitive DNA sensor for sensitive detection of *Escherichia coli* O157:H7 in potable water based on the z3276 genetic marker: fabrication and analytical performance. *Analyst* 145 (6): 2267–2278. https://doi.org/10.1039/c9an02291k.

89 Gupta, A., Bhardwaj, S.K., Sharma, A.L. et al. (2019). Development of an advanced electrochemical biosensing platform for *E. coli* using hybrid metal-organic framework/polyaniline composite. *Environ. Res.* 171: 395–402.

90 Li, Z., Fu, Y., Fang, W., and Li, Y. (2015). Electrochemical impedance immunosensor based on self-assembled monolayers for rapid detection of *Escherichia coli* O157:H7 with signal amplification using lectin. *Sensors* 15 (8): 19212–19224.

91 Dos Santos, M.B., Azevedo, S., Agusil, J.P. et al. (2015). Label-free ITO-based immunosensor for the detection of very low concentrations of pathogenic bacteria. *Bioelectrochemistry* 101: 146–152.

92 Pandey, C.M., Tiwari, I., Singh, V.N. et al. (2017). Highly sensitive electrochemical immunosensor based on graphene-wrapped copper oxide-cysteine hierarchical structure for detection of pathogenic bacteria. *Sensors Actuators B Chem.* 238: 1060–1069.

93 Zarei, S.S., Soleimanian-Zad, S., and Ensafi, A.A. (2018). An impedimetric aptasensor for *Shigella dysenteriae* using a gold nanoparticle-modified glassy carbon electrode. *Microchim. Acta* 185 (12): 538.

94 Qian, X., Qu, Q., Li, L. et al. (2018). Ultrasensitive electrochemical detection of *Clostridium perfringens* DNA based morphology-dependent DNA adsorption properties of CeO_2 nanorods in dairy products. *Sensors* 18 (6): 1878.

95 Wu, J., Wang, R., Lu, Y. et al. (2018). Facile preparation of a bacteria imprinted artificial receptor for highly selective bacterial recognition and label-free impedimetric detection. *Anal. Chem.* 91 (1): 1027–1033.

96 Pathania, P., Sharma, A., Kumar, B. et al. (2017). Selective identification of specific aptamers for the detection of non-typhoidal salmonellosis in an apta-impedimetric sensing format. *Microchim. Acta* 184 (5): 1499–1508.

97 Ranjbar, S., Shahrokhian, S., and Nurmohammadi, F. (2018). Nanoporous gold as a suitable substrate for preparation of a new sensitive electrochemical aptasensor for detection of *Salmonella typhimurium. Sensors Actuators B Chem.* 255: 1536–1544.

98 Primiceri, E., Chiriacò, M.S., de Feo, F. et al. (2016). A multipurpose biochip for food pathogen detection. *Anal. Methods* 8 (15): 3055–3060.

99 Sheikhzadeh, E., CHamsaz, M., Turner, A. et al. (2016). Label-free impedimetric biosensor for *Salmonella Typhimurium* detection based on poly [pyrrole-co-3-carboxyl-pyrrole] copolymer supported aptamer. *Biosens. Bioelectron.* 80: 194–200.

100 Bagheryan, Z., Raoof, J.B., Golabi, M. et al. (2016). Diazonium-based impedimetric aptasensor for the rapid label-free detection of *Salmonella typhimurium* in food sample. *Biosens. Bioelectron.* 80: 566–573. https://doi.org/10.1016/j.bios.2016.02.024.

101 Jia, F., Duan, N., Wu, S. et al. (2016). Impedimetric *Salmonella aptasensor* using a glassy carbon electrode modified with an electrodeposited composite consisting of reduced graphene oxide and carbon nanotubes. *Microchim. Acta* 183 (1): 337–344.

102 Liu, J., Jasim, I., Shen, Z. et al. (2019). A microfluidic based biosensor for rapid detection of *Salmonella* in food products. *PLoS One* 14 (5): e0216873. https://doi.org/10.1371/journal.pone.0216873.

103 Liu, X., Marrakchi, M., Xu, D. et al. (2016). Biosensors based on modularly designed synthetic peptides for recognition, detection and live/dead differentiation of pathogenic bacteria. *Biosens. Bioelectron.* 80: 9–16. https://doi.org/10.1016/j.bios.2016.01.041.

104 Tam, P.D. and Thang, C.X. (2016). Label-free electrochemical immunosensor based on cerium oxide nanowires for *Vibrio cholerae* O1 detection. *Mater. Sci. Eng. C Mater. Biol. Appl.* 58: 953–959. https://doi.org/10.1016/j.msec.2015.09.027.

105 Ward, A., Hannah, A., Kendrick, S. et al. (2018). Identification and characterisation of *Staphylococcus aureus* on low cost screen printed carbon electrodes using impedance spectroscopy. *Biosens. Bioelectron.* 110: 65–70.

106 Sotero, A.F., Wang, R., Tao, W. et al. (2018). A portable impedance aptasensing system for rapid detection of *Salmonella typhimurium* in poultry products. *2018 ASABE Annual International Meeting*, 2018, 1801147. American Society of Agricultural and Biological Engineers. doi:10.13031/aim.201801147.

107 Helali, S., Alatawi, A.S.E., and Abdelghani, A. (2018). Pathogenic *Escherichia coli* biosensor detection on chicken food samples. *J. Food Saf.* 38 (5): e12510.

108 Mutreja, R., Jariyal, M., Pathania, P. et al. (2016). Novel surface antigen based impedimetric immunosensor for detection of *Salmonella typhimurium* in water and juice samples. *Biosens. Bioelectron.* 85: 707–713. https://doi.org/10.1016/j.bios.2016.05.079.

109 Shaibani, P.M., Etayash, H., Jiang, K. et al. (2018). Portable nanofiber-light addressable potentiometric sensor for rapid *Escherichia coli* detection in orange juice. *ACS Sens.* 3 (4): 815–822. https://doi.org/10.1021/acssensors.8b00063.

110 Silva, N.F.D., Almeida, C.M.R., Magalhaes, J. et al. (2019). Development of a disposable paper-based potentiometric immunosensor for real-time detection of a foodborne pathogen. *Biosens. Bioelectron.* 141: 111317. https://doi.org/10.1016/j.bios.2019.111317.

111 Silva, N.F., Magalhães, J.M., Barroso, M.F. et al. (2019). In situ formation of gold nanoparticles in polymer inclusion membrane: application as platform in a label-free potentiometric immunosensor for *Salmonella typhimurium* detection. *Talanta* 194: 134–142.

112 Shaibani, P.M., Jiang, K., Haghighat, G. et al. (2016). The detection of *Escherichia coli* (*E. coli*) with the pH sensitive hydrogel nanofiber-light addressable potentiometric sensor (NF-LAPS). *Sensors Actuators B Chem.* 226: 176–183.

113 Hua, R., Hao, N., Lu, J. et al. (2018). A sensitive potentiometric resolved ratiometric photoelectrochemical aptasensor for *Escherichia coli* detection fabricated with non-metallic nanomaterials. *Biosens. Bioelectron.* 106: 57–63. https://doi.org/10.1016/j.bios.2018.01.053.

114 Lv, E., Ding, J., and Qin, W. (2018). Potentiometric detection of *Listeria monocytogenes* via a short antimicrobial peptide pair-based sandwich assay. *Anal. Chem.* 90 (22): 13600–13606. https://doi.org/10.1021/acs.analchem.8b03809.

115 Tian, Y., Liang, T., Zhu, P. et al. (2019). Label-free detection of *E. coli* O157:H7 DNA using light-addressable potentiometric sensors with highly oriented ZnO nanorod arrays. *Sensors* 19 (24): 5473.

116 Bozal-Palabiyik, B., Gumustas, A., Ozkan, S.A., and Uslu, B. (2018). Biosensor-based methods for the determination of foodborne pathogens. In: *Foodborne Diseases* (ed. A.M. Holban and A.M. Grumezescu), 379–420. Elsevier.

117 Razmi, N., Hasanzadeh, M., Willander, M., and Nur, O. (2020). Recent progress on the electrochemical biosensing of *Escherichia coli* O157:H7: material and methods overview. *Biosens. (Basel)* 10 (5): 54. https://doi.org/10.3390/bios10050054.

118 Mishra, A., Tyagi, M., Pilloton, R. et al. (2020). Evolving techniques for the detection of *Listeria monocytogenes*: underlining the electrochemical approach. *Int. J. Environ. Anal. Chem.* 100 (4): 507–523.

119 Boukamp, B.A. (2004). Electrochemical impedance spectroscopy in solid state ionics: recent advances. *Solid State Ionics* 169 (1-4): 65–73.

120 Chiriacò, M.S., Parlangeli, I., Sirsi, F. et al. (2018). Impedance sensing platform for detection of the food pathogen *Listeria monocytogenes*. *Electronics* 7 (12): 347.

6

Zinc Oxide-Based Miniature Sensor Networks for Continuous Monitoring of Aqueous pH in Smart Agriculture

Akshaya Kumar Aliyana[1], Aiswarya Baburaj[1], Naveen Kumar S. K.[1], and Renny Edwin Fernandez[2]

[1] Department of Electronics, Mangalore University, Konaje, India
[2] Department of Engineering, Norfolk State University, Norfolk, VA, USA

6.1 Introduction

The world has lost one-third of its arable land due to erosion or pollution in the past 40 years. As the world population is expected to exceed 9.6 billion people by the year 2050, Food and Agriculture Organization (FAO) reports that the agricultural industry will have to resort to novel ways to increase agricultural crop yield in available arable lands [1]. The quality, health, and resilience of soil and water bodies play an important role in the agricultural crop yield and environmental safety [2, 3]. A critical factor that affects crop yield is the availability and concentration levels of the nutrients in the water and soil resources. The nutrient retainment and supplying capacity of water bodies are heavily influenced by pH levels. The pH is a critical parameter in regulating the reactivity of many chemical and biological species. In an aqueous medium, the following equilibrium exists between the water (H_2O), the acid (H^+), and the alkali (OH^-).

$$H_2O = H^+ + OH^- \tag{6.1}$$

A pH value is defined as the negative logarithm of the activity of hydrogen ions that is $pH = -\log \alpha H$ and monitoring of pH level is a fundamental requirement of agriculture to food processing [4, 5], environmental science to chemical engineering [6], and biomedical engineering to health applications [7]. Soil nutrients are best soluble and bio-available at pH values between 6.5 and 7.5 [8]. This range

Sensing Technologies for Real Time Monitoring of Water Quality, First Edition.
Libu Manjakkal, Leandro Lorenzelli, and Magnus Willander.

of soil pH is generally compatible to plant root growth because essential nutrients are optimally available form. Nitrogen (N), Phosphorus (P), Potassium (K), Iron (Fe), Manganese (Mn), Copper (Cu), Zinc (Zn), Boron (B), Molybdenum (Mo), and Chlorine (Cl) are affected directly by soil pH than many other factors [9]. At higher alkaline levels, Phosphate ions are disposed to react with Calcium (Ca) and Magnesium (Mg) to form less soluble compounds and even acidic conditions; Phosphate ions react with Aluminum (Al) and Iron (Fe) to again form less-soluble compounds. The solubility of Al, Mn, and Fe, which are toxic to plants in excess, increases as pH concentration of soil decreases. Nitrogen (N) level is critical to plant growth and yield; Ammonium (NH_3^+) holds a major part of the N level in the soil and it stimulates root branching, flowering time, leaf expansion, and seed dormancy [8]. Phosphorus enhances the overall root system, seed development, crop maturation, and plant photosynthesis and respiration levels. K^+ is responsible for the regulation of water usage in plants, disease resistance, stem strength, and protein synthesis [8]. Hence, pH levels and nutrients show the interrelated response and their direct effects on plant growth and production [10, 11]. The monitoring of physicochemical properties of water soil has been an open research problem as it has implications in areas such as wastewater management, biomedical, environmental, and soil-agriculture applications [12–15].

This chapter reports our recent efforts in building nanostructured ZnO-based miniature sensor networks for continuous monitoring of pH in smart agriculture. The elaborate discussion in this chapter includes various ZnO nanostructured (NRs, NWs, NFs, and Nfs) active layer-based pH sensors, and the analysis of the sensitivity of the various nanostructures. Our sensors show a sensitivity of ~1.06 nF/pH in the range of pH 4–10. The pH sensor is highly selective to hydrogen ions and possesses good stability for repetitive detection of pH with an acceptable error margin [16]. We also describe the sensor network architecture by an internet of things (IoT)-enabled Node MCU microcontroller platform for continuous monitoring of pH [17].

6.2 Metal Oxide-Based Sensors and Detection Methods

Electrochemical sensing mechanisms are centered on the surface potential changes due to the analyte interactions with the active layer. Metal oxides (MOx)-based sensors with excellent response time and stability are promising candidates for the next-generation IoT-enabled real-time sensing. Analytical signals from a chemical sensor are dominated by the target concentration. Besides the target molecule, interfering ions, process parameters like solution pH, conductivity, temperature, electrode parameters such as geometry, and material properties also contribute to the sensor signals [17–20]. In addition, nanostructured MOx materials offer a large surface-to-volume ratio to the specific ionic reactions and improve the sensor performance. MOx nanostructures were fabricated through facile synthesis methods to have varying electrical and chemical sensing properties [21–24]. Depending on the type of ions (oxidizing or reducing), an interaction occurs on the surface, which alters the electrical/chemical/optical properties of the MOx layers. The MOx was initially recognized in 1984 as a sensitive material for pH sensors [25]. A further variety of MOx has been utilized to fabricate pH sensors. Among them, ZnO [26], SnO_2 [27], TiO_2 [28], RuO_2 [29, 30], IrO_2 [31, 32], and CuO [33] are most studied active materials with their wide range of nanostructures and their impacts on sensitivity. In addition to this, conjugated M/MOx-based composites were sensitive to several redox agents and reported consistent departure from predicted behavior. Some of the disadvantages of the MOx include their buffering drift on the conductivity, instability at higher working frequency, and low resistance to alkaline solutions, which affects their life. The design of thin and thick/thin MOx films on nonconductive bottom layers overcomes many of these difficulties [34].

In general, various electrochemical, optical, and enzymatic methods have been developed specifically for ion sensing in agriculture [35–39]. The glass probe-based pH electrodes [40], ion-selective electrodes (ISEs) [41, 42], ion-selective field-effect transistor (ISFET) [43], potentiometric [44, 45], conductimetric [46], and chemiresistive [30] methods are used to fabricated pH sensors. Interdigitated electrodes (IDEs)-based electrochemical pH sensors have advantageous features such as small size, ease of operation, high sensitivity, fast response, and low cost of fabrication [29, 33, 47, 48]. IDEs have simple two-electrode configuration, operate without reference electrodes, and are widely used for miniaturized sensor applications [49–52]. Nanostructures can be modified with ionophore active layer that enables the detection of specific ions [53, 54].

6.3 pH Sensor Fabrication

6.3.1 Detection of pH: Materials and Method

Zinc oxide (ZnO) has been explored in a variety of sensing applications such as solar cells [63], electronic devices [64], and biomedical applications [65, 66]. The enhanced sensitivity of the ZnO is attributed to its surface-to-volume ratio, electrical conductivity (EC), fast response, wide band-gap (3.37 eV), large exciton binding energy (60 meV), and piezoelectricity [67–69]. The various ZnO nanostructures such as Nanorods (NRs), Nanowires (NWs), Nanoflowers (NFs), and Nanoflakes (Nfs) and their impacts on pH sensing performance are discussed in detail (Table 6.1).

Table 6.1 Metal oxide (MOx)-based electrochemical pH sensors.

Materials	Fabrication method	Structure of the material	Sensitivity	pH range	References
ZnO	Low temperature Hydrothermal method	Nanorods (NRs) Thin film	44.56 mV/pH 34.82 mV/pH	4, 6, 7, 8, 10	[55]
TiO_2	Sol–gel method	Thin film	58.73 mV/pH	1–11	[28]
HfO_2	Atomic layer deposition (ALD)	Thin film	54.5 mV/pH	3–9	[56]
SnO_2	Low-temperature hydrothermal method	Nanorods (NRs)	55.18 mV/pH	1–13	[27]
WO_3	Hydrothermal method	Nano slab-type shape	56.7 ± 1.3 mV/pH	5–9	[57]
IrO_2	Sol–gel method	Thin films	51.7 mV/pH	1.5–12	[31]
RuO_2	Screen printing technology	Thick film	57 mV/pH	2–10	[30]
CuO	Low-temperature chemical bath	Nanoflowers (NFs)	28 mV/pH	2–11	[58]
	Low-temperature hydrothermal method	Nanorods (NRs)	0.64 µF/pH	5–8.5	[33]
Ta_2O_5	Screen printing technology	Thick film	45.92 mV/pH	2–10	[59]
	Screen printing technology	Thick film	56.17 mV/pH	1–10	[60]
Co_2O_3	Thermal decomposition method	Thick film	54.9–60.3 mV/pH	1–12	[61]
PbO_2	Microfabrication	Thin film	84 ± 1 mV/pH	0.25–13	[62]

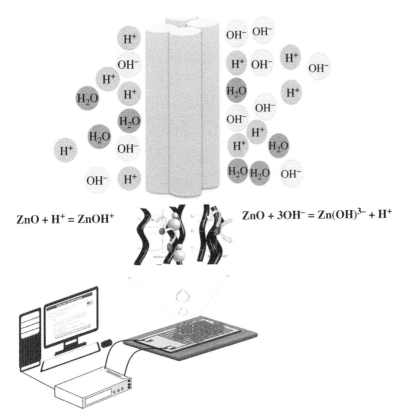

$$ZnO + H^+ = ZnOH^+$$

$$ZnO + 3OH^- = Zn(OH)^{3-} + H^+$$

Figure 6.1 Schematic representation of the pH sensor electrochemical mechanism.

The different electrochemical mechanisms are considered to calculate the sensitivity of the device. Simple ion exchange to oxygen intercalation mechanisms of electrodes and active materials cause changes in electrochemical signals and surface potentials (Figure 6.1). The NRs have gained more attention for pH sensing due to chemical stability, nontoxicity, electrochemical activity, fast response, and low costs [70]. Al-Hilli et al. discussed the electrochemical potential response of ZnO NRs and explored the sensitivity of $-59\,mV/pH$ [71]. But, Fulati et al. reported a pH sensor based on ZnO nanotubes and nanorod structures. ZnO nanotubes have double the sensitivity of nanorods ($45.9\,mV/pH$ vs. $28.4\,mV/pH$), due to subsurface oxygen vacancies, as well as a larger effective surface area and higher surface-to-volume ratio than ZnO NRs [42]. The selectivity influence of interfering ions on pH is typically unimportant in most MOx active layer nanostructures [72]. However, the ZnO active layer-based sensor's sensitivity, stability, and reaction time have depended on its varied nanostructure layers, and the

(a) (b)

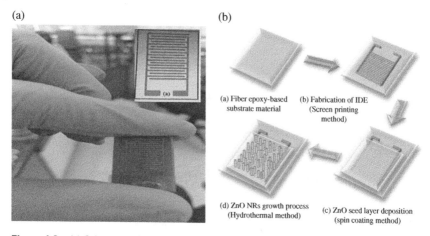

(a) Fiber epoxy-based
substrate material

(b) Fabrication of IDE
(Screen printing
method)

(d) ZnO NRs growth process
(Hydrothermal method)

(c) ZnO seed layer deposition
(spin coating method)

Figure 6.2 (a) Schematic illustration of pH sensor fabrication process, (b) prototype of screen-printed IDEs, and. *Source:* Reproduced with permission from Ref. [73], © 2019, IOP Publishing/Public Domain CC BY 4.0.

sensor's performance in terms of selectivity morally depends on the concentration of interfering ions in the test solution.

Figure 6.2 illustrates the fabrication process of IDEs modified ZnO-NRs-based pH sensing arrays using a low-temperature hydrothermal growth process. Nonconducting fiber epoxy resins were used as the bottom layer of the sensor. IDEs have been screen printed on the bottom layer with a finger length of 11 mm, electrode width of 0.5 mm, and spacing of 0.5 mm. The fabricated IDEs consist of a total of 18 individual electrodes, and they will provide a total sensing area of about $18 \times 16 \, \text{mm}^2$ for the pH sensing layer. A temperature-controlled spin coating unit was used to deposit the ZnO seed solution on the IDEs with a four-step programming process.

6.3.2 Detection of pH: Surface Morphology of the Nanostructured ZnO and IDEs

The surface morphology of the various ZnO nanostructured active layers such as (a) Seed particles, (b, c) ZnO-NRs, (d) ZnO-NFs, (e) ZnO-NCs, (f) ZnO-Nfs, and (g–i) IDEs were evaluated using field emission scanning electron microscopy (FeSEM) (Figure 6.3).

The majority of the synthesized ZnO-nanostructures were grown uniformly in high compactness with proper alignment and perpendicular to the IDEs. The top view of the FeSEM image shows a hexagonal faceted morphology of the nanostructures. The average length and diameters of the ZnO-NRs are 1 μm and

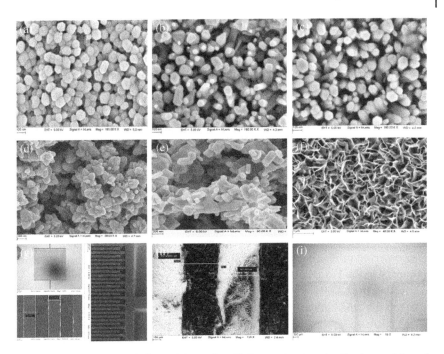

Figure 6.3 Surface morphology of the various ZnO nanostructured active layers: (a) seed particles, (b, c) NRs, (d) NFs, (e) NCs, (f) Nfs, and (g–i) IDEs [73–75].

$\approx 60 \pm 10$ nm, respectively. The average diameter and thickness of nanoflakes around 4.13 µm and 90 nm, respectively, were lying horizontally on the substrate [74]. Most of these nanostructures show branched structure, parent nanostructures further grew into smaller and narrow nanostructures. The grown ZnO NFs were well distributed on the surface of the IDEs, indicating a large-scale fabrication of ZnO NFs and it exhibited the petals with $\approx 150 \pm 10$ nm in length and $\approx 80 \pm 10$ nm in width [75]. The size of the ZnO nanostructures is strongly dependent on the seed size of the prepared ZnO seed layer. As the grain size of the particle layer decreases, smaller sizes of ZnO-NRs in diameter are grown and vice versa. The screen-printed IDEs pattern was modeled and designed for the dimension of 500 µm for the finger gap of the electrodes.

6.3.3 Detection of pH: Electrochemical Sensing Performance

To measure the electrochemical characteristics of the ZnO nanostructure-based pH sensor, the two contact electrodes of the sensor were connected to the probe station using a semiconductor characterization system. Standard pH buffer solutions are available in the pH 2–12 range. Capacitance-Voltage (C-V) and

Capacitance-Frequency (C-F) characteristics of the sensor were performed with increasing pH concentrations to assess the sensing performance of the device (Figure 6.4a,b). The device's sensitivity was determined by plotting a graph of pH level (pH) vs capacitance (F) at a frequency of 1 kHz. We have discussed a two-contact electrode arrangement with the IDEs platform for the pH sensor. When minimum voltage is applied across the contact electrodes of the IDEs, a local electrical field is generated between the electrodes, leading to a change in the electrical properties of the active layer. While varying the pH concentration of the solution, the positive or negatively charged surface groups of the electrical double layer (EDL) formed at the interface of ZnO nanostructures, and electrolyte gets disturbed which further alters the electrical properties of the active layer. According to the theory of conductivity [76], the interaction of more ions across the active layer will increase the conductivity and the low frequency of voltage signals will lead to additional electrons movement across electrodes [77].

As observed in Figure 6.4c, capacitance was found to drastically decrease with frequency in the 1–3 kHz range. However, the capacitance value did not decrease any further from 4 kHz. It was also observed that the capacitance values were also influenced by the buffer pH, as seen in Figure 6.4d. Maximum capacitance value (C_{max}) decreased from 9.65 nF to 13.07 pF during buffer pH values varying from pH 2–12 at selected frequencies (1–5 kHz). Here, ZnO-NRs-based pH sensor shows linear response and displays a sensitivity of ~1.06 nF/pH. Zhang et al. adopted Au-assisted chemical vapor deposition (CVD) to fabricate ultrafine ZnO-NWs and NFs hybrid structures as functional pH sensors. The pH sensitivity of the ZnO-NWs to the NFs sensor was 36.65–34.74 mV/pH [78]. Similarly, Sheng-Joue et al. reported pH sensitivity of the ZnO NRs to thin film (low-temperature hydrothermal technique) was 44.56–34.82 mV/pH [55]. The ZnO NRs have higher sensitivity and stability toward target ions than other nanostructures (Table 6.2). The dangling bonds and surface states density induced on the surface of ZnO thin-film and the sidewall surface of the NRs. The Fermi level pinning effect creates a small drift in the sensing parameters. As a result, the sensitivity of the un-passivated i-ZnO NRs to intrinsic-ZnO NRs was enhanced from 42.37 to 49.35 mV/pH [71, 79].

The response time of the fabricated pH device was found to be about 1–10 seconds for acidic and alkaline solutions. Sensors showed a faster response for acid solutions than alkaline (Figure 6.4e). The device was seen to be highly stable in pH 6–8. The stability of the fabricated device was tested for longer periods. Even after four weeks, sensors displayed highly stable results. The reproducibility test of the devices confirms the errors margins for four sensors were with 0.8%, 1.3% and 0.7%, which is an acceptable error margin of <2% (Figure 6.4f). Major advantages of ZnO-NRs-based pH sensors which make them more efficient than other pH sensors include fast response, high stability, high accuracy and low interference to other ions, low cost, miniaturized size, good performance in extreme conditions, and more.

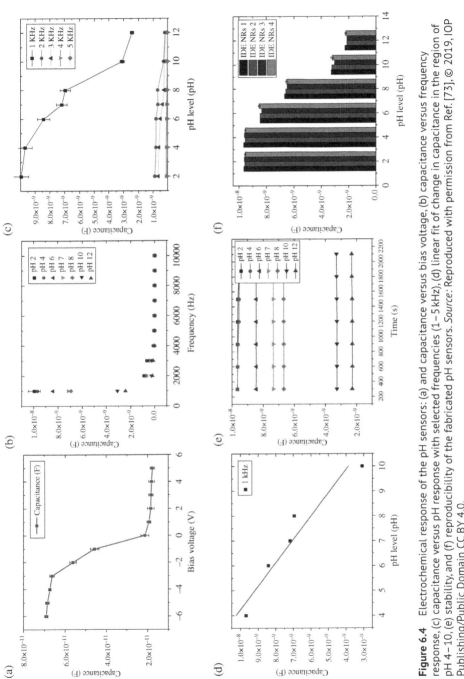

Figure 6.4 Electrochemical response of the pH sensors: (a) and capacitance versus bias voltage, (b) capacitance versus frequency response, (c) capacitance versus pH response with selected frequencies, (d) linear fit of change in capacitance in the region of pH 4–10, (e) stability, and (f) reproducibility of the fabricated pH sensors. *Source:* Reproduced with permission from Ref. [73], © 2019, IOP Publishing/Public Domain CC BY 4.0.

Table 6.2 ZnO nanostructured-based pH sensors.

Fabrication method	Structure of the material	Sensitivity (mV/pH)	pH range	References
Vapor cooling condensation method	Thin film Nanorods (NRs) array	42.37 mV/pH 49.35 mV/pH	4–12	[79, 80]
Au-assisted chemical vapor deposition	Ultrafine nanowires (NWs) Nanoflakes (NFs)	34.74 mV/pH 36.65 mV/pH	4–9	[78]
Low-temperature growth method	Nanorods (NRs)	51.881 mV/pH	4–11	[71]
Low-temperature Hydrothermal method	Nanorods (NRs) Thin film	44.56 mV/pH 34.82 mV/pH	4, 6–8, 10	[55]
Low-temperature growth method	Nanotube (NTs) Nanorods (NRs)	45.9 mV/pH 28.4 mV/pH	4–12	[42]
Sol–gel method	Thin films	38 mV/pH	2–12	[26]
Self-catalyst growth method	Nanorods (NRs)	—	2–12	[81]

Even though this sort of sensor has numerous advantages, its application is limited due to the possible side effect of material deterioration from long-term operation in various biological and chemical environments [34, 69, 82–84]. In a dynamic setting, several parameters contribute heavily to signal erraticism [85]. The response of such intricate systems cannot be interpreted using calibration curves or modeled using mathematical equations. When trying to electrochemically analyze complex mixtures of structurally similar analytes in real matrices, various interferences occur, leading to faulty estimation of the target concentration. Instability in electrochemical sensing is caused by structurally similar molecules/ ions (Ca^{2+}, Mg^{2+}, K^+, H^+, and Na^+) [86], sensor fouling, temperature, etc. [87].

6.3.4 Detection of Real-Time pH Level in Smart Agriculture: Wireless Sensor Networks and Embedded System

The real-time monitoring of the aqueous pH conditions in agriculture field will provide key information not only to improve resource utilization to maximize farming outputs and minimize environmental side effects but also to build site-specific databases of relationships between aqueous-soil conditions and plant growth for intelligent and sustainable agriculture systems. The fertigation process is impacted by the arbitrary variation of aqueous-soil pH, EC, organic carbon (OC), and available NPK concentration, which in turn influences crop output [88, 89]. Hence, pH is one of the important parameters to be assessed. Laterite soil samples were obtained from the Joint Director Agriculture Office, Dakshina Kannada District, Karnataka, India, and test samples are prepared using standard protocol for the pH measurement. The pH 7 carried laterite dry soil named sample 1 was added to the standard pH buffer solution and samples were prepared to correspond to the target range pH 2–8. Elemental compositions of the five different laterite soil samples collected from the Dakshina Kannada district, Karnataka, India, are listed in Table 6.3 [90]. Initially, the sensor is placed under normal ambient conditions and the impedance of the sensor is monitored. Further, the sensor is deployed on the surface of the prepared samples (pH 2–8) and the change in impedance magnitude is monitored. All the measurements are taken at room temperature ($25 \pm 2\,°C$). The sensing

Table 6.3 Dakshina Kannada District laterite soil samples-based elemental composition.

Soil samples			Soil compositions						
			N	K	Ca	Mn	Fe	EC	OC
Sample	Soil type	pH			wt.%			kg/H	
1		7	0.40	0.06	0.01	0.03	0.99	0.070	1.45
2		6.95	0.57	0.29	0.03	0.02	1.5	0.078	1.53
3		7.08	0	0.06	0.02	0.03	0.92	0.070	1.45
4		7.06	2.18	0	0.02	0.04	0.95	0.066	1.40
5		6.82	0	0.06	0.02	0.02	0.98	0.074	1.42

EC, electrical conductivity; kg, kilogram; H, hectare; OC, organic content. *Source:* Reproduced with permission from Ref. [90], © 2021, IEEE.

features of the sensor toward pH levels are confirmed by comparing the calibrated pH level with obtained pH level, which is estimated from the data collected.

Wireless sensor networks (WSNs) have enabled the promising hardware and software protocols for various real-time monitoring applications. The internet-connected microcontroller platforms allow multiple nanosensors to communicate with each other with minimal human intervention. The total WSNs monitoring system consists of three main sub-systems (sensor nodes, base-station, and cloud-based system) [91]. Figure 6.5 shows the architecture of the IoT-enabled embedded system (Node MCU ESP8266 microcontroller unit) for real-time sensor data measurement applications. The hardware part of the system, which enables the data processing and wireless communication, is facilitated using ESP8266 as the main processor. The tuned properties of the active layers of the device alter the impedance/capacitance of the sensors due to the various sensing ion interaction with the active layer. Most of the system is powered by a 3–5 V supply and associated with an inbuilt frequency generator. An individual set of sensors were assembled into sensor arrays and connected to a signal condition circuit through the analog pins of the Node MCU unit [92].

A 10 kΩ resistor and a 1 kHz frequency generator circuit were linked in series with the sensors in the signal conditioning circuit to provide a voltage signal for the Node MCU development kit (Figure 6.6). The repeatability and stability data of the change in impedance measurements of the calibrated pH sensor are among the system design factors used to predict the corresponding pH level in an aqueous and soil environment [93].

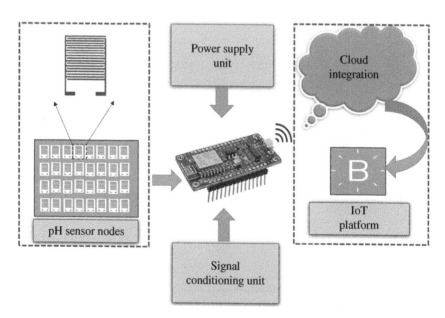

Figure 6.5 Block diagram of IoT-enabled embedded system for real-time sensor data monitoring.

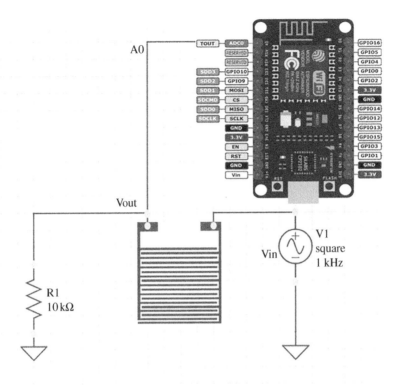

Figure 6.6 Impedance characteristic measurement electronic circuit with 1 kHz frequency generator.

The software part of this work is classified into three major sections such as initialization, setup, and loop function. Network elements are initialized by prompting for a user ID and auth token of the android applications (Blynk and ThingSpeak). The setup function enables serial communication and associates the internet connectivity. The loop function evaluates the sensing data and displays all monitored sensor data values in the serial monitor. Further pushes the data to the corresponding channels and displays the data in the modern android smartphone applications.

6.4 Conclusion

This chapter reports the fabrication and deployment of nanostructured metal oxide-based miniature sensor networks for continuous monitoring of pH in smart agriculture. We reported the in-depth view of various ZnO nanostructured materials that have been used to develop pH sensors in smart agriculture applications.

Individual sensors are fabricated by different deposition methods and low-temperature hydrothermal growth leads to a wide range of ZnO nanostructured patterns (NRs, NWs, NFs, and Nfs) and impacts on pH sensing performance are also discussed in detail. Sensors possess good stability for repetitive detection of pH within an acceptable error margin. Deployment of a low-cost sensor network using an IoT-enabled Node MCU ESP8266 microcontroller platform to facilitate continuous pH monitoring in a field was also discussed.

Acknowledgments

We would like to thank the Department of Science and Technology (DST) for the INSPIRE FELLOWSHIP (Registration Number: IF170544) and University Grant Commission (194-3/2016(IC)) sponsored the Indo-US Bilateral Research project for funding and instruments facilities provided to this research work.

References

1 Cervantes-Godoy, D., Dewbre, J., Amegnaglo, C. J. et al. (2014). The future of food and agriculture: trends and challenges. Technical Report. Food Agriculture Organization, United Nations.

2 Jahany, M. and Rezapour, S. (2020). Assesment of the quality indices of soils irrigated with treated wastewater in a calcareous semi-arid environment. *Ecol. Indic.* 109 (May 2019): 105800. https://doi.org/10.1016/j.ecolind.2019.105800.

3 Lal, R. (2020). Managing soils for negative feedback to climate change and positive impact on food and nutritional security. *Soil Sci. Plant Nutr.* 66 (1): 1–9. https://doi.org/10.1080/00380768.2020.1718548.

4 Sanches, G.M., Magalhães, P.S.G., Remacre, A.Z., and Franco, H.C.J. (2018). Potential of apparent soil electrical conductivity to describe the soil pH and improve lime application in a clayey soil. *Soil Tillage Res.* 175 (March 2017): 217–225. https://doi.org/10.1016/j.still.2017.09.010.

5 Simic, M., Manjakkal, L., Zaraska, K. et al. (2016). TiO_2 based thick film pH sensor. *IEEE Sens. J.* 17 (2): 248–255. https://doi.org/10.1109/JSEN.2016.2628765.

6 Pan, T.M., Wang, C.W., Mondal, S., and Pang, S.T. (2018). Super-Nernstian sensitivity in microfabricated electrochemical pH sensor based on $CeTi_xO_y$ film for biofluid monitoring. *Electrochim. Acta* 261: 482–490. https://doi.org/10.1016/j.electacta.2017.12.163.

7 Dong, Z., Wejinya, U.C., and Elhajj, I.H. (2013). Fabrication and testing of ISFET based pH sensors for microliter target solutions. *Sensors Actuators A Phys.* 194: 181–187. https://doi.org/10.1016/j.sna.2013.02.008.

8 Kumar, A.A. and Kumar, S.K.N. (2019). Comprehensive review on pH and nutrients detection sensitive materials and methods for agriculture applications. *Sens. Lett.* 17 (9): 663–670. https://doi.org/10.1166/sl.2019.4126.

9 Grusak, M.A., Ars, U., Broadley, M.R., and White, P.J. (2016). Plant macro- and micronutrient minerals. *eLS* (December): https://doi.org/10.1002/9780470015902. a0001306.pub2.

10 Xie, K., Lu, Z., Pan, Y. et al. (2020). Leaf photosynthesis is mediated by the coordination of nitrogen and potassium: the importance of anatomical-determined mesophyll conductance to CO_2 and carboxylation capacity. *Plant Sci.* 290: 110267. https://doi.org/10.1016/j.plantsci.2019.110267.

11 Sattar, A., Naveed, M., Ali, M. et al. (2018). Perspectives of potassium solubilizing microbes in sustainable food production system: a review. *Appl. Soil Ecol.* 133: 146–159. https://doi.org/10.1016/j.apsoil.2018.09.012.

12 Burton, L., Jayachandran, K., and Bhansali, S. (2020). Review — the 'Real-Time' revolution for in situ soil nutrient sensing review. *J. Electrochem. Soc.* 167: 037569. https://doi.org/10.1149/1945-7111/ab6f5d.

13 Mohammad, M.J. and Mazahreh, N. (2011). Changes in soil fertility parameters in response to irrigation of forage crops with secondary treated wastewater. *Commun. Soil Sci. Plant Anal.* 34 (May 2013): 37–41. https://doi.org/10.1081/ CSS-120020444.

14 Tadesse, T., Dechassa, N., Bayu, W., and Gebeyehu, S. (2013). Effects of farmyard manure and inorganic fertilizer application on soil physico-chemical properties and nutrient balance in rain-fed lowland rice ecosystem. *Am. J. Plant Sci.* 4 (2): 309–316.

15 Agbeshie, A.A., Abugre, S., Adjei, R. et al. (2020). Impact of land use types and seasonal variations on soil physico-chemical properties and microbial biomass dynamics in a tropical climate. *Ghana* 21 (1): 34–49. https://doi.org/10.9734/ AIR/2020/v21i130180.

16 Akshaya Kumar, A., Naveen Kumar, S.K., Aniley, A.A. et al. (2019). JES focus issue on 4D materials and systems hydrothermal growth of zinc oxide (ZnO) nanorods (NRs) on screen printed IDEs for pH measurement application. *J. Electrochem. Soc.* 166 (9): 3264–3270. https://doi.org/10.1149/2.0431909jes.

17 Akshaya Kumar, A., Naveen Kumar, S.K., and Fernandez, R.E. (2020). Real time sensing of soil potassium levels using zinc oxide-multiwall carbon nanotube based sensors. *IEEE Trans. NanoBiosci.* 20: 50–56.

18 Fazio, E., Spadaro, S., Corsaro, C. et al. (2021). Metal-oxide based nanomaterials: synthesis, characterization and their applications in electrical and electrochemical sensors. *Sensors* 21 (7): https://doi.org/10.3390/s21072494.

19 Chavali, M.S. and Nikolova, M.P. (2019). Metal oxide nanoparticles and their applications in nanotechnology. *SN Appl. Sci.* 1 (6): https://doi.org/10.1007/ s42452-019-0592-3.

20 Hatchett, D.W. and Josowicz, M. (2008). Composites of intrinsically conducting polymers as sensing nanomaterials. *Chem. Rev.* 108 (2): 746–769. https://doi.org/10.1021/cr068112h.

21 Franco, F.F., Manjakkal, L., Shakthivel, D., and Dahiya, R. (2019). ZnO based screen printed aqueous ammonia sensor for water quality monitoring. *Proceedings of the IEEE Sensors*, Montreal, QC (27–30 October 2019), 1–4. https://doi.org/10.1109/SENSORS43011.2019.8956763.

22 Vinoth, E. and Gopalakrishnan, N. (2020). Fabrication of interdigitated electrode (IDE) based ZnO sensors for room temperature ammonia detection. *J. Alloys Compd.* 824: 153900. https://doi.org/10.1016/j.jallcom.2020.153900.

23 Punetha, D. and Pandey, S.K. (2020). Enhancement and optimization in sensing characteristics of ammonia gas sensor based on light assisted nanostructured WO_3 thin film. *IEEE Sens. J.* 20 (24): 14617–14623. https://doi.org/10.1109/JSEN.2020.3009661.

24 Tohidi, S., Parhizkar, M., Bidadi, H., and Mohamad-Rezaei, R. (2020). Electrodeposition of polyaniline/three-dimensional reduced graphene oxide hybrid films for detection of ammonia gas at room temperature. *IEEE Sens. J.* 20 (17): 9660–9667. https://doi.org/10.1109/JSEN.2020.2991128.

25 Unwerslty, T. (1984). Electronic semiconducting oxides as pH sensors. *Sensors Actuators* 5: 137–146.

26 Batista, P.D. and Mulato, M. (2005). ZnO extended-gate field-effect transistors as pH sensors. *Appl. Phys. Lett.* 87 (14): 1–3. https://doi.org/10.1063/1.2084319.

27 Li, H.H., Dai, W.S., Chou, J.C., and Cheng, H.C. (2012). An extended-gate field-effect transistor with low-temperature hydrothermally synthesized SnO_2 nanorods as pH sensor. *IEEE Electron Device Lett.* 33 (10): 1495–1497. https://doi.org/10.1109/LED.2012.2210274.

28 Liao, Y.H. and Chou, J.C. (2009). Preparation and characterization of the titanium dioxide thin films used for pH electrode and procaine drug sensor by sol-gel method. *Mater. Chem. Phys.* 114 (2–3): 542–548. https://doi.org/10.1016/j.matchemphys.2008.10.014.

29 Manjakkal, L., Cvejin, K., Kulawik, J. et al. (2014). Electrochemical interdigitated conductimetric pH sensor based on RuO_2 thick film sensitive layer. *Proceedings of the 2014 International Conference and Exposition on Electrical and Power Engineering (EPE)*, Iasi, Romania (16–18 October 2014), 797–800. https://doi.org/10.1109/ICEPE.2014.6970020.

30 Manjakkal, L., Cvejin, K., Kulawik, J. et al. (2013). A low-cost pH sensor based on RuO_2 resistor material. *Nano Hybrids* 5: 1–15. https://doi.org/10.4028/www.scientific.net/nh.5.1.

31 Huang, W.D., Cao, H., Deb, S. et al. (2011). A flexible pH sensor based on the iridium oxide sensing film. *Sensors Actuators A Phys.* 169 (1): 1–11. https://doi.org/10.1016/j.sna.2011.05.016.

32 Marsh, P., Manjakkal, L., Yang, X. et al. (2020). Flexible iridium oxide based pH sensor integrated with inductively coupled wireless transmission system for wearable applications. *IEEE Sens. J.* 20 (10): 5130–5138. https://doi.org/10.1109/JSEN.2020.2970926.

33 Manjakkal, L., Sakthivel, B., Gopalakrishnan, N., and Dahiya, R. (2018). Printed flexible electrochemical pH sensors based on CuO nanorods. *Sensors Actuators B Chem.* 263: 50–58. https://doi.org/10.1016/j.snb.2018.02.092.

34 Manjakkal, L., Szwagierczak, D., and Dahiya, R. (2020). Metal oxides based electrochemical pH sensors: current progress and future perspectives. *Prog. Mater. Sci.* 109: 100635.

35 Alahi, M.E.E., Xie, L., Mukhopadhyay, S., and Burkitt, L. (2017). A temperature compensated smart nitrate-sensor for agricultural industry. *IEEE Trans. Ind. Electron.* 64 (9): 7333–7341. https://doi.org/10.1109/TIE.2017.2696508.

36 Li, Y., Yang, Q., Chen, M. et al. (2019). An ISE-based on-site soil nitrate nitrogen detection system. *Sensors* (3): https://doi.org/10.3390/s19214669.

37 Ali, A., Dong, L., Dhau, J. et al. (2020). Perspective – electrochemical sensors for soil quality assessment. *J. Electrochem. Soc.* 167: 037550. https://doi.org/10.1149/1945-7111/ab69fe.

38 Li, D., Xu, X., Li, Z. et al. (2020). Detection methods of ammonia nitrogen in water: a review. *TrAC Trends Anal. Chem.* 127: 115890. https://doi.org/10.1016/j.trac.2020.115890.

39 Ryu, H., Thompson, D., Huang, Y. et al. (2020). Electrochemical sensors for nitrogen species: a review. *Sensors Actuators Rep.* 2 (1): 100022. https://doi.org/10.1016/j.snr.2020.100022.

40 Kahlert, H., Steinhardt, T., Behnert, J., and Scholz, F. (2004). A new calibration free pH-probe for in situ measurements of soil pH. *Electroanalysis* 16 (24): 2058–2064. https://doi.org/10.1002/elan.200403059.

41 Liao, Y.-H. and Chou, J.-C. (2009). Fabrication and characterization of a ruthenium nitride membrane for electrochemical pH sensors. *Sensors* 9 (4): 2478–2490. https://doi.org/10.3390/s90402478.

42 Fulati, A., Usman Ali, S.M., Riaz, M. et al. (2009). Miniaturized pH sensors based on zinc oxide nanotubes/nanorods. *Sensors* 9 (11): 8911–8923. https://doi.org/10.3390/s91108911.

43 Moises, M., Velasco, M., Mart, E., and Mart, H.L. (2012). ISFET sensor characterization. *Procedia Eng.* 35: 270–275. https://doi.org/10.1016/j.proeng.2012.04.190.

44 Manjakkal, L., Dervin, S., and Dahiya, R. (2020). Flexible potentiometric pH sensors for wearable systems. *RSC Adv.* 10 (15): 8594–8617. https://doi.org/10.1039/d0ra00016g.

45 Manjakkal, L., Dang, W., Yogeswaran, N., and Dahiya, R. (2019). Textile-based potentiometric electrochemical PH sensor for wearable applications. *Biosensors* 9 (1): 1–12. https://doi.org/10.3390/bios9010014.

46 Simic, M., Manjakkal, L., Zaraska, K. et al. (2017). TiO$_2$-based thick film pH sensor. *IEEE Sens. J.* 17 (2): 248–255. https://doi.org/10.1109/JSEN.2016.2628765.

47 Manjakkal, L., Cvejin, K., Kulawik, J. et al. (2016). X-ray photoelectron spectroscopic and electrochemical impedance spectroscopic analysis of RuO$_2$-Ta$_2$O$_5$ thick film pH sensors. *Anal. Chim. Acta* 931: 47–56. https://doi.org/10.1016/j.aca.2016.05.012.

48 Zhong, M.L., Zeng, D.C., Liu, Z.W. et al. (2010). Synthesis, growth mechanism and gas-sensing properties of large-scale CuO nanowires. *Acta Mater.* 58 (18): 5926–5932. https://doi.org/10.1016/j.actamat.2010.07.008.

49 Arshak, K., Gill, E., Arshak, A., and Korostynska, O. (2007). Investigation of tin oxides as sensing layers in conductimetric interdigitated pH sensors. *Sensors Actuators B Chem.* 127 (1): 42–53. https://doi.org/10.1016/j.snb.2007.07.014.

50 Rosli, A.B., Awang, Z., Shariffudin, S.S., and Herman, S.H. (2019). Fabrication of integrated solid state electrode for extended gate-FET pH sensor. *Mater. Res. Express* 6 (1): 1–8. https://doi.org/10.1088/2053-1591/aae739.

51 Young, S.-J., Yang, C.-C., and Lai, L.-T. (2016). Review—growth of Al-, Ga-, and in-doped ZnO nanostructures via a low-temperature process and their application to field emission devices and ultraviolet photosensors. *J. Electrochem. Soc.* 164 (5): B3013–B3028. https://doi.org/10.1149/2.0051705jes.

52 Ji, L., Wu, C., Lin, C. et al. Characteristic improvements of ZnO-based metal – semiconductor – metal photodetector on flexible substrate with ZnO cap layer. *Jpn. J. Appl. Phys.* 49: 052201. https://doi.org/10.1143/JJAP.49.052201.

53 Day, C., Søpstad, S., Ma, H. et al. (2018). Impedance-based sensor for potassium ions. *Anal. Chim. Acta* 1034: 39–45. https://doi.org/10.1016/j.aca.2018.06.044.

54 Lima, A.S., Bocchi, N., Gomes, H.M., and Teixeira, M.F.S. (2009). An electrochemical sensor based on nanostructured hollandite-type manganese oxide for detection of potassium ions. *Sensors* 9 (9): 6613–6625. https://doi.org/10.3390/s90906613.

55 Young, S.J., Lai, L.T., and Tang, W.L. (2019). Improving the performance of pH sensors with one-dimensional ZnO nanostructures. *IEEE Sens. J.* 19 (23): 10972–10976. https://doi.org/10.1109/JSEN.2019.2932627.

56 Fredj, Z., Baraket, A., Ali, M.B. et al. (2021). Capacitance electrochemical pH sensor based on different hafnium dioxide (HfO$_2$) thicknesses. *Chemosensors* 9 (1): 1–13. https://doi.org/10.3390/chemosensors9010013.

57 Santos, L., Neto, J.P., Crespo, A. et al. (2014). WO$_3$ nanoparticle-based conformable pH sensor. *ACS Appl. Mater. Interfaces* 6 (15): 12226–12234. https://doi.org/10.1021/am501724h.

58 Zaman, S., Asif, M.H., Zainelabdin, A. et al. (2011). CuO nanoflowers as an electrochemical pH sensor and the effect of pH on the growth. *J. Electroanal. Chem.* 662 (2): 421–425. https://doi.org/10.1016/j.jelechem.2011.09.015.

59 Manjakkal, L., Cvejin, K., Bajac, B. et al. (2015). Microstructural, impedance spectroscopic and potentiometric analysis of Ta$_2$O$_5$ electrochemical thick film pH sensors. *Electroanalysis* 27 (3): 770–781. https://doi.org/10.1002/elan.201400571.

60 Chen, M., Jin, Y., Qu, X. et al. (2014). Electrochemical impedance spectroscopy study of Ta$_2$O$_5$ based EIOS pH sensors in acid environment. *Sensors Actuators B Chem.* 192: 399–405.

61 Qingwen, L., Guoan, L., and Youqin, S. (2000). Response of nanosized cobalt oxide electrodes as pH sensors. *Anal. Chim. Acta* 409 (1–2): 137–142. https://doi. org/10.1016/S0003-2670(99)00850-8.

62 Arida, H. (2015). Novel pH microsensor based on a thin film gold electrode modified with lead dioxide nanoparticles. *Microchim. Acta* 182 (1–2): 149–156. https://doi.org/10.1007/s00604-014-1311-9.

63 Salam, S., Islam, M., and Akram, A. (2013). Sol-gel synthesis of intrinsic and aluminum-doped zinc oxide thin films as transparent conducting oxides for thin film solar cells. *Thin Solid Films* 529: 242–247. https://doi.org/10.1016/ j.tsf.2012.10.079.

64 Kumar, R., Al-Dossary, O., Kumar, G., and Umar, A. (2015). Zinc oxide nanostructures for NO2 gas–sensor applications: a review. *Nano-Micro Lett.* 7 (2): 97–120. https://doi.org/10.1007/s40820-014-0023-3.

65 Cruz, D.M., Mostafavi, E., Vernet-Crua, A. et al. (2020). Green nanotechnology-based zinc oxide (ZnO) nanomaterials for biomedical applications: a review. *J. Phys. Mater.* 3 (3): https://doi.org/10.1088/2515-7639/ab8186.

66 Hatamie, A., Khan, A., Golabi, M. et al. (2015). Zinc oxide nanostructure-modified textile and its application to biosensing, photocatalysis, and as antibacterial material. *Langmuir* 31 (39): 10913–10921. https://doi.org/10.1021/ acs.langmuir.5b02341.

67 Naveen Kumar, S.K., Akshaya Kumar, A., Aniley, A.A. et al. (2019). Hydrothermal growth of zinc oxide (ZnO) nanorods (NRs), structural, and chemical composition studies for pH measurement sensor applications. *ECS Trans.* 88 (1): 437–447.

68 Xiang, H.J., Yang, J., Hou, J.G., and Zhu, Q. (2006). Piezoelectricity in ZnO nanowires: a first-principles study. *Appl. Phys. Lett.* 89 (22): 223111–223111-3. https://doi.org/10.1063/1.2397013.

69 Wei, A., Pan, L., and Huang, W. (2011). Recent progress in the ZnO nanostructure-based sensors. *Mater. Sci. Eng. B Solid State Mater. Adv. Technol.* 176 (18): 1409–1421. https://doi.org/10.1016/j.mseb.2011.09.005.

70 Ghoneim, M.T., Nguyen, A., Dereje, N. et al. (2019). Recent progress in electrochemical pH-sensing materials and configurations for biomedical applications. *Chem. Rev.* 119 (8): 5248–5297. https://doi.org/10.1021/acs. chemrev.8b00655.

71 Al-Hilli, S.M., Willander, M., Öst, A., and Strlfors, P. (2007). ZnO nanorods as an intracellular sensor for pH measurements. *J. Appl. Phys.* 102 (8): 0–5. https://doi. org/10.1063/1.2798582.

72 Manjakkal, L., Szwagierczak, D., and Dahiya, R. (2020). Progress in materials science metal oxides based electrochemical pH sensors: current progress and future perspectives. *Prog. Mater. Sci.* 109 (December 2019): 100635. https://doi.org/10.1016/j.pmatsci.2019.100635.

73 Akshaya Kumar, A., Naveen Kumar, S.K., Aniley, A.A. et al. (2019). Hydrothermal growth of zinc oxide (ZnO) nanorods (NRs) on screen printed IDEs for pH measurement application. *J. Electrochem. Soc.* 166 (9): B3264–B3270. https://doi.org/10.1149/2.0431909je.

74 Kumar, A.A., Kalappa, S., and Kumar, N. (2020). Integration of interdigitated electrodes (IDEs) with ZnO nanoflakes (Nfs) active layers. *AIP Conf. Proc.* 2244 (June): 070008.

75 Kumar, A.A. and Kumar, S.K.N. (2019). Fabrication of interdigitated electrodes (IDEs) modified with sensitive layers of ZnO nanorods and nanoflowers. *AIP Conf. Proc.* 2162 (1): 1–7.

76 Schreiber, M.A., Moyer, K.L., Mueller, B.J. et al. (2001). Development and validation of a cholate binding capacity method for DMP 504, a bile acid sequestrant. *J. Pharm. Biomed. Anal.* 25 (3–4): 343–351. https://doi.org/10.1016/S0731-7085(00)00521-5.

77 Lin, T.K. (2014). Fabrication of interdigitated electrodes (IDE's) by conventional photolithography technique for pH measurement using micro-gap structure. Proceedings of the 2014 IEEE Conference on Biomedical Engineering and Sciences (IECBES), Kuala Lumpur, Malaysia (8–10 December 2014). 1570016883(8): 8–9.

78 Zhang, Q., Liu, W., Sun, C. et al. (2015). On-chip surface modified nanostructured ZnO as functional pH sensors. *Nanotechnology* 26 (35): 355202. https://doi.org/10.1088/0957-4484/26/35/355202.

79 Lee, C.T., Chiu, Y.S., Lou, L.R. et al. (2014). Integrated pH sensors and performance improvement mechanism of ZnO-based ion-sensitive field-effect transistors. *IEEE Sens. J.* 14 (2): 490–496. https://doi.org/10.1109/JSEN.2013.2285488.

80 Chiu, Y.S., Tseng, C.Y., and Lee, C.T. (2012). Nanostructured EGFET pH sensors with surface-passivated ZnO thin-film and nanorod array. *IEEE Sens. J.* 12 (5): 930–934. https://doi.org/10.1109/JSEN.2011.2162317.

81 Kang, B.S., Ren, F., Heo, Y.W. et al. (2005). pH measurements with single ZnO nanorods integrated with a microchannel. *Appl. Phys. Lett.* 86 (11): 1–3. https://doi.org/10.1063/1.1883330.

82 Kurzweil, P. (2009). Metal oxides and ion-exchanging surfaces as pH sensors in liquids: state-of-the-art and outlook. *Sensors* 9 (6): 4955–4985.

83 Shaba, E.Y., Jacob, J.O., Tijani, J.O., and Suleiman, M.A.T. (2021). A critical review of synthesis parameters affecting the properties of zinc oxide nanoparticle and its application in wastewater treatment. *Appl. Water Sci.* 11: 48.

84 Rong, P., Ren, S., Yu, Q., and Review, D. (2018). Fabrications and applications of ZnO nanomaterials in flexible functional devices-a review. *Crit. Rev. Anal. Chem.* 49: 336–349. https://doi.org/10.1080/10408347.2018.1531691.

85 Hu, Y., Tan, O.K., Pan, J.S. et al. (2005). The effects of annealing temperature on the sensing properties of low temperature nano-sized $SrTiO_3$ oxygen gas sensor. *Sensors Actuators B Chem.* 108: 244–249. https://doi.org/10.1016/j.snb.2004.10.053.

86 Silvester, D.S. (2011). Recent advances in the use of ionic liquids for electrochemical sensing. *Analyst* 136 (23): 4871–4882. https://doi.org/10.1039/c1an15699c.

87 Wang, J. (1991). Modified electrodes for electrochemical sensors. *Electroanalysis* 3 (4–5): 255–259. https://doi.org/10.1002/elan.1140030404.

88 Aniley, A.A., Naveen Kumar, S.K., Akshaya Kumar, A. et al. (2019). Thin film dual probe heat pulse (DPHP) micro heater network for soil moisture content estimation in smart agriculture. *J. Electrochem. Soc.* 166 (2): B63–B67. https://doi.org/10.1149/2.0511902jes.

89 Bhat, R. and Sujatha, S. (2009). Soil fertility and nutrient uptake by arecanut (*Areca catechu* L.) as affected by level and frequency of fertigation in a laterite soil. *Agric. Water Manag.* 96 (3): 445–456. https://doi.org/10.1016/j.agwat.2008.09.007.

90 Kumar, A.A., Kumar, S.K.N., and Fernandez, R.E. (2021). Real time sensing of soil potassium levels using zinc oxide-multiwall carbon nanotube-based sensors. *IEEE Trans. Nanobioscience* 20 (1): 50–56. https://doi.org/10.1109/TNB.2020.3027863.

91 Pule, M., Yahya, A., and Chuma, J. (2017). Wireless sensor networks: a survey on monitoring water quality. *J. Appl. Res. Technol.* 15 (6): 562–570. https://doi.org/10.1016/j.jart.2017.07.004.

92 Aliyana, A.K., Ganguly, P., Beniwal, A. et al. (2022). Disposable pH sensor on paper using screen-printed graphene-carbon ink modified zinc oxide nanoparticles. *IEEE Sens. J.* 22 (21): 1–1. https://doi.org/10.1109/jsen.2022.3206212.

93 Aliyana, A.K., Kumar, S.K.N., Marimuthu, P., and Baburaj, A. (2021). Machine learning-assisted ammonium detection using zinc oxide/multi-walled carbon nanotube composite based impedance sensors. *Sci. Rep.* 11: 1–10.

Section II

Readout Electronic and Packaging

7

Integration and Packaging for Water Monitoring Systems

Muhammad Hassan Malik and Ali Roshanghias

Silicon Austria Labs GmbH, Heterogeneous Integration Technologies, Villach, Austria

7.1 Introduction

In this chapter, system integration and packaging of the multisensory system for water quality monitoring are addressed. In the previous chapters of this book, a variety of sensors in a wide range of configurations was introduced, together with the corresponding CMOS/ASIC development and communication platforms. Here, the integration of all blocks in the form of a package with a small footprint is aimed. There are several commercial system-level packaged sensors available for water quality monitoring. These sensors have essentially a large form-factor, are rigid, bulky, and costly. The application of these sensors varies from monitoring aquarium water quality to the salinity of seawater. Hydrolab HL7 – Multiparameter Sonde (Figure 7.1) is an example of a multiparameter water quality monitoring system that can monitor pH, ammonium, nitride conductivity, and temperature of the water [1].

In agricultural areas, real-time water quality monitoring is essential for the high efficiency and growth of crops [2]. As an example, at China Agricultural University, a multiparameter monitoring sensor was developed which can monitor pH, dissolved oxygen (DO), temperature, and ammonia. Figure 7.2 (top) illustrates the principle of the system. The whole system can be divided into application layer, process layer, and perception layer. The perception layer consists of the sensor objects. The user interface is within the application layer and the process layer acting as an interconnect between the two layers. The perception layer collects six parameters of water quality. The data is then processed in the processing layer and

Sensing Technologies for Real Time Monitoring of Water Quality, First Edition.
Libu Manjakkal, Leandro Lorenzelli, and Magnus Willander.
© 2023 The Institute of Electrical and Electronics Engineers, Inc.
Published 2023 by John Wiley & Sons, Inc.

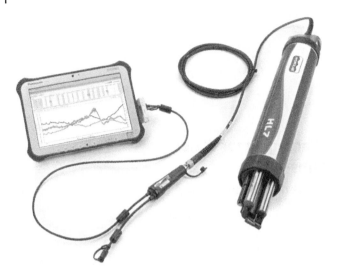

Figure 7.1 Hydrolab HL7 – Multiparameter Sonde. *Source:* From Ref. [1], OTT HydroMet.

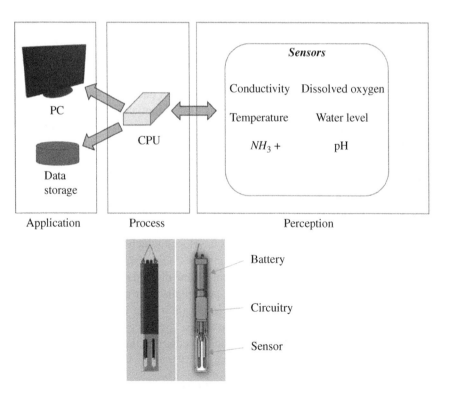

Figure 7.2 Overall principle of a multiparameter sensor system (top) and water quality sensor system (bottom). *Source:* From Ref. [2], Springer Nature.

further sent to the application layer where the user can configure the data accordingly. Figure 7.2 (bottom) shows the multiparameter integrated water quality sensors system. It can be seen that the system was packaged in a cylindrical form, with a hook on its top. PCB is placed in the middle of the system while the batteries are put near the top of the system. Sensors and water pump connected with PCB are fixed at the lower place. Signals are collected through sensors from the bottom of the system.

Another example of a multisensory system package was demonstrated by WET labs and Sea-Bird Electronics, where a water quality sensing system with a continuous monitoring feature was developed [3]. The sensor package provides the measurement of physical and biogeochemical measurements like conductivity, temperature, pressure, DO, and chlorophyll fluorescence. The package is 65.4 cm in length and 18.5 cm in diameter. The package is made of copper and durable plastic. Sensor is shrouded in a copper ring guard assembly and additionally protected from biofouling with two EPA-approved antifoulant cartridges. The system features a processor and storage drive onboard which can measure data at 1 Hz frequency [3] (Figure 7.3).

The Wireless Water Sentinel project (WaterWiSe@SG) demonstrated a generic wireless sensor network's capabilities to enable real-time monitoring of a water distribution network [4] (Figure 7.4a). As shown in Figure 7.4b, the sensor node is

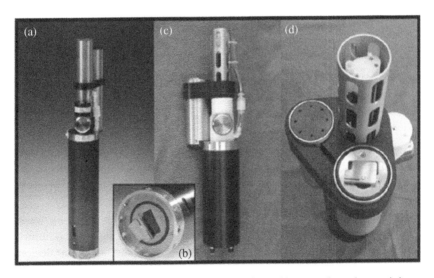

Figure 7.3 (a) Water quality monitoring sensor package, (b) copper face plate and ring guard, (c) initial WQM prototype with perforated ring-guard assembly covering the temperature and conductivity sensor, and (d) WQM prototype used during the San Luis Bay, CA and Chesapeake Bay, MD deployments with ECO-FLNTU oriented facing upward. *Source:* From Ref. [3], IEEE.

(a)

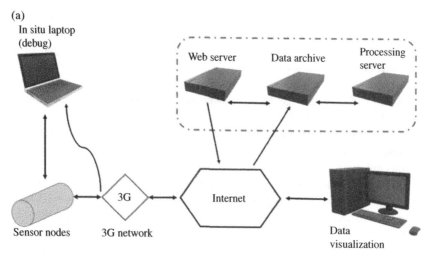

In situ laptop (debug)

Web server   Data archive   Processing server

3G

Internet

Sensor nodes   3G network

Data visualization

(b)

Figure 7.4 (a) WaterWiSe@SG communication structure and (b) WaterWiSe@SG sensor node in its water-resistant packaging. *Source:* From Ref. [4], American Society of Civil Engineers.

packaged in a clear plastic acrylic tube, with PVC caps at each end. These end caps house waterproof debugging ports, sensor ports, and attention buttons. The water-resistant packaging facilitates the deployment of the sensors inside a manhole if needed.

Given those examples, one can infer that the current technologies for water quality monitoring are rather confined to system-level packaging which is limited to the printed circuit board (PCB) technology and rigid surface mount devices (SMD). In fact, the sensor package manufacturing as well as other electronics manufacturing industries are primarily PCB based and any new system integration technology has to compete with this already established technology. A famous example of the successful shift from PCB-based system integration to another manufacturing platform (flexible, roll-to-roll, and film-based) is radio frequency identification (RFID) labels. Further in this chapter, state-of-the-art system integration technologies, as well as innovative packaging concepts that can be used for water quality monitoring, will be discussed.

7.2 Advanced Water Quality Monitoring Systems

7.2.1 Multi-sensing on a Single Chip

It is well known that the relative distributions and chemical species of toxic contaminants are highly influenced by environmental parameters such as pH and oxidation–reduction potential (ORP). Many of these measurements are currently made using chemical electrodes of various types, including those for measuring pH, DO, phosphates, and heavy metals such as lead, cadmium, and mercury. These devices are valuable tools for direct detection of pollutants in streams or lakes, wastewater treatment reactors, and water distribution systems due to the speed of analysis, low maintenance cost, and availability of necessary equipment.

Recently, there has been a growing interest in the application of microelectromechanical systems (MEMS) technologies to monitor and measure contaminants in waters. Lab-on-a-chip (LOC) MEMS devices are currently the major trend in research for water quality monitoring. In a LOC device, several laboratory functions are integrated on a single chip with a size of few millimeters to centimeters. The basis for most LOC fabrication processes is photolithography. Initially, most processes were in silicon, as these well-developed technologies were directly derived from semiconductor fabrication. The demand for cheap LOC, however, pushed the technology toward the fabrication of PDMS microfluidic devices as shown in Figure 7.5. Generally, a LOC device can consist of passive microfluidic components, a biochemical sensor array, on-chip pressure sources, and a sample injection interface [5].

The use of the LOC concept enables miniaturized instruments for detecting nitrate, pH, ORP, and heavy metals in surface water and groundwater at shallow depths. Furthermore, miniaturized on-chip electrochemical sensors with planar microelectrodes reduce the sensor cost due to mass production, additionally due to its batch fabrication methods and the use of low-cost polymers. It also permits the use of low-cost components with high accuracy and reliability [5, 6]. For the development of potentiometric nitrate and pH LOC sensor package, the nanobead packing technique was applied to fabricate self-assembled nanobead hetro (nBH) columns in a polymer chip cartridge.

As shown in Figure 7.6, the hydrophilic or hydrophobic nanobeads were partially packed in a designated region by capillary electrophoresis. As a result, the developed nBH column is composed of a hydrophilic silica beadpacked area for

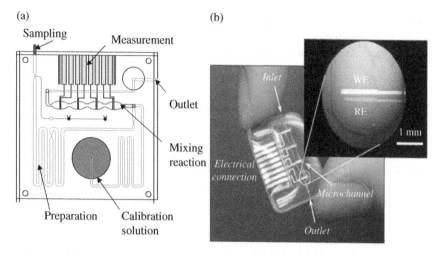

Figure 7.5 Schematic diagram of a state-of-the-art LOC chip (a) and the fabricated device (b). *Source:* From Ref. [5], IEEE.

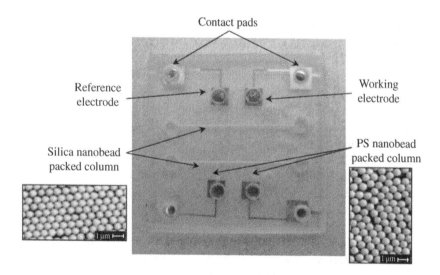

Figure 7.6 Photograph of the dual ion-selective lab chip and the ESEM images of the heterogeneous nanobead packed column. *Source:* From Ref. [6], IOP Publishing.

electrolyte loading and a hydrophobic polystyrene bead-packed area for a LIX membrane which is permeable for target ions but impermeable for interfering species. The nBH column provides a new electrochemical sensing platform with high sensitivity and excellent ion selectivity for environmental monitoring [6].

7.2.2 Heterogeneous Integration

The trends toward the Internet of Things (IoT) and the internet of everything along with advances in software, data mining, machine learning, and artificial intelligence (AI) require intelligent sensor systems that are smart, environmental friendly, and affordable. For the realization of such sensor systems in the era of IoT, a system is to be developed with a comparatively small form-factor compared to the examples mentioned in the introduction. LOC devices are much more compacter compared to conventional sensors due to the integration of several functionalities in a small area as well as massive parallelization and low fluid volumes consumption. On the other hand, LOC devices are still not yet fully developed for the widespread use and are complex and require expensive equipment and specialized personnel.

Another aspect of IoT sensor packages is to keep the final cost minimal. It is noteworthy to mention that 50–80% of the production cost of a sensor device is attributed to packaging. The incorporation of intelligence, data storage, microprocessing, and connectivity into sensor packages at low cost necessitates hybrid integration of semiconductor bare dies into inexpensive platforms such as polymers, e.g. PET or paper [7]. Since the final product is designed to be flexible, cheap, thin, and manufacturable on a large scale, innovative packaging techniques are required which are applicable in large scales such as sheet-to-sheet or roll-to-roll manufacturing [8]. Nowadays, different integration configurations can be found for electronic-based systems such as a full rigid system, flexible system, or hybrid (rigid-flex) system. All have their own advantages and disadvantages. Figure 7.7 illustrates two conceptual demonstrations of simplified sensor systems with encapsulated circuitry and exposed sensing areas which can be realized in either rigid format, flex, or hybrid.

7.2.3 Case Study: MoboSens

Wang demonstrated a mobile phone sensing platform, MoboSens, with integrated plug-n-play microelectronic ionic sensor that performs electrochemical measurement by using audio jack of a smartphone [9]. This platform was used to measure nitrate concentration using few microliter liquid samples on the field along with providing geospatial map locations through wireless networks. A photograph of the complete package of MoboSens including the microfluidics, mobile apps, and the controlling circuits is shown in Figure 7.8. Here, the nitrate sensor is fabricated on a glass substrate by a photolithography process. To facilitate portability, a small circuit box was made attached with a light fixture frame fabricated using rapid prototyping manufacturing process. The fixture mechanism will make the circuit box stable when attached to a mobile phone.

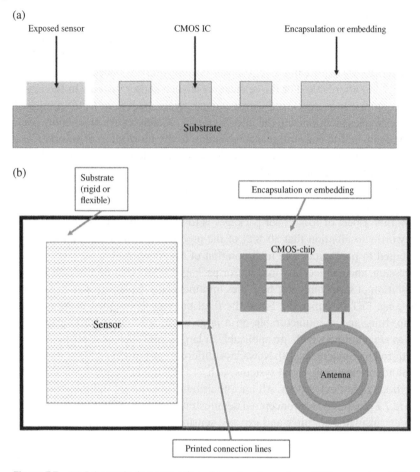

Figure 7.7 (a) Schematic demonstration of a system in package (SiP) concept and (b) top view of sensor package.

A microfluidic PDMS chamber was bonded to the active region of the electrochemical sensor on the glass slide. Under the capillary design of the test chamber, the sample liquid was dragged into the microfluidic chamber by capillary action, when a sample is placed at the sensor opening to perform testing. By making use of the CPU, ADC, and DAC from the phone itself, instead of external electronic components, the form-factor and cost (~$10) of the package were significantly reduced to achieve one handheld platform to do electrochemical testing [9].

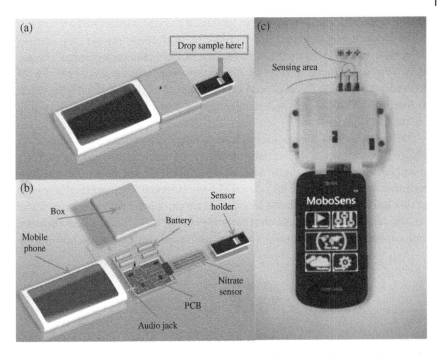

Figure 7.8 Concept design and a fabricated package of MoboSens. (a) Assembly view of MoboSens. (b) Detailed components of MoboSens. (c) Photograph showing the complete MoboSens system. *Source:* From Ref. [9], Elsevier.

7.3 Basics of Packaging

An electronics package primarily provides an electrical interconnection scheme among various active and passive components in a system. It also serves as a protective enclosure and the interface to sensing media. An advanced packaging can also provide different thermal management solutions by integrating diel-level, board level, or system-level cooling such as by adding thermal vias, heat sinks, heat pipes, thermoelectric refrigeration, etc. In the last decades, the trend of electronic packages has moved toward miniaturization, higher interconnects density, low power consumption, and higher speed. As to fulfill these trends, innovative technologies such as System in Package (SiP), wafer-level packaging, and more recently 2.5 and 3D integrations were introduced. Also, fan-out wafer-level packaging (FOWLP) and panel-level packaging (FOPLP) have spurred increasing interest due to significant cost advantages over competitive technologies, increased interconnect density, as well as enhanced electrical and thermal package performance [10]. These technologies have been successful in following the trends of the Moore's law, according to which the number of devices per chip doubles every two years [11]. Since the classification of electronic packages has become more complex, different levels of integration were suggested as one way of differentiation. These levels depend on the number of elements and subsystems in the package. A schematic demonstration of these levels can be observed in Figure 7.9 [12].

The lowest level of this package is level 0. This mostly consists of a chip or die which is made commonly of silicon. There are some interconnects that are already made on the chip here. Also, wafer-level capping of sensor dies and hermetic encapsulation can be introduced already in level 0. The next level is level 1. In this level, the chip is placed in a protective container. These containers can carry several chips or a single chip according to the application. These chips are then connected to the electronic components outside this protective housing via

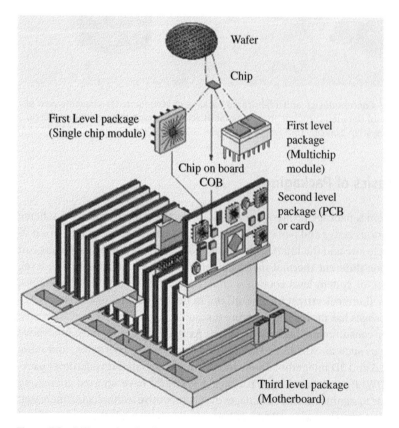

Figure 7.9 Different levels of packaging. *Source:* From Ref. [12]/Springer Nature.

lead frames, through vias or protruded bumps. For sensors, based on their applications, procedures such as over-molding, exposed die molding, or cavity housings are utilized to generate the packages. In level 2, the packaged dies are connected to other components. Mostly this is done on an interposer that can be organic (such as a PCB, flexible printed foil, or even 3D printed board), silicon, or glass interposers [13]. In level 3, these interposers are connected to a physical interface that can consist of several boards connected providing a supporting medium.

7.4 Hybrid Flexible Packaging

A visionary approach for the future of advanced water quality monitoring devices is the hybrid integration of active and passive components in one system (SiP). With time there has been significant development in hybrid packaging as the application of electronics packaging has become wider and different materials such as plastic foils, papers, ceramics, glasses, metals, and semiconductors are to be integrated into one package. Sensor hybrid packages are divided into two groups: hermetic (metal and ceramics) and semi-hermetic (plastic) packages. In low-end sensory applications, plastic packages are still popular as they are most cost-effective and more compact compared to the metallic or ceramic packages [14]. Here, different passivation layers and encapsulating coatings can be applied to provide requisite hermeticity to polymer packages. As the application of hybrid packages has widened in the last decades toward the IoT, flexible packages have gained more attention. The specific advantages of flexible packages are as follows [15]:

- Can be foldable, deformable, stretchable
- Can be disposable and environmentally friendly
- Freedom in design for manufacturers and ease of use for consumers
- Cost-effectiveness
- No need for sophisticated semiconductor manufacturing processes
- Compatibility of mass production in the form of a roll-to-roll or sheet-to-sheet fabrication

Low-cost materials can be easily adapted for flexible sensor systems. There are several examples of printed sensor systems in the literature which were realized inside inexpensive flexible materials. Paper has been recently used as a preferred substrate material for flexible sensor systems, especially for disposable medical applications. However, using paper as an electronic substrate requires hermetic sealing against water absorption. An illustration of this can be observed in Figure 7.7, as encapsulation or embedding can be used to create this sealing.

Polymers such as Polyethylene terephthalate (PET) and polyamide (PI) have been also extensively used for hybrid packaging [16]. By employing PET or paper, the cost of the substrate can be cut down with a factor of 5–10 [17].

7.4.1 Interconnects

Inexpensive hybrid packaging opens the opportunity for integrating multi-sensor systems at a low cost which is crucial for mass production. As the technologies are different from conventional packaging, alternative interconnects are required which are applicable at low temperatures and low pressures. These processes also should be adaptable with roll-to-roll or sheet-to-sheet processing. With regards to interconnect materials, adhesives are exhibiting considerable advantages over wire bonding or soldering. Similar interconnecting technologies to RFID tags are currently employed in flexible sensor production, in which isotropic conductive adhesives (ICA) are dispensed and cured at low temperatures. Anisotropic conductive adhesive (ACA) either in the form of a film (ACF) or paste (ACP) is another popular material in the fabrication of flexible packages. These adhesives consist of the polymer-based matrix with conductive particles which provide a vertical conductive path between the pads on the chip and the solder [18, 19]. Figure 7.10a and b show the two techniques employed for hybrid bonding.

An advantage of ACAs compared to conventional soldering is that they can cater to the need for fine pitch assembly and can be used for flex on flex assembly [7, 20, 22]. As the flexible substrate is also different from PCBs, printing technologies can be used in this. Concerning the circuitry, priming technologies (such as screen-printing, gravure, and inkjet printing) are the well-established low-cost solutions in comparison to etched metal structures. These flexible substrates also open the way of using biodegradable substrates like paper [23, 24]. In the literature, numerous flexible sensor packages were reported for pH sensing. For these pH sensors, substrates like polyimide (Pi) are used and the sensing electrode is printed using a screen printing method [25]. Another application for these types of printed sensors is measuring human body temperature. Inkjet-printed electrodes have also shown good results in this perspective [26].

The second part of this hybrid integration is the bonding of dies or chips to the substrate. In flip-chip bonding, the main possibilities for attaching the die to the substrate are solder bumps, Cu pillars, Au stud-bumps, and adhesives [14]. An example of Flip-chip configuration is illustrated in Figure 7.11. Flip-chip bonding also has some advantages over wire bonding [14]:

- low lead inductances, as it has a shorter length compared to wire,
- reduced need to attach precious metals, and
- high productivity, as a greater number of bonds can be made simultaneously.

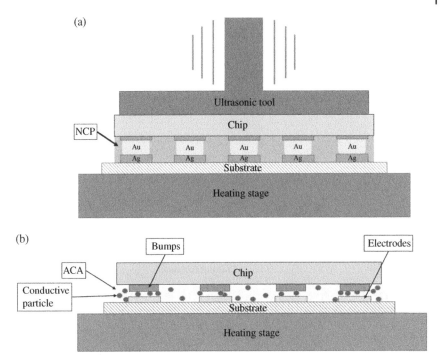

Figure 7.10 (a) Thermosonic flip chip bonding. *Source:* Adapted from Ref. [20]. (b) ACA illustration. *Source:* Adapted from Ref. [21].

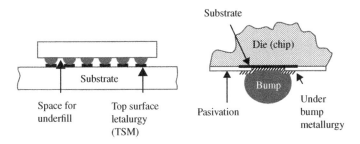

Figure 7.11 Flip chip package. *Source:* From Ref. [14].

Thermocompression, thermosonic, and pressure-less flip-chip bonding techniques have been used to bond bare dies to paper and PET substrate [23]. A simplified sketch of thermosonic bonding of Au-bumped silicon dies to screen-printed substrates can be observed in Figure 7.10a.

Materials like non-conductive paste (NCP), ACF, and ACP can be used in this method to make conductive joints. Another advantage of using flip-chip bonding

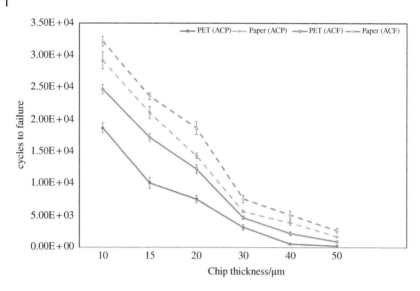

Figure 7.12 The results of the cyclic bending test; the effects of die thickness, adhesives, and substrates on the failure cycle. *Source:* From Ref. [7]/Elsevier/CC BY-4.0.

with ACP is, it can be performed at low temperatures and low pressures. As an example, Silicon chips with 100-µm pitch were bonded with an NCP and 170 °C temperature and showed good resistance results [27]. Furthermore, this technique provides us the gateway to bond ultrathin chips to flexible substrates. In literature, silicon dies with a minimum thickness of 10 µm were bonded to paper and PET substrates utilizing both ACP and ACF followed by their mechanical stability being investigated by performing a dynamic bending test [7]. The effect of die thickness on the bending reliability can be observed in Figure 7.12.

These adhesives can also be used for thin chip stacking. A four-layer flip-chip stack with 170 µm has been produced and showed promising results; this is visible in Figure 7.13 [28].

7.4.2 Thin Die Embedding

Most recently, electronic products have emerged with a whole system integrated into a flexible foil [15]. These flexible systems are the most promising choice for structural health monitoring such as on a robotic arm in which the sensor must be bent for installation in a specific orientation. This requires both the interposers and the components to be thin and flexible [29]. To facilitate this flexibility, even the bare silicon die should be thinned down to sub 100 µm. This provides some flexibility in the final configuration. For example, a 25 µm die can be bent to a radius of 1 cm.

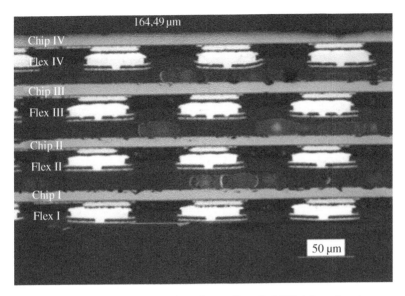

Figure 7.13 Stacked thin chip on Flex. *Source:* From Ref. [28], International Microelectronics Assembly & Packaging Society.

Most commonly the interposer materials used for this type of process are Pi foil, as it possesses properties like good thermal and dimensional stability. To make this process, cost-effective materials like polyethylene naphthalate (PEN) and PET have gained more attention [30]. But there is a disadvantage of thermal stability compared to Pi. The glass transition temperatures of PEN and PET (155 °C and 110 °C) are comparatively low to PI (350 °C) [15, 31].

A process flow for thin die integration in a flexible substrate is shown in Figure 7.16. The first step of this process is the thinning of the silicon die. This is done with grinding and polishing of the excess silicon on the die. Simultaneously the substrate is also prepared, which starts with the deposition of the release layer on a glass substrate. This substrate is used as a processing carrier. A layer of Pi is then spin-coated and further cured. As the Pi has a high curing temperature, this limits the material selection for the release layer. The next process step is to glue the thinned die to this substrate using the thermosetting polymer BCB and coated with a second layer of Pi. The next step is to pattern the Pi (top layer) to expose the contact pads. The final step is to release the package from the substrate as shown in Figure 7.14 [33].

Conventionally for flexible printed electronics, printing technologies such as inkjet, aerosol, gravure, and screen printing have been readily utilized to make the required circuitry. However, they are still limited in resolution and yielding poor conductivity. Die embedding and topside metallization, on the other hand, render

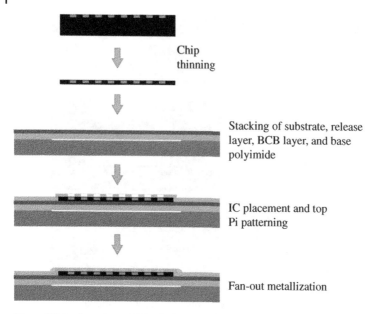

Chip
thinning

Stacking of substrate, release
layer, BCB layer, and base
polyimide

IC placement and top
Pi patterning

Fan-out metallization

Figure 7.14 Overview of thin die integration in foil. *Source:* From Ref. [33],
© (2011), IEEE.

a robust system in foil methodology with lithographic resolution [34]. Here, physical vapor deposition (PVD) is the technology used for topside metallization and encapsulation where the material is evaporated and condensed to form a thin film coating over an object (substrate). This technique has been successfully adapted for flexible sensor systems. An example of a flexible system in foil with topside PVD metallization is shown in Figure 7.15 which was fabricated as an implantable neural probe. The package consists of a CMOS chip encapsulated in Pi resulting in a thickness of 75 μm [35].

7.4.3 Encapsulation and Hermeticity

For biological applications, the sensor package must be also hermetic, safe, and compatible with the human body. Polydimethylsiloxane (PDMS) has been considered one of the most biocompatible polymers. It was employed as the sensor package interface to the body leading to minimum tissue reaction among other polymeric materials [36]. PDMS, however, has a high permeability to moisture which is a major drawback. Therefore, it cannot be used for encapsulation of electrodes such as Platinum (Pt). The moisture uptake from delaminated interfaces between PDMS and Pt deteriorates the hermeticity and long-term reliability of the sensor package [32]. To cater to this need, a further layer of Hafnium(IV) oxide

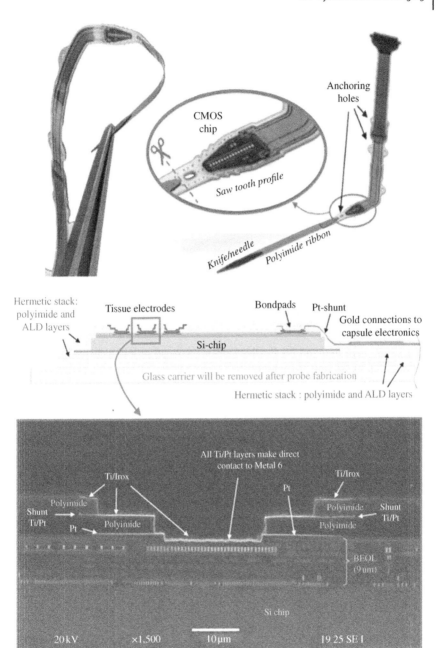

Figure 7.15 System in foil via die embedding (top) and PVD routings (bottom).
Source: From Ref. [35], © (2017), IEEE.

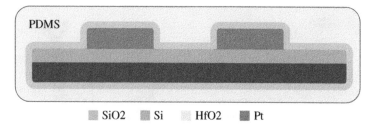

PDMS

⬛ SiO2   ⬛ Si   ⬜ HfO2   ⬛ Pt

Figure 7.16 Cross-section of material stacking. *Source:* From Ref. [32], © (2020), IEEE.

(HfO$_3$) was deposited via atomic layer deposition (ALD) between the Pt and PDMS. In this way, the Pt is protected from the moisture. Figure 7.16 gives an overview of the material stacking in the package.

The main advantage of using this bilayer package is HfO$_3$ provides good stability in the environment and prevents any delamination or moisture seepage. PDMS provides a long-lasting adhesion with the ALD layer.

7.4.4 Roll to Roll Assembly

Due to the high demand for flexible sensor packages at a low cost, high volume production is essential. To achieve that, Roll to Roll or Reel to Reel (R2R) manufacturing lines are gradually adapted to printed, flexible electrics. These large-scale, continuous manufacturing lines consist of deposition, coating, patterning, micro-assembly, lamination, and encapsulation of the electronic devices in one go [37]. R2R has been already proven in the manufacturing of RFID labels for some time. Figure 7.17 provides an overview of the process flow.

Recently, R2R has also been implemented for heterogeneous assembly and integration. The Two compatible micro-assembly techniques with R2R lines are fan-out die embedding in foil and flip-chip integration of dies on the foil. With regards to flip-chip bonding, two requirements must be fulfilled: (i) high alignment

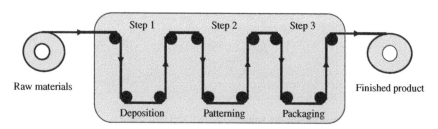

Raw materials   Finished product

Step 1   Step 2   Step 3

Deposition   Patterning   Packaging

Figure 7.17 Roll-to-roll manufacturing process flow. *Source:* From Ref. [8].

accuracy of the dies to substrate and (ii) fast running die handling systems to achieve high throughput. Optical alignment systems are necessary to meet the alignment requirements. For the fan-out embedding, a chip-foil package approach was proposed by Palavesam et al. [38], where the thinned chips are firstly embedded in a fan-out foil. Then, the foil with relaxed large pitch sizes is transferred to R2R line. As an alternative to these classical pick and placement technologies, disruptive approaches such as self-alignment of dies on foils were also proposed in the literature. In this method, the thinned chips are placed on the polar assembly liquid. The surface tension of the assembly liquid forces the chip to align with the target substrate [38].

7.5 Conclusion

Water quality monitoring systems are already present in the market. The shortcomings of these systems are large form-factor, rigidity, and cost effectiveness. To realize a small form-factor system, integrating a whole system on a single substrate utilizing die-level packaging can be a proposed solution. For this, already established techniques such as thermosonic and thermocompression flip-chip techniques can be used. Adhesives like ACAs are also available in the market which can aid in the manufacturing of a system. Furthermore, using substrates such as polymers and roll-to-roll manufacturing techniques can reduce the price of these systems.

References

1 Ott HydroMet (2020). Hydrolab HL7 – Multiparameter Sonde. https://www.ott.com/ products/water-quality-2/hydrolab-hl7-multiparameter-sonde-2338/ (accessed 13 June 2020).

2 Mingli, L., Daoliang, L., Ya, C., and Chengfei, G. (2013). A multi-parameter integrated water quality sensors system. *IFIP Adv. Inf. Commun. Technol.* 392: 260–270.

3 Cristina, O., Casey, M., Dave, R., et al. (2007). WQM: a new integrated water quality monitoring package for long-term in-situ observation of physical and biogeochemical parameters. *Oceans*, Vancouver (29 September to 4 October 2007).

4 Andrew Whitttle, L.G. (2011). WaterWiSe@SG: a testbed for continuous monitoring of the water distribution system in Singapore. *12th Annual Conference on Water Distribution Systems Analysis (WDSA)* Tucson (12–15 September 2010).

5 Ahn, C.H. (2004). Disposable smart lab on a chip for point-of-care clinical diagnostics. *Proc. IEEE* 1: 154–173.

6 Jang, A., Zou, Z., Lee, K.-K.P. et al. (2011). State-of-the-art lab chip sensors for environmental water monitoring. *Meas. Sci. Technol.* 22 (3): 032001.

7 Malik, M.H., Grosso, G., Zangl, H. et al. (2021). Flip chip integration of ultra-thinned dies in low-cost flexible printed electronics; the effects of die thickness, encapsulation and conductive adhesives. *Microelectron. Reliab.* 123: 114204.

8 Schwartz, E. (2006). *Roll to Roll Processing for Flexible Electronics*. Ithaca, New York: Cornell University.

9 Wang, X. (2015). Audio jack based miniaturized mobile phone electrochemical sensing platform. *Sensors Actuators B Chem.* 209: 677–685.

10 Roshanghias, A., Dreissigacker, M., Scherf, C. et al. (2020). On the feasibility of fan-out wafer-level packaging of capacitive micromachined ultrasound transducers (CMUT) by using inkjet-printed redistribution layers. *Micromachines* 11: 564.

11 Sakuma, K., Andry, P.S., Cornelia, T. et al. (2008). 3D chip-stacking technology with TSV's and low-volume lead-free interconnections. *IBM J. Res. Dev.* 52 (6): 611–622.

12 Suhling, J.C. and Lall, P. (2008). *Electronic Packaging Applications*. Springer Handbook of Experimental Solid Mechanics.

13 Roshanghias, A. (2019). Additive-manufactured organic interposers. *J. Electron. Packag.* 7 (26): 014501–014506.

14 Ivan, S.Z.E.N.D.I.U.C.H. (2011). Development in electronic packaging. *Radioengineering* 20: 214–220.

15 Burghartz, J. (ed.) (2011). *Ultra-Thin Chip Technology and Applications*. New York: Springer.

16 Zikulnig, J., Roshanghias, A., Rauter, L., and Hirschl, C. (2020). Evaluation of the sheet resistance of inkjet-printed Ag-layers on flexible, uncoated paper substrates using Van-der-Pauw's method. *Sensors* 20 (8): 2398.

17 van den Brand, J., Kusters, R., Heeren, M., et al. (2010). Flipchip bonding of ultrahin Si dies onto PEN/PET substrates with low cost circuitry. *Proceedings of the 3rd Electronics System Integration Technology Conference ESTC*, Berlin (13–16 September 2010).

18 Yoon, D.-J., Malik, M.H., Yan, P. et al. (2021). ACF bonding technology for paper- and PET-based disposable flexible hybrid electronics. *J. Mater. Sci. Mater. Electron.* 32 (2): 2283–2292.

19 Wong, C.P., Lu, D., and (Grace) Li, Y. (2010). *Electrical Conductive Adhesives with Nanotechnologies*. New York: Springer.

20 Yoon, D.-J., Lee, S.-H., and Paik, K.-W. (2019). Effects of polyacrylonitrile anchoring polymer layer solder anisotropic conductive films on the solder ball movement for fine-pitch flex-on-flex (FOF) assembly. *IEEE Trans. Compon. Packag. Manuf. Technol.* 9 (5): 830–835.

21 Yim, M.J. and Paik, K.W. (2006). Recent advances on anisotropic conductive adhesives (ACAs) for flat panel displays and semiconductor packaging applications. *Int. J. Adhes. Adhes.* 26 (5): 304–313.

22 Malik, M.H., Tsiamis, A., Zangl, H. et al. (2022). Die-level thinning for flip-chip integration on flexible substrates. *Electronics* 11 (6): 849.

23 Roshanghias, A., Rodrigues, A.D., and Holzmann, D. (2020). Thermosonic fine-pitch flipchip bonding of silicon chips on screen printed paper and PET substrates. *Microelectron. Eng.* 228.

24 Malik, M.H., Rauter, L., Zangl, H., et al. (2022). Ultra-Thin Chips (UTC) integration on inkjet-printed papers. *Proceedings of the 2022 IEEE International Conference on Flexible and Printable Sensors and Systems (FLEPS)*, Vienna (10–13 July 2022).

25 Dobie, A. (2019). Flexible PET substrate for high-definition printing of polymer thick-film conductive pastes. *J. Microelectron. Electron. Packag.* 16 (2): 103–116.

26 Ali, S., Khan, S., and Bermak, A. (2019). Inkjet-printed human body temperature sensor for wearable electronics. *IEEE Access* 7: 163981–163987.

27 Roshanghias, A. and Rodrigues, A.D. (2019). Low-temperature fine-pitch flip-chip bonding by using snap cure adhesives and Au stud bumps. *Proceedings of the 22nd European Microelectronics and Packaging Conference & Exhibition (EMPC)*, Pisa, Italy (16–19 September 2019).

28 Haberland, J., Lütke-Notarp, D., Aschenbrenner, R. (2010). Ultrathin 3D ACA flipchip-in-flex technology. *Proceedings of the Device Packaging Conference*, SCOTTSDALE, AZ (8–11 March 2010).

29 Elsobky, M., Alavi, G., Albrecht, B., et al. (2018). Ultra-thin sensor systems integrating silicon chips with on-foil passive and active components. *Eurosensors Conference*, Graz (9–12 September 2018).

30 Fjelstad, J. (2007). *Flexible Circuit Technology*, 3e. Seaside: BR Publishing.

31 Demirel, B., Yaraş, A., and ELÇIÇEK, H. (2011). *Crystallization Behavior of PET Materials*. Balıkesir: Balıkesir Üniversitesi Fen Bilimleri Enstitü Dergisi.

32 Nanbakhsh, K., Ritasalo, R., Serdijn, W.A., and Giagka, V. (2020). Long-term encapsulation of platinum metallization. *Proceedings of the EEE Electronic Components and Technology Conference (ECTC)*, Orlando, FL (3–30 June 2020).

33 Sterken, T., Vanfleteren, J., Torfs, T., et al. (2011). Ultra-Thin Chip Package (UTCP) and stretchable circuit technologies for wearable ECG system. *Annual International Conference of the IEEE Engineering in Medicine and Biology Society. IEEE Engineering in Medicine and Biology Society*, Boston (30 August to 3 September 2011).

34 van den Brand, J., Kusters, R., Barink, M., and Dietzel, A. (2010). Flexible embedded circuitry: a novel process for high density, cost effective electronics. *Microelectron. Eng.* 87 (10): 1861–1867.

35 Op, M. de Beeck, R. Verplancke, D.S, et al. (2007). Ultra-thin biocompatible implantable chip for bidirectional communication with peripheral nerves. *IEEE Biomedical Circuits and Systems Conference (BioCAS)*, Turin (19–21 October 2017).

36 Vanhoestenberghe, A. and Donaldson, N. (2011). The limits of hermeticity test methods for micropackages. *Artif. Organs* 35 (3): 242–244.

37 Gregg, A., York, L., and Strnad, M. (2005). *Flexible Flat Panel Displays*, 409–445. Hoboken: Wiley.

38 Palavesam, N., Marin, S., Hemmetzberger, D. et al. (2018). Roll-to-roll processing of film substrates for hybrid integrated flexible electronics. *Flex. Print. Electron.* 3 (1): 014002.

8

A Survey on Transmit and Receive Circuits in Underwater Communication for Sensor Nodes

Noushin Ghaderi and Leandro Lorenzelli

Fondazione Bruno Kessler, Center for Sensors and Devices (FBK-SD), Trento, Italy

8.1   Introduction

In recent years, wireless underwater communication has been done through various technologies. The primary method is acoustic technology, commonly used for telecommunications and first observed in ancient Greece. Dolphins and whales, for example, use sound to communicate over long distances in ocean waters. Jean-Daniel Colladon, a physicist, and Charles-Francois Sturm, a mathematician, performed the first direct experimental proof of underwater sound propagation and velocity measurement in 1826. In their experiment, an underwater bell rang simultaneously with a flash of light. At the same time, an observer at a considerable distance measured the time interval between the arrival of light and sound. The speed they measured for sound in water was 1435 m/s, which is close to today's 1438 m/s at 8 °C [1]. Today, sound waves are widely used for underwater communication and monitoring due to their low attenuation in water. However, there are some limitations for using this method, especially in shallow water applications, which are limited bandwidth, high latency, and ambient noise. In addition, sound waves cannot easily cross the water/air border. Therefore, acoustic wave transmission is unsuitable for high data rates and low latency applications.

The other method is optical technology which performs well at short distances. Besides being low cost, optical communication systems can provide high data rates (up to Gigabits per second) and low latency, which is its main advantage over acoustic technology. However, at the same time, they need precise alignment of nodes and lines, and environmental conditions influence them due to the intense

Sensing Technologies for Real Time Monitoring of Water Quality, First Edition.
Libu Manjakkal, Leandro Lorenzelli, and Magnus Willander.
© 2023 The Institute of Electrical and Electronics Engineers, Inc.
Published 2023 by John Wiley & Sons, Inc.

absorption that occurs at the optical frequencies and the strong scattering due to the presence of suspended particles in the environment [2]. Also, similar to sound waves, they cannot easily cross the water/air border.

Researchers have also studied the use of underwater radio frequency (RF) electromagnetic (EM) [3, 4] waves for years. These waves can reach high bandwidth and have low channel latency. They are not affected by turbidity, salinity, and pressure gradient and are protected against acoustic noise. Finally, these EM waves can pass through water/air and water/seabed borders, making transmission much easier in shallow water, and are suitable for underwater-air communication. However, due to the highly conductive nature of water, which causes RF waves to fade rapidly during underwater transmission, they are not a popular solution for long distances. Therefore, despite severe attenuation in seawater, EM waves can be used in communications in the range of 5–10 m.

This chapter aims to review the design of existing transmission and receiving circuits that can be included in the final product.

8.2 Sensor Networks in an Underwater Environment

8.2.1 Acoustic Sensor Network

In the early days of underwater acoustic communications, the main focus was developing systems for military applications such as active and passive SOund Navigation and Ranging (SONAR).

Today, Underwater Acoustic Sensor Networks (UW-ASNs) are considered the latest technological achievement in terms of communication. These networks consist of a set of underwater communication sensors, UW-ASNs, intended for the observation and exploration of lakes, rivers, seas, and oceans. Recently, they have received special attention due to their great potential in terms of promising applications in various fields (military, environmental, scientific, etc.).

A typical UW-ASN consists of several network nodes scattered in the target area, as shown in Figure 8.1. Each network node can receive and send data to a neighboring network node [5].

In recent years, researchers have begun to develop technology in underwater applications based on significant advances in integrated circuits, battery technology, digital signal processing, low-cost piezoelectric converters, and underwater channel modeling. There are many applications involved with this acoustic sensor network.

The first application is monitoring. Underwater sensor networks (USNs) can monitor various cases such as water quality, temperature, and ocean currents. These systems can monitor the detection of hazardous substances that affect marine

Figure 8.1 Underwater Acoustic Sensor Network.

ecosystems. These systems can also be used to track and understand the activities of various fish and other marine life. Using these USNs, monitoring of underwater seismic activity can be performed from remote locations, leading to early tsunami warnings. Other applications include monitoring offshore oil and gas pipelines to identify problems such as oil spills and to evaluate extraction sites. An underwater acoustic sensor can also help detect dangerous objects on the seabed and water for shipping lines. Adjusting maritime traffic can also be done by them.

Despite the many applications of these sensor networks, there are many challenges in designing them, as mentioned below.

8.2.1.1 Energy Sink-Hole Problem

One of the characteristic features of sensor networks is that all deployed sensors must transmit the collected data to the sink. Therefore, all transmitted data necessarily passes through sensors close to the sink. As shown in [6], the sensors closest to the sink are susceptible to depleting their provided energy faster than other sensors. Therefore, unlike sensors that are further away, sensors near the sink, especially those that are 1-hop away from the sink, have a significant packet load on the relay, in addition to their collected data, which may quickly exhaust their limited energy. In [7], the authors proved that when the nearest sensors exhaust their provided energy, the distant nodes still have 93% of their original energy. This unbalanced energy consumption can drastically reduce network lifetime because it creates energy holes that prevent reports from reaching the sink, resulting in potential network outages. This effect, one of the major concerns in the underwater network, is called the sink-hole problem.

In [8], the authors concluded that the linear sparsely sensor networks are most affected by energy holes. They proposed an energy-balanced hybrid data propagation (EBH) algorithm based on the residual energy of nodes to reduce the early energy depletion at the nearest sink.

Accordingly, each node in distributed linear networks uses the first multi-hop mode and sends data to its next nearest neighbor toward the sink. Then, the node sends data directly to the sink if its residual energy is greater than the next hop energy. Based on EBH, nodes should exchange control messages that contain their remaining energy information. The authors also proposed a different initial battery assignment (DIB) strategy that uses super nodes to form clusters with basic nodes. Having a higher battery power and longer transmit distance, the super node collects data from basic nodes in its cluster and forwards them to other super nodes or directly to the sink.

In [9], an energy-saving mobicast routing protocol is proposed to reduce the energy hole problem caused by ocean currents and the nonuniform deployment of sensor nodes. This protocol relies on a mobile autonomous underwater vehicle (AUV) to collect time-varying data from sensors in a series of 3-D reference zone (3D ZOR) while traveling along a user-defined path. To save energy, the sensors are usually in sleep mode and are only notified by the mobicast message from the AUV to deliver the data. To ensure the continuity of the delivery paths to the AUV and circumvent the hole problem in each 3D ZOR, the authors introduced an adaptive "apple-slicing" technique to construct routing segments around the created coverage hole.

In [10], the authors proposed a spherical hole repair technique (SHORT) to repair the problem of coverage and energy holes in underwater wireless sensor networks (UWSN). In this method, higher overlapping nodes are responsible for hole reparation. The authors also suggested using multiple transmission ranges while sending data to the sinks to save more energy while considering residual energy and depth.

8.2.1.2 Acoustic Sensor Design Problems

There are some challenges related to water properties and wireless communication inside the water [5, 11]. The first challenge is the pass loss of the waves through the water. As sound waves move through the water, they are gradually attenuated due to converting acoustic energy into heat, which increases with distance and frequency. Another reason for this loss is the geometric expansion of power due to the development of the wavefront, which increases with the propagation distance and is independent of frequency.

The next problem is underwater noise, mainly produced by two sources. The first is artificial noise due to human machinery noise and shipping activity. The other ambient noise source is primarily because of hydrodynamics, like the noise from water movement.

Another critical challenge in underwater communication is the propagation of multiple paths, leading to severe degradation of the acoustic communication signal and creating inter-symbol interference (ISI). In addition, the speed of propagation in the underwater channel is much slower than in the radio channel. Therefore, this high propagation delay reduces the channel coherence bandwidth. Furthermore, the delay in underwater sound propagation depends on the temperature, water depth, and salinity.

Another design challenge in underwater systems is the Doppler frequency propagation, which can be significant in the underwater channel, causing many adjacent symbols to interfere at the receiver side, which requires complex signal processing to mitigate the ISI generated.

The last challenge is the battery power limitation which leads the designers to design very low-power circuits inside the nodes [12].

Figure 8.2 indicates the typical features of an underwater acoustic node. It consists of four main building blocks described in the following sections.

8.2.1.3 The Underwater Transducer

The underwater transducer, the costliest component, converts mechanical energy (sound) to electrical energy or transforms electrical energy into mechanical energy. They are typically made from piezoelectric materials and are encapsulated in a potting compound to prevent contact with any conductive fluids. The commercially available all-purpose underwater converters [13, 14] range in price from $2000 to 3000. Therefore, there is a lot of focus on designing an inexpensive modem as an alternative to a custom commercial transducer. Jurdak et al. replaced the converter with cheap speakers and microphones but could only obtain data rates of 42 bps for the 17-m transmission range [15]. Benson et al. replaced a custom transducer with a commercially available fish finder transducer (which costs $50) but could only obtain an 80-bps data rate for a 6-m transmission range [16].

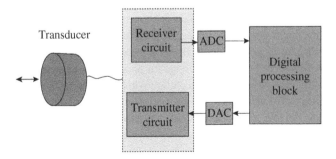

Figure 8.2 Main modules of an underwater acoustic sensor node.

8.2.1.4 Amplifier Design

One of the challenging parts is the power amplifier stage located between the transducer and the analog-to-digital converter (ADC) and amplifies signals before delivering them to the other blocks. The challenge related to the amplifier design includes suitable bandwidth, power budget requirement, and an acceptable amount of distortion which is discussed here briefly.

Bandwidth   Two critical factors in the path loss of a signal in an acoustic channel are the distance and frequency of the movement. An important factor in selecting the frequency is the absorption loss, which affects the channel's capacity concerning space [17]. The absorption of sound waves according to the frequency in seawater can be determined by an experimental equation of absorption coefficient presented by Thorp's formula [18, 19]. Fisher and Simmons [20] calculated the curves of the sound absorption in seawater from 6 to 350 kHz frequency. Based on their turns, the optimum wave bandwidth, considering the absorption rate, is up to 30 kHz. The second important parameter in selecting the frequency band is the ambient noise. Four ocean noise sources are shipping, turbulence, waves, and thermal noise. The minimum noise occurs at 20–200 kHz, and the lowest absorption occurs below 50 kHz. Based on the above observations, it can be concluded that the optimal bandwidth for sound generation for short to medium-range USNs is up to 30 kHz [12].

Power Budget Consideration   In a battery-powered sensor, with applications such as wireless underwater monitoring, the output power required for the amplifier can vary greatly depending on the distance and position of the adjacent nodes. Therefore, the amplifier should be designed to be efficient at a wide range of output power levels to improve the battery life.

Load Compatibility   Usually, the piezoelectric transducer is used for underwater sound generation because of its higher efficiency. Using piezoelectric converters as amplifier loads pose two significant challenges for circuit designers. The first challenge is the supply voltage. High drive voltages (in the range of 10–100 V) are required to activate such transducers effectively. Because these battery nodes have a limited supply voltage, the need for the high voltage for piezoelectric transducers necessitates using a dc–dc converter or boosting techniques in the circuit design part. The second challenge associated with driving piezoelectric transducers is due to the capacitive nature of these devices, which has a significant effect on the amplifier.

Amplifier Design   The designed amplifier should have high linearity, low EM interface, and low power consumption. For generating ac signals for transducers, usually, class A, B, or class AB linear amplifiers are used. A typical class B output stage amplifier is shown in Figure 8.3. Since NMOS and PMOS transistors conduct alternatively in positive and negative cycles of the input signal, the quiescent current from supply voltage will be minimum in the presence of a resistive load. Therefore, the class B amplifier can reach an efficiency of around 78%. However, in a capacitance load, such as a capacitive transducer [21], the efficiency of a class B amplifier can be reduced to 44% [21]. This is due to the phase difference between the voltage across a capacitive load and its current. Since in the capacitance load, the output current will reach its maximum when the output voltage is zero, the high voltage drop across the transistors causes a decrease in the efficiency value. Some methods for improving the efficiency in the presence of a capacitance load were reported in the

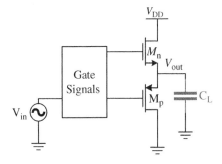

Figure 8.3 A typical class B stage.

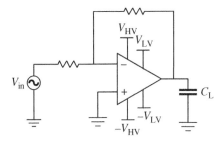

Figure 8.4 Simplified block diagram of a class G amplifier.

papers [22]. They try to make some charge recovery in the pass between the input and output, but the efficiency remains almost low.

A class G amplifier is proposed to increase the efficiency in [18], as shown in Figure 8.4. It applies a high voltage and low voltage supply rail to provide suitable instantaneous output voltage levels to increase efficiency. When a low voltage level is required at the output, the circuit will switch to the low power supply to save power consumption.

While this idea increases efficiency, switching the power sources is not easy and complicates the circuit. In addition, the overlap between the voltage across the transistors and their currents, due to the capacitive load and the continuous conduction state of the class B output stage, still exists, which causes a drop in efficiency.

Class D sound amplifiers are also used [23, 24] to improve efficiency. The idea of a class D amplifier is to convert the input sinusoidal signal to a pulse width modulation signal (PWM) and drive the power circuit to this PWM signal to obtain the maximum efficiency and retrieve the sine signal by applying a low-pass filter (Figure 8.5).

If the switching of the output transistors is extremely fast or ideal, the V-I overlap is almost zero. So, an efficiency of 100% can be expected in the perfect case.

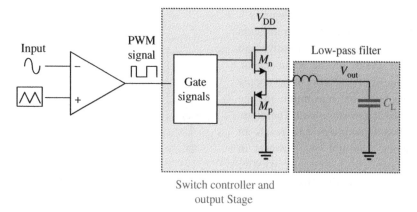

Figure 8.5 A typical class D amplifier.

If the switches are excellent, the output voltage will be a linear function of the duty cycle of input (D). It means that it can be used in an application with high linearity. The non-idealities in the PWM signal, like dead zone problems of the comparator, delays, and noise, can be compensated by adding a feedback loop (not shown in the figure).

Despite several advantages, the output voltage of a class D amplifier is limited to the supply voltage, while, as mentioned earlier, driving a piezoelectric transducer needs a higher voltage supply. This problem is usually solved by adding a dc–dc converter circuit to the class D amplifier to increase its supply voltage. In this case, the total efficiency of the circuit will be limited to the efficiency of the dc–dc boost circuit.

The Voltage-Boosting Output Stage The idea of the voltage boosting circuit is to combine the class D output stage and the boost converter circuit into a single circuit. In this case, the efficiency will improve due to the decrease in the component. The concept of the circuit is shown in Figure 8.6.

When M_p is on, the inductor L will charge through the power supply (V_{DD}). In the next phase, by turning off M_p and turning on M_n, the output voltage (V_{out}) will be boosted by taking energy from L and delivering it into the output load through M_n. At the same time, the low pass filtering is also performed to extract the final sine waveform, similar to the class D stage. But, here, the maximum amount of the output signal is not limited to the overall power supply of the circuit. So, this circuit is suitable for applying piezoelectric transducers in USNs.

The disadvantage of this amplifier is its poor linearity. In addition, adding a resistive load will add a Half-Plane Zero (RHPZ) to the circuit, which causes a decrease in the circuit bandwidth.

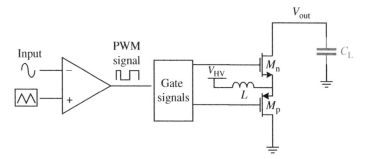

Figure 8.6 Concept of voltage-boosting output stage.

Considering that linearity is not a critical parameter in underwater circuit design and the range of frequency in acoustic sensors underwater is limited to 30 kHz, this circuit can be a good choice as the amplifier in applying the acoustic sensor network. Notice that in the case of a piezoelectric transducer, the load is capacitive and primarily pushes (RHPZ) to a higher frequency.

To improve the linearity of the above circuit, several techniques were proposed in the papers [25, 26]. In [27], a new control strategy is presented in a Boost dc–ac inverter [28] to control each Boost with a double loop control scheme with an additional inductor current control inner loop and an output control outer loop. The concept of the Boost inverter is shown in Figure 8.7. This circuit consists of two individual Boost converter circuits, which are driven by two dc-biased sinusoidal voltage references with 180° phase differences. An ac output voltage will be produced in the output, which can be greater or lower than the dc input voltage.

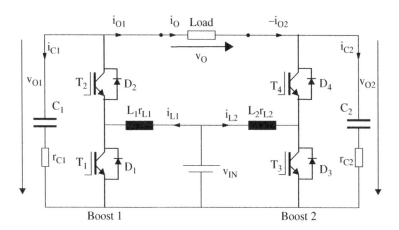

Figure 8.7 The concept of the Boost inverter circuit.

The advantage of this circuit is that it can produce a greater dc output voltage from a lower dc input voltage. The quality of the sinusoidal output waveform and the reduced number of switches are also additional advantages of this circuit. Using the differential structure also causes a cancellation of the second harmony, increasing the circuit's linearity.

Canceling the nonlinear sections in the circuit's transfer function as a dynamic cancellation is another suggested literature method [26]. In [29], a full bridge is used with a bidirectional boost converter to achieve a high output voltage swing in a piezo actuator application. However, the bandwidth and output power are very limited in this circuit.

8.2.1.5 Analog-to-Digital Converter

Usually, a successive approximation analog-to-digital converter (SAR-ADC) [11] converts the analog signal to digital due to low power consumption. In SAR-ADC, several comparison cycles are used to complete one conversion. Therefore, it is mainly applied in low-power and low-speed applications and is a good candidate for underwater sensor applications. The fundamental blocks of a SAR-ADC are sample and hold (S/H) circuit, comparator, capacitor network digital to analog (DAC), and SAR logic. The main power consumption parts are the switched capacitor DAC and comparator parts. Several switching schemes have been proposed to reduce the capacitor size of the DAC literature [30, 31]. However, developing a complicated SAR control logic causes an increase in the number of capacitors and networks, increasing power consumption. Another switching method is the monolithic switching method [32], which reduces the power consumption by 81% and the capacitor network by 50%. However, the input signal in this method should be in the differential. In addition, the comparator design faces some challenges due to significant variations in common-mode voltage. In [33], a particular split capacitor structure is merged with the monotonic switching to considerably reduce the capacitor size and switching energy to the half value of the split capacitor method.

SAR-ADC Architecture　Figure 8.8 shows a parallel array of a binary-waited capacitor that makes a charge redistribution DAC. First, all capacitors are discharged, and then Di switches from the digital control section to connect each capacitor to the ground or V_{ref}. Therefore, the output voltage, V_{out}, is a function of the voltage distribution between the capacitors.

In an architecture based on charge redistribution, the capacitor can be used as an S/H circuit. Therefore, there is no need to use a separate S/H circuit in this architecture. For N-bit DAC, the total capacitance in this method is $2^N C$. The size of the MSB capacitor becomes a significant concern by increasing the resolution. It causes an increase in the chip area and reduces the sample and hold block

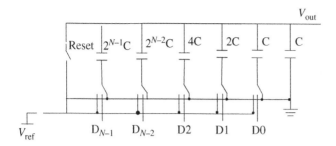

Figure 8.8   Charge redistribution DAC architecture.

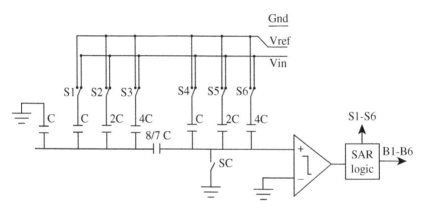

Figure 8.9   Split capacitor array method in a 6-bits SAR-ADC.

speed, which causes an increase in the settling time of the DAC output voltage. To minimize the above problems, the split array technique is used to reduce the size of the capacitors. Figure 8.9 shows a 6-bits SAR-ADC based on a split capacitor array. An additional attenuation capacitor separates the array into an LSB array and an MSB array. In this architecture, MSB is connected to the rightmost switches and LSB to the leftmost switches, and the attenuation capacitor is located between the two sides. The value of this capacitor must be selected so that the combination of the series of this capacitor and the LSB array is equal to C. Since, in this method, the capacitors are divided into two separate categories, the giant capacitor is turned into two separate capacitors with a value of $2^{(N/2)-1}C$. Therefore, the sum of the capacitors is reduced from the value of $2^{N}C$ to $2^{(N/2)+1}C$. This reduction is more significant in high-resolution ADCs.

The modification of the split capacitor array structures [29] is shown in Figure 8.10. In the sampling step, when the sampling switches are turned on, the

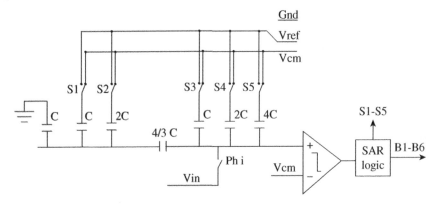

Figure 8.10 Modification of the split capacitor array.

input signal is sampled through the bootstrap switches on the top plates of the capacitors of the MSB array. At the same time, the capacitor's bottom plates are connected to the common-mode voltage reference. This means that the first bit of the MSB is obtained by the first comparison. Therefore, compared to the traditional method of split capacitor array in Figure 8.9, no switching energy is consumed. It means that the largest capacitor in LSB array capacitors is not necessary anymore and can be removed. As can be seen from the comparison between Figures 8.9 and 8.10, capacitor $4C$ is omitted from the LSB array of the circuit. The total number of capacitors in this method is $3 \times 2^{N/2-1}C$, while in the typical division capacitor array method, it is $2^{N/2+1}C = 4 \times 2^{N/2-1}C$ (Figure 8.9). Therefore, the percentage of capacitor reduction by this method in comparison to the split capacitor is equal to:

$$\frac{4 \times 2^{\frac{N}{2}-1} - 3 \times 2^{\frac{N}{2}-1}}{4 \times 2^{\frac{N}{2}-1}} \times 100 = 25\% \tag{8.1}$$

Due to the presence of smaller capacitors in this node compared to the structure of Figure 8.9, the settling speed and the input bandwidth increase. Overall, this scheme reduces the switching energy by 50% compared to a conventional split capacitor array scheme.

The other power-consuming part is the comparator circuit. Usually, in the low-power design, a dynamic comparator [30] circuit, as shown in Figure 8.11, is used to omit the static current consumption. In this circuit, at the low level of the clock signal, the output nodes are disconnected from inputs and are charged to V_{DD} It means that the ON transistors operate in

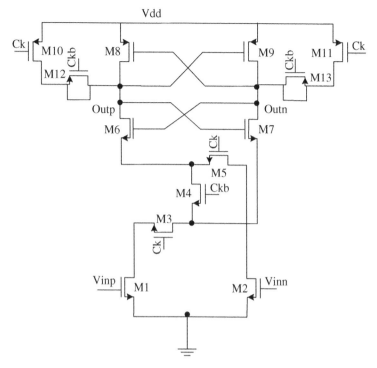

Figure 8.11 Low kickback dynamic comparator.

the triode region and do not consume energy. Also, the kickback noise is decreased by isolating input transistors from output nodes through the switched M3 and M5.

8.2.2 Electromagnetic (EM) Waves Underwater Sensors

EM waves in the RF range are also suitable for underwater wireless communication systems. They are less sensitive to refraction and reflection effects than acoustic waves. In addition, suspended particles have minimal impact on them. So, EM sensors are more suitable in shallow water environments than acoustic sensors. The main parameters of EM waves include conductivity (σ), permittivity (ε), permeability (μ), and volume charge density (ρ) changes with the type of water. The absorption coefficient and propagation speed values also vary depending on the water. One of the main problems with underwater EM communication is the high attenuation due to the conductivity of the seawater. The attenuation increases with increasing the frequency [34]. It is why the range of EM sensor systems is usually in the field of tens of meters.

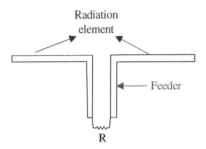

Figure 8.12 Dipole antenna.

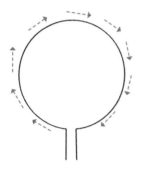

Figure 8.13 Circular loop antenna.

8.2.2.1 Antenna Design

Antenna design should be based on some requirements like low cost and easy fabrication. In addition, it needs to be compact and reliable at very low frequencies. The most common antenna is a dipole antenna. It is made of two conductors of equal length like metal wires or rods fed in the center through a transmission line, as shown in Figure 8.12 [35].

The circular loop antenna is the other standard used as an antenna in underwater sensor applications. In this antenna, the conductive wire can be in different shapes, including square, triangular, or circular, which allows the perimeter to be more than the wavelength. The circular is the most popular among them, as indicated in Figure 8.13.

Both dipole and loop antennas are low-cost, easy to fabricate, and very flexible. The size of the antennas is quite strict due to the frequency range requirement of the underwater application. The goal is to design an antenna less than 15 cm in diameter. Therefore, the antennas will be short in comparison to the wavelength. It means that its performance is degraded in contrast with the larger antenna. To increase the performance of the loop, it is possible to create a multi-loop antenna. In this method, the loop's magnetic field increases by adding more loops. In addition, the compactness of the antenna is not affected by increasing the number of loops. A ferrite element can also be added at the center of the coil to increase the transmitted magnetic field by the loop antenna even more. Since the transmitter has to deal with high-power signals, a class D amplifier is usually used to amplify the generated signal, as described in Section 8.2.1.5 The SAR-ADC block, as mentioned in Section 8.2.1.5, is the most suitable choice due to its low power consumption in comparison to the other types of ADCs.

8.2.2.2 Multipath Propagation

In studying the behavior of EM waves underwater, the influence of water–air and water–seabed boundaries is also important. Due to the high permittivity, the wave will have a large refraction angle, which leads to the transmission of the signal in a path almost parallel to the water. In this case, the signal appears to be radiating from a patch of water just above the transmitter. This effect allows transmission

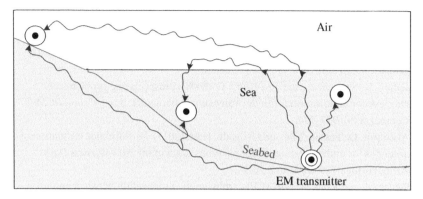

Figure 8.14 Multipath propagation of EM waves in seawater.

from a submerged station to the land or between shallow stations without the need for a surface repeater. The seabed can produce a similar effect. Seabed conductivity is much lower than water, so it can provide an alternative communication path with low loss and low noise. Figure 8.14 shows this effect. In many deployments, the single propagation path with the least resistance will dominate. If an air route or seabed route is chosen, the transmission range can be relatively longer. Therefore, multipath propagation of EM waves can be used for signal transmission in shallow water [36].

8.3 Conclusion

The objective of this chapter was to explain various underwater communications through different technologies.

Acoustic technology, which is commonly used for underwater communication and monitoring due to its low attenuation in water, was described. In addition, its limitations, especially in shallow water applications, due to limited bandwidth, high latency, and ambient noise, were discussed. The challenges related to acoustic sensor design were investigated, and suitable circuits for each block were introduced.

The EM wireless transmission was also described as a replacement for shallow water and underwater–air communication.

Acknowledgment

This work was supported by the European Commission in the framework of the AQUASENSE (H2020-MSCA-ITN-2018-813680) project.

References

1 Beyer, R.T. and Raichel, D.R. (1999). Sounds of our times, two hundred years of acoustics. *J. Acoust. Soc. Am.* 106 (1): 15–16.

2 Lanbo, L., Shengli, Z., and Jun-Hong, C. (2008). Prospects and problems of wireless communication for underwater sensor networks. *Wirel. Commun. Mob. Comput.* 8 (9): 977–994.

3 Aboderin, O., Pessoa, L.M., and Salgado, H.M. (2017). Performance evaluation of antennas for underwater applications. *Proceedings of the 2017 Wireless Days*, Porto, Portugal (29–31 March 2017), 4.

4 Siegel, M. and King, M. (1973). Electromagnetic propagation between antennas submerged in the ocean. *IEEE Trans. Antennas Propag.* 21 (4): 507–513.

5 Akyildiz, I.F., Pompili, D., and Melodia, T. (2005). Underwater acoustic sensor networks: research challenges. *Ad Hoc Netw.* 3 (3): 257–279.

6 Lian, J., Naik, K., and Agnew, G.B. (2004). Modeling and enhancing data capacity in wireless sensor networks. *IEEE Monograph on Sensor Network Operations* (January 2004).

7 Wadaa, A., Olariu, S., Jones, K. et al. (2005). Training a sensor. *Proceedings of Monet*, Orlando, FL, USA (January 2005).

8 H. Luo, Z. Wu, F. Hong and Y. Feng, Energy balanced strategies for maximizing the lifetime of sparsely deployed underwater acoustic sensor networks, *Sensors J.*, 9, 6626–6651 2009.

9 Chen, Y.-S. and Lin, Y.-W. (2013). Mobicast routing protocol for underwater sensor networks. *IEEE Sensors J.* 13 (2): 737–749.

10 Latif, K., Javaid, N., Ahmad, A. et al. (2016). On energy hole and coverage hole avoidance in underwater wireless sensor networks. *IEEE Sensors J.* 16 (11): 4431–4442.

11 Heidemann, J., Ye, W., Wills, J. et al. (2006). Research challenges and applications for underwater sensor networking. *Proceedings of the IEEE Wireless Communications and Networking Conference (WCNC 2006)*, Las Vegas, NV (3–6 April 2006).

12 Amir, S. (2020). *Voltage Boosting Amplifier Techniques for Underwater Sensor Networks: Theory and Design*. Twente: Integrated Circuit Design, University of Twente.

13 Freitag, L., Grund, M., Singh, S., et al. (2005). The WHOI micro-modem: an acoustic communications and navigation system for multiple platforms. *Proceedings of the IEEE OCEANS Conference*, Washington, DC (17–23 September 2005).

14 Yan, H., Zhou, S., Shi, Z., and Li, B. (2007). A DSP implementation of OFDM acoustic modem. *WUWNet '07: Proceedings of the 2nd Workshop on Underwater Networks*, Montreal (September 2007).

15 Jurdak, R., Aguiar, P., Baldi, P., and lopes, C.V. (2007). Software modems for underwater sensor networks. *IEEE OCEANS-Europe*(June 2007), pp. 1–6.

16 Benson, B., Chang, G., Manov, D. et al. (2006). Design of a low-cost acoustic modem for moored oceanographic applications. *ACM International Workshop on Underwater Networks*, Los Angeles, CA (January 2006).

17 Stojanovic, M. (2007). On the relationship between capacity and distance in an underwater acoustic communication channel. *ACM SIGMOBILE Mob. Comput. Commun. Rev.* 11 (4): 34–43.

18 Blom, K.C.H., Wester, R., Kokkeler, A.B.J., and Smit, G.J.M. (2012). Low-cost multi-channel underwater acoustic signal processing testbed. *IEEE 7th Sensor Array and Multichannel Signal Processing Workshop (SAM)* (17–20 June 2012).

19 Ainselie, M.A. (2010). *Principles of Sonar Performance Modelling*. Springer.

20 Fisher, F.H. and Simmons, V.P. (1977). Sound absorption in sea water. *J. Acoust. Soc. Am.* 62 (3): 558–564.

21 Nielsen, D. (2014). Class d audio amplifiers for high voltage capacitive transducers. PhD dissertation, Department of Electrical Engineering, Denmark.

22 Wallenhauer, C., Gottlieb, B., Zeichfusl, R., and Kappel, A. (2010). Efficiency-improved high-voltage analog power amplifier for driving piezoelectric actuators. *IEEE Trans. Circuits Syst. Regul. Pap.* 57 (1): 291–298.

23 Berkhout, M. (2003). An integrated 200-w class-d audio amplifier. *IEEE J. Solid-State Circuits* 38 (7): 1198–2003.

24 Agbossou, K., Dion, J.L., Carignan, S. et al. (2000). Class D amplifier for a power piezoelectric load. *IEEE Trans. Ultrason. Ferroelectr. Freq. Control* 47 (4): 1036–1041.

25 Jha, K. and Mishra, S. (2010). Large-signal linearization of a boost converter. *IEEE Energy Conversion Congress and Exposition*, Atlanta, GA (12–16 September 2010).

26 Michal, V. (2012). Modulated-ramp PWM generator for linear control of the boost converter's power stage. *IEEE Trans. Power Electron.* 27 (6): 2958–2965.

27 Sanchis, P., Ursæa, A., Gubía, E., and Marroyo, L. (2005). Boost DC–AC inverter: a new control strategy. *IEEE Trans. Power Electron.* 20 (2): 343–353.

28 Caceres, R. and Barbi, I. (1995). A boost DC–AC converter: operation, analysis, control and experimentation. *Proceedings of IECON '95 – 21st Annual Conference on IEEE Industrial Electronics,* Orlando (6–10 November 1995).

29 Chaput, S., Brooks, D., and Wei, G.Y. (2017). 21.5 A 3-to-5V input 100Vpp output 57.7mW 0.42% THD+N highly integrated piezoelectric actuator driver. *IEEE International Solid-State Circuits Conference (ISSCC)*, San Francisco, CA (5–9 February 2017).

30 Rahim, E. and Yavari, M. (2014). Energy-efficient high-accuracy switching method for SAR ADCs. *Electron. Lett.* 50 (7): 499–501.

31 Akbari, M., Nazari, O., and Hashemipour, O. (2019). Energy-efficient and area-efficient switching schemes for SAR ADCs. *2019 IEEE 62nd International Midwest Symposium on Circuits and Systems (MWSCAS)*, Dallas, TX (4–7 August 2019).

32 Liu, C.C., Chang, S.J., Huang, G.Y., and Lin, Y.Z. (2010). A 10-bit 50-MS/s SAR ADC with a monotonic capacitor switching procedure. *IEEE J. Solid State Circuits* 45 (4): 731–740.

33 Ghaderi, N., Adami, A., and Lorenzelli, L. (2021). A low-power 6-bit successive approximation register ADC using a new split capacitor array method. *Proceedings of the 2021 IEEE Asia Pacific Conference on Circuit and Systems (APCCAS)*, Penang, Malaysia (22–26 November 2021).

34 Massaccesi, A. (2015). *Analysis and Design of Underwater Antennas for Diving Applications*. Politecnico di Torino.

35 Aboderin, O., Pessoa, L.M., and Salgado, H.M. (2017). Performance evaluation of antennas for underwater applications. *Wireless Days*, Porto, Portugal (29–31 March 2017).

36 Wells, I., Dickers, G., and Gong, X. (2011). Re-evaluation of RF electromagnetic communication in underwater sensor networks. *IEEE Commun. Mag.* 48 (12): 143–151.

Section III

Sensing Data Assessment and Deployment Including Extreme Environment and Advanced Pollutants

9

An Introduction to Microplastics, and Its Sampling Processes and Assessment Techniques

Bappa Mitra[1], Andrea Adami[1], Ravinder Dahiya[2], and Leandro Lorenzelli[1]

[1] Fondazione Bruno Kessler, Center for Sensors and Devices (FBK-SD), Trento, Italy
[2] Bendable Electronics and Sensing Technologies Group, Department of Electrical and Computer Engineering, Northeastern University, Boston, USA

9.1 Introduction

Plastics are one of the major wastes found in seas and oceans. This is due to the widespread use of plastic produced for human needs, which started around six decades ago [1]. Since then, it has grown continuously till date with a global production of more than 348 million tons and 368 million tons in the year 2017 and 2019, respectively [2, 3]. Figure 9.1 shows the plastic production worldwide in the year 2017 and Figure 9.2 shows the plastic production by types and applications in Europe in the year 2017. While the presence of microplastic pellets was reported as early as the 1970s, microplastics started gaining attention after a seminal paper was published by [1]. Microplastics are tiny, ubiquitous particles or fragments of plastics. Microplastics are generally considered as plastic particles having a size less than 5 mm as their largest dimension. Although a standard definition of microplastics is still lacking, various authors have considered different size distributions to define and classify microplastics. Microplastics are present in the environment in different shapes, sizes, and densities. They can be found diversely dispersed in areas such as marine sediments, air, and water bodies and even in the internal bodies of aquatic animals. Microplastics originate from different sources which are generally classified as primary sources and secondary sources. Industry-manufactured microplastics that enter the environment from industry applications such as paint industry, sandblasting, pharmaceutical vectors, and

Sensing Technologies for Real Time Monitoring of Water Quality, First Edition.
Libu Manjakkal, Leandro Lorenzelli, and Magnus Willander.
© 2023 The Institute of Electrical and Electronics Engineers, Inc.
Published 2023 by John Wiley & Sons, Inc.

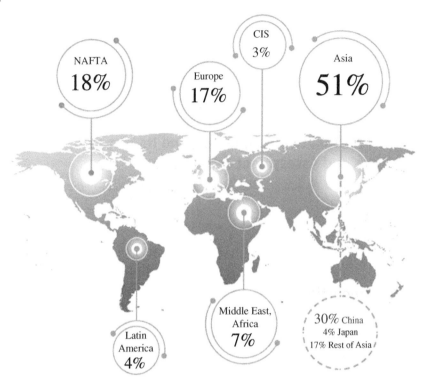

Figure 9.1 Global plastic production percentage based on regions. *Source:* From Ref. [2]/Plastics Europe Market Research Group.

domestic applications such as exfoliating facial scrubs in cosmetics, toothpastes, detergents, etc., are examples of primary sources. Personal care products such facial scrubs containing exfoliates and abrasives for skincare were reported to be one of the major primary sources of microplastic pollution as these products are commonly used and enter into water systems as wastewaters. Similarly, micro-plastics from medical applications, paint and polish industries, drilling fluids used for oil and gas exploration, and 3D printing by-products are other contributors as a primary source of microplastics. Efforts have been taken to limit the microplas-tics entering the environment. This includes banning the use of microspheres in facial scrubs.

The major challenge, however, lies in plastics that have already existed in the environment for several decades. In addition to the unmanaged littering and dumping of plastic wastes, these plastics act as a secondary source of microplastics by undergoing slow chemical or biological degradation to form smaller plastic particles. Once discarded, microplastics and their sources end up as marine litter

(a)

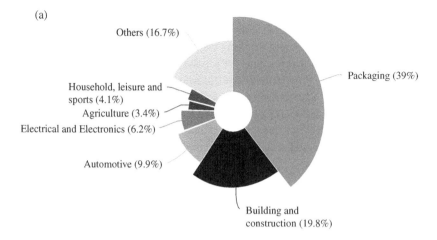

(b)

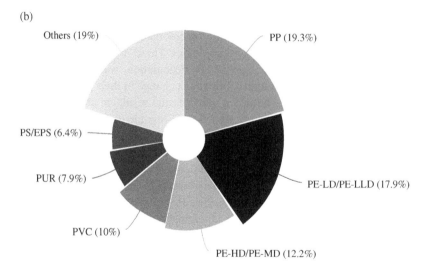

Figure 9.2 (a) Plastic consumption in different sectors. (b) Global plastic demand by resin. *Source:* Adapted from Ref. [2].

and remain in water bodies before breaking down into further smaller components. A global analysis done on the production and fate of plastics produced so far estimates that the cumulative plastic waste generated till 2015 amounted to 6300 metric tonnes out of which 79% was discarded to the natural environment [4]. With this assumption, they predicted an accumulation of 12 000 metric tonnes of plastic waste in the environment. Plastics, which are secondary sources of microplastics, pose a major threat alongside the primary sources that contribute to the release of tonnes of microplastics in water bodies annually on its own. In a review published by [5], authors assessed microplastics from 50 studies done in various freshwaters and drinking water. They found that polymers such as polyethylene (PE), polypropylene (PP), polystyrene (PS), polyvinyl chloride (PVC), and polyethylene terephthalate (PET) were present in higher numbers in decreasing order than other polymers. This also matched the trend of EU plastic demand in the year 2018; PE-29.7%, PP-19.3%, PS/EPS-6.4%, PVC-10%, and PET-7.7%. These plastics that have accumulated in the water bodies for several years are undergoing degradation to produce microplastics.

9.1.1 Properties of Microplastics

Microplastics have small sizes (<5 mm), varying shapes, and a higher surface area to volume ratio. These properties play an important role in absorption and adsorption of organic and inorganic compounds on their surfaces, which help chemical ions, biofilms, and biological entities accumulate on the plastics. In addition to that, colonization of microplastics with microorganisms has been shown to change the physical properties of microplastics. These microplastics present in the environment undergo processes of UV degradation, weathering, abrasion, chemical modification, etc., thereby, bringing a change in their properties. Apart from these, the hydrophobic nature of microplastics results in attracting organic pollutants. Some examples include attracting of organic pollutants such as aromatic hydrocarbons, organochlorine pesticides, etc. [6]. Similarly, microplastics are known to interact well with metallic toxicants such as cadmium, nickel, mercury, etc., which can modulate autophagy. Microplastics such as polyethylene and polystyrene which are reported in higher amount than others are shown to have a higher absorption and adsorption properties in laboratory conditions [7]. Thus, microplastics can act directly or indirectly impacting the aquatic ecosystem and get into human body forming a route through food chain.

Weathering plays an important role in altering the physicochemical properties of microplastics and the changes in properties of microplastics due to weathering are likely to show different toxic effects from the pristine ones [8]. The change in properties includes physical changes such as change in size, specific surface area, color, density, and mechanical properties, while chemical changes would include changes in sorption properties, crystallinity, and oxygen containing groups and release of additives during the weathering process. The increase in sorption capacity due to weathering results in increased surface polarity and charges, and, therefore, increased sorption of hydrophilic organic pollutants and metals. In addition to this, weathering also favors the formation of biofilms on the surfaces which further affects the sorption behavior by decreasing surface hydrophobicity and increasing the heterogeneity. Microplastics can be found acting as a vector of transport of toxic compounds and pathogenic organisms. Organisms such as bacteria, cyanobacteria, bryozoans, fungi, etc., have been found on microplastic surfaces in marine environments. Presence of microbial biofilms and adhering of organisms on microplastic surfaces result in change in size and buoyancy distribution of microplastics in different zones, vertical depths of water level, and horizontal transport to new environments through wind and ocean currents.

9.1.2 Microplastics in Food Chain

Microplastics are known to be ingested both by invertebrates such as Polychaeta worms, crustaceans, echinoderms, ciliates, bryozoans, and bivalves and by vertebrates such as fish and sea birds [9–12]. Microplastics have also been detected in planktonic organisms such as zooplankton, chaetognatha, larval fish, etc. [12]. Negative effects of microplastics in these animals include complications such as lacerations and inflammatory response to consumed microplastics, resulting in reduced feeding activity [13, 14]. These also have been reported to cause several complications that include pathological stress, false satiation, reproductive complications, blocked enzyme production, etc., in these organisms [15, 16]. Similarly, the accumulation of microplastics in the larvae, gut, gill, liver, and in rare cases the brain of larger marine vertebrates such as fish can be found in the literature [14, 17]. A summarized table on effects of micro/nanoplastics on mouse models has also been presented by [7] where the presence of micro/nanoplastics were reported in the gut, liver, and kidney inducing pathological changes. In general, however, studies have shown that the degree of toxicity of micro/nanoparticles was less severe in mice than fish [7]. This raises a general question regarding the effects of microplastics in human health.

9.1.3 Human Consumption of Microplastics and Possible Health Effects

Human beings are exposed to microplastics through inhalation, drinking water, and food products. Microplastics, which can be confused as a prey for aquatic animals when consumed by them can act as a possible route to human beings through the food chain. Presence of microplastics in air exposes us to its entry in our respiratory system; consumption of microplastic containing food and liquids transports them into our digestive track. A report presented by [18] reported that an average human consumes about 39 000–52 000 particles person year. A further breakdown of microplastic consumption paper showed that humans can consume 1.48 MPs/g of seafood, 0.44 MPs/g of sugar, 0.10 MPs/g of honey, 0.11 MPs/g of salt, 32.27 MPs/l of alcohol, and tap water consumption has been assumed to expose us to 4000 particles per year. Similarly, mineral water from bottles would result in an additional exposure of approximately 90 000 MP particles per year [19]. The risk of consumption of microplastics through fruits and vegetables is also possible since treated wastewater and bio-solids used in irrigation channels are reported to contain microplastics [20]. A study carried out by [21] to determine the presence of microplastics showed all the eight human stool samples examined for the presence of microplastics tested positive. A median of 20 MPs (50–500 μm in size) per 10 g of human stool was identified out of which nine plastic types were detected, with polypropylene and polyethylene terephthalate being the most abundant. Similarly, the presence of microplastics in human placentae was also reported [22]. The toxicity and possible negative health effects were also examined by [7] to determine the effect of PS particles to human cells at an experimental dosage of approximately 500 μg/ml. The study, however, reveals that PS particles with diameters of 10–100 μm were not significantly cytotoxic but smaller PS particles with diameters of 460 nm and 1 μm affected red blood cells [23]. It is possible that the transfer of nanoplastics into the bloodstream after ingestion could lead to local inflammation or induce allergic reactions in tissues. A greater concern arises from the additives present in the microplastics which can be harmful when they undergo fragmentation or leaching inside human bodies. Microplastics, originating from secondary sources mostly contain a wide range of plasticizer chemicals to give them specific physical properties such as elasticity, rigidity, UV, stability, flame retardants, and heavy metals for colorings. Additives such as bisphenol-A, phthalates such as di-*n*-butyl phthalate, di-(2-ethylhexyl) phthalate, polybrominated diphenyl ethers (PBDEs), and metals have been reported to be either toxic or endocrine disruptors [24, 25]. A critical report provided by [6] provides an overview on how bio persistence of microplastics can be fatal for humans giving rise to inflammation, oxidative stress, genotoxocity, etc. Although no direct effects of microplastics in human health have been reported so far, the amount of research done to understand the microplastics in human health is relatively poor. The amount of information available till date is not enough to fully understand the risks and potential threats of microplastics on human health.

9.1.4 Overview

The abovementioned concerns have led to increasing attempts to collect and process selective, bulk, and volume-reduced environmental samples for the identification and analysis of microplastics. The amount of microplastics present in the selective sampling is the representation of microplastics in the limited sample collected. For covering large volumes or areas, volume-reduced sampling approach is quicker as a fixed volume of sample is collected from a certain area. The amount of samples collected is representative of the overall volume from which the sample is collected. Since volume-reduced approach discards most of the sample, it often leads to wrong estimation of microplastics present in an area. Thus, some studies report sample collection using bulk sampling approach whereby entire sample is assessed. Figure 9.3 shows the overall process involved in microplastic

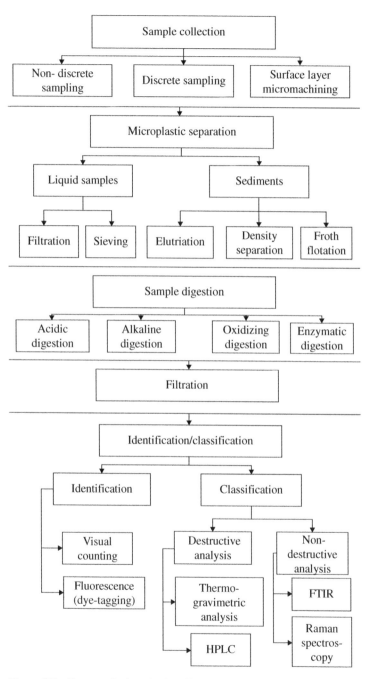

Figure 9.3 Process of microplastic collection and identification process.

collection, processing, and analysis. The rest of this chapter discusses each step of this process in more detail.

9.2 Microplastic Sampling Tools

To identify and quantify the presence of microplastics in water, samples must be collected from the surface layer as well as at various depths due to varying densities of microplastics. Different approaches have been employed for the sampling of microplastics in water. Manta trawls and neuston nets are the most commonly used equipment for surface water sampling. In case of water column sampling, plankton nets, bongo nets, continuous plankton recorders (CPR), etc., are widely used [15]. There are some alternative tools that are occasionally used in surface water or water column sampling for microplastics, such as plankton traps, water collection bottles, or water intake pumps. Mesh size of the sampling tools varies from tens of microns to millimeters, with the most common aperture size being 333 mm. The employment of sampling tools with different mesh sizes, therefore, makes it difficult to compare the available monitoring data. Sampling devices can be divided into three categories: non-discrete sampling devices (nets and pumping systems), discrete sampling devices (Niskin bottles, bucket, bottle, and steel sampler), and sampling devices of the surface microlayer (sieves and rotating drum sampler).

9.2.1 Non-Discrete Sampling Devices

9.2.1.1 Nets

Microplastic sample collection in water bodies is done in a dynamic way and stationary ways. Dynamic way refers to collection of water samples by towing different types of nets or trawls attached tied to a boat using long ropes. While stationary sampling is done by attaching floating nets to the banks. Microplastic sampling on the sea surface by manta nets is a widely used method for the sampling of microplastics on the sea surface. A large volume of water can be filtered through the manta net, thus the possibility of trapping a relevant number of microplastics is high and the results are perceived to be reliable. Manta net is most often fixed from the side of the vessel. A flow meter attached to the nets records the volume of filtered water. This enables the normalization of results per volume of sample. If a flow meter is not attached, the net opening size and the length of the transect can be used to estimate the amount of water filtered at a time. The most frequently used manta nets have around 300 μm mesh size and are 3–4.5 m long. A 333-μm net mesh is most commonly used. Nets are towed horizontally, within a superficial layer or at greater depths. It can also be used to collect the water column entirely by placing it vertically or obliquely from the bottom to the surface of the water column. Most often, nets are used for sampling up to a depth of 0.5 m. In this case, the trawling is stopped immediately to prevent nets from damage or losing microplastics [26].

Nets such as neuston nets, plankton nets, manta nets, continuous nets, etc., are the most used devices for MP sampling. Abundances of microplastics recovered from the water matrix are directly influenced by the mesh size of the sampling tools. Under circumstances of higher waves, neuston catamarans are used [27]. The mesh size dictates the efficiency and minimum size of microplastics that can be captured. It has been suggested that the efficiency of 100 μm net is 10 times more than 500 μm nets. Meanwhile, the efficiency of 333 μm net is only 2.5 times less than 100 μm net. While found out that when an 80-μm mesh was used, the results were 100 000 times higher than that by a 450-μm mesh [28]. Figure 9.4 shows the schematic of various nets that have been used for microplastic sample collection.

9.2.1.2 Pump Tools

Pumps allow for sampling a large volume of seawater. Pumps provide an advantage that devices can work straight for several hours along the same transect and in different sampling stations for minutes. Pumps are usually lowered from the

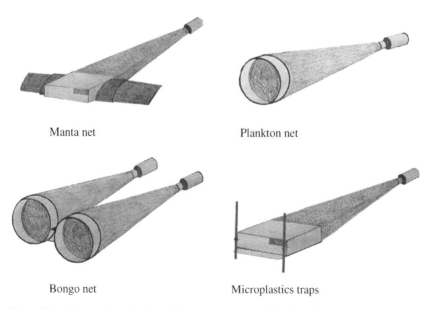

Manta net   Plankton net

Bongo net   Microplastics traps

Figure 9.4 Schematics of net used for water sample collection. *Source:* From Ref. [28]/ MDPI/CC BY 4.0.

vessel using a sideways winch or stern of the ship. Pumping systems, however, are less commonly used than nets. Also, pumping systems lack standardization in terms of sampling time. The filters/sieves of the pumping systems determine the size of microplastics that can be retained, which allow for selecting plastic of the size of interest. Several filters with a varying mesh size can be used as layers to remove larger plastic particles, then a second filter or a series of sieves with a different mesh size to separate microplastics according to their sizes. Figure 9.5 shows a pump system with a series of meshes used for separation of particles of various sizes from water samples.

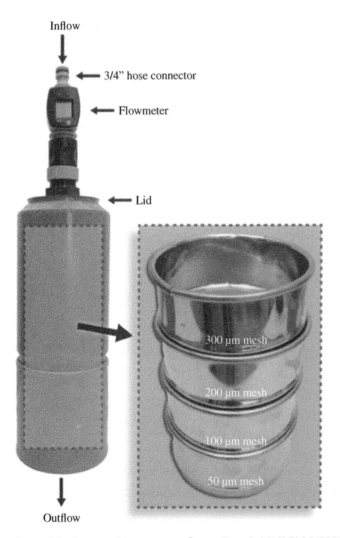

Figure 9.5 Pump underway system. *Source:* From Ref. [29]/PLOS/CC BY 4.0.

9.2.2 Discrete Sampling Devices

Niskin bottle, rosette, and Integrating Water Sampler (IWS) are commonly used discrete sampling devices for water sampling for different purposes. An example would be analysis of total suspended solids [30]. In addition to these, buckets, bottles, and steel samplers are also used. Sample collection is done by lowering the device manually to a sampling depth or through a winch from the vessel. Such devices are pretty straightforward to use and sample collection could be done quickly allowing sampling of smaller microplastics in a known water volume. However, such devices should be attached or held on a boat and transportation of bulky samples to the laboratory with possible risk of contamination. As different mesh sizes and depths are used, major issues of such systems are the non-consistent data and comparability of such study. Along with that, the reporting units and the method of preparation also results in nonuniformity in reporting of data. Thus, a standardization of discrete sampling and preparation methods is still lacking.

9.2.3 Surface Microlayer Sampling Devices

Devices for surface microlayer sampling include simple stainless-steel sieves or rotating drum which does not require further specialized equipment for sample collections. The size of the mesh used in such devices can be chosen as per requirement and on-field separation of microplastics can be done. However, such devices are prone to sample contamination and allow only low volume of water collection which then needs to be transported manually to laboratory.

9.3 Microplastics Separation

Microplastics present in environmental samples – sediments, sands, beach, river, etc., need to be separated for its identification. Therefore, based on the type of environment, different separation techniques are used which have been discussed below. Although several techniques discussed in this chapter have been presented for liquid samples and for sediments, the techniques used for sediment samples can also be used for separating microplastics for liquid samples.

9.3.1 Separating Microplastics from Liquid Samples

9.3.1.1 Filtration

Filtration is an effective method which is commonly used to separate microplastic particles from liquids. The process involves usage of a filter medium that allows only liquid to pass through while the solid particles are retained in the filter. The filtration medium that is most commonly used for filtration of microplastics is made up of glass fibers. Several other filters such as nitro cellulose, polycarbonate membranes, zooplankton filters, or iso-pore filters are also used for microplastic filtration process [31]. Filtration being the easiest procedure to separate microplastics from liquids is often met with challenges of particles of different sizes which can clog the filtration medium. This leads to the requirement of frequently changing the filters. Various strategies such as volume-reduced sampling, pre-filtration steps, or addition of chemicals such as ferrous sulphate have been used to increase the effectiveness. Filtration is the simplest process for separating particles from liquids; however, major drawback arises due to clogging of filter due to microscopic particles or debris which lowers the effectiveness of the process [15].

Figure 9.6 is a schematic representation of a filtration system used for filtration of solid particles from water samples.

9.3.1.2 Sieving

Sieves are used to capture solid materials physically that are larger than the mesh size by allowing water and smaller particles to be removed from the sample. The mesh size of sieves can be chosen depending upon the desired size range of microplastics to be collected. The majority of size ranges from 0.035 to 4.75 mm [15]. Figure 9.7 shows a sieving device and its internal construction along with the sample collected.

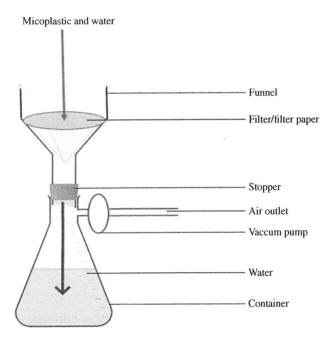

Micoplastic and water

— Funnel

— Filter/filter paper

— Stopper

— Air outlet

— Vaccum pump

— Water

— Container

Figure 9.6 Vacuum filtration system schematic.

(a)

(b)

(c)

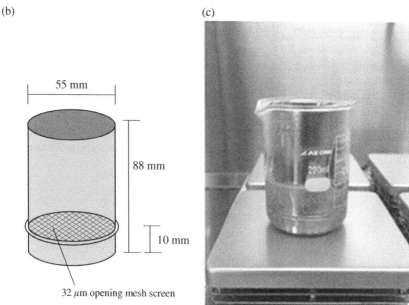

Figure 9.7 (a) Sieving device. *Source:* From Ref. [29]/Elsevier/CC BY 4.0. (b) Schematic of sieving container. *Source:* From Ref. [29]/with permission of Elsevier. (c) Sample collected using container. *Source:* From Ref. [29]/Elsevier/CC BY 4.0.

9.3.2 Separating Microplastics from Sediments

9.3.2.1 Density Separation

Density separation makes use of the difference of densities between materials of interest and other unwanted materials. As shown in Figure 9.8, using a liquid with an intermediate density, a binary separation of particles lighter and denser than the liquid can be done by shaking and allowing the mixture of particles and liquid to settle. In case of sediments, sand particles and other particulates with a density of $2.65\,g/cm^3$ are typically present. Since the specific densities for most plastics range from 0.8 to $1.70\,g/cm^3$, plastic particles and sand can be effectively separated by using the density separation method. Sodium chloride (NaCl) solution ($1.202\,g/cm^3$) being inexpensive and eco-friendly is the most frequently used salt solution for the density separation process. Using NaCl solution, low-density microplastics such as polyethylene (PE, 0.917–$0.965\,g/cm^3$), polypropylene (PP, 0.85–$0.94\,g/cm$), and polystyrene (PS, 1.04–$1.1\,g/cm^3$) can be separated from sediments and high-density microplastics. However, for complete separation of microplastics which also includes plastics such as polyvinylchloride (PVC, 1.3–$1.7\,g/cm^3$), polyethylene terephthalate (PET, 1.4–$1.6\,g/cm^3$), and the saturated NaCl solution is not the best solution. To address this issue, some high-density salt solutions, such as solution of sodium bromide (NaBr, $1.38\,g/cm^3$), sodium polytungstate ($3Na_2WO_4 \cdot 9WO_3 \cdot H_2O$, $1.4\,g/cm^3$), zinc chloride (ZnCl$_2$, $1.7\,g/cm^3$), or sodium iodide (NaI, $1.8\,g/cm^3$), are used [27]. Nevertheless, high-density salts are generally expensive, and some are environmentally hazardous.

9.3.2.2 Elutriation

Elutriation is a process in which particles are separated based upon their size, shape, and density by using a stream of gas or liquid flowing in a direction opposite to the direction of sedimentation. In terms of microplastics, the technique was

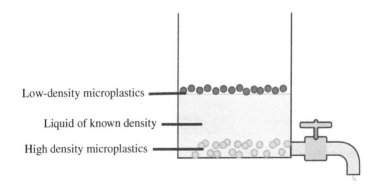

Figure 9.8 Density separation schematic.

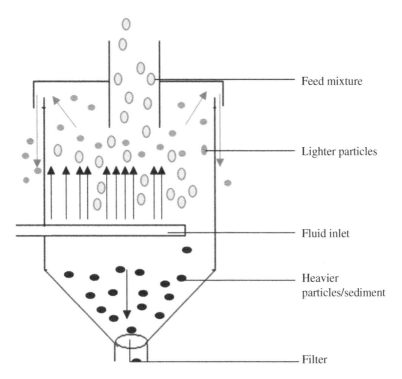

Figure 9.9 Elutriation.

first used to separate microplastics from sediment by directing an upward flow of water through a column, thereby inducing fluidization of the sediment. In this process, microplastics are first collected in a mesh through the upward flow of water and then separated using solutions of higher density such as NaCl solution. A schematic of the process has been shown in Figure 9.9. Elutriation provides a cheap and efficient alternative to separation of microplastics from large volumes of sediments. This process also acts as sample volume reduction for density separation. The drawback of elutriation for microplastic separation, however, comes at a cost of higher time consumption and additional pre-steps of sieving steps by size range [32].

9.3.2.3 Froth Floatation
Froth flotation is a process that selectively separates materials based upon whether they are water repelling (hydrophobic) or have an affinity for water (hydrophilic). Thus, the process of froth flotation is not solely dependent upon the density of the material; it is also dependent upon its hydrophobic nature. In this technique, particles of interest are physically separated from a liquid phase because of

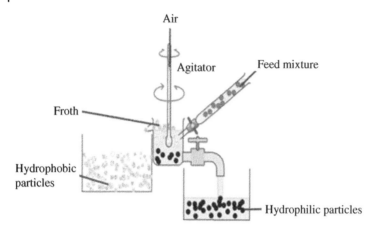

Figure 9.10 Froth flotation.

differences in the ability of air bubbles to selectively adhere to the surface of the particles, based upon their hydrophobicity. The hydrophobic particles with the air bubbles attached are carried to the surface, thereby forming a froth which can be removed, while hydrophilic materials stay in the liquid phase. Since plastics are generally hydrophobic materials, froth flotation has been used for the separation of plastic materials in millimeter range. However, there are many factors to consider in the process of froth flotation, such as the surface free energy of the microplastics and the surface tension of the liquid in the flotation bath, as well as the critical surface tension, which defines the surface tension at which the liquid completely wets the solid microplastics. Froth flotation, however, is not widely used for particles which are below millimeter range. This process is incompatible with the smaller fraction of microplastics due to low buoyancy and change of particle density due to surface fouling. In addition to that, attachment of nonplastic particles and denser particles makes it incompatible. Further, the unpredictability of bubble formation and recovery of plastics from the air–liquid interface complicates the process [33]. Therefore, the froth flotation process is not well suited for particles of smaller sizes as its selectivity and recovery is not efficient. A schematic of froth flotation process is shown in Figure 9.10.

9.4 Microplastic Sample Digestion Process

Environmental samples can be contaminated with organic and inorganic materials. In case of attachment of biological materials or biofilms in the samples, such entities result in overestimation of concentration of microplastics from water samples and sometimes from sediments. Therefore, it is necessary that samples must undergo pretreatment with a chemical or enzymatic digestion procedure for removal the organic matter. However, this adds a risk of damaging microplastics due to friction, degradation, or heating of the sample.

9.4.1 Acidic Digestion

Acidic digestion is used for microplastic sample treatment to degrade organic matter present over its surface. However, an optimum concentration, temperature, and duration of exposure to acids is required to degrade the organic matter as polymers have low resistance to acids and thus might start digesting the microplastic. HCl and HNO_3 are used for organic matter digestion in microplastic sample treatment. In comparison, HCl is not effective when compared to HNO_3 as it does not destroy all organic matter. HNO_3 is widely used in case of acidic digestion as it is quicker and has a better digestion under periodic heating in a timely manner during exposure of sample in the acidic solution [34]. However, acidic digestion can easily lead to underestimation of microplastics.

9.4.2 Alkaline Digestion

An alternate option for sample digestion is the utilization of bases such as NaOH or KOH solution. In comparison to acidic digestion, digestion of organic matter is better and has a good recovery of plastics. Also, with increase of molarity and the temperature, a more effective digestion can be achieved. However, the usage of 10 M NaOH degrades several polymers like PC, cellulose acetate (CA), PET, and PVC [35]. Alkaline digestion often leaves oily residues and residues on plastic surfaces which complicates characterization by spectroscopic techniques. Sequential usage of acid and alkali digestion can also be done to obtain good digestion of organic matter with high recovery rates.

9.4.3 Oxidizing Digestion

Oxidizing digestion performs more efficiently in digestion of organic matter and shows better performance when compared with acidic and alkaline digestion. Also, such digestion methods results in very less to no harm in microplastic integrity. H_2O_2 is widely used as it is very effective in removal of organic material. The polymers only change slightly getting more transparent, smaller, or thinner when using H_2O_2. This method causes production of foams which may result in reduction of microplastic retrieval. However, concentrated H_2O_2 solutions is highly reactive against plastics; therefore, it is recommended to use 10% H_2O_2 with an exposure time of 18 hours [36].

9.4.4 Enzymatic Degradation

Enzymes have also been used in some studies for removing organic matters. The advantage of enzymatic degradation is that polymers treated with enzymes are neither dissolved nor degraded. Therefore, they are more likely to damage microplastics. Moreover, enzymes are not hazardous. Enzymes such as cellulase, lipase, chitinase, protease, proteinase K, etc., have been used for microplastic digestion. The disadvantage of such digestion technique is the time needed for processing samples as each enzyme has its optimal temperature and pH [37]. An extensive review can be found in the literature [38].

9.5 Microplastic Identification and Classification

Several techniques have been used for identification and classification of microplastics. Since there is no standardization, the size, and the type of microplastics that can be identified are also reliant on the technique used. The techniques have been discussed below.

9.5.1 Visual Counting

Visual identification is the preliminary step applied for microplastic sorting and identification. For microplastics of bigger size (>1 mm), identification by naked eye is the simplest and quickest method. Microplastics of size lesser than 1 mm is often counted by using stereomicroscope or an optical microscope with magnification. Some reports on microplastic assessment using visual identification in river water and lake, and wastewater treatment plants can be found in the literature. The data quality presented due to visual counting relies heavily upon the person responsible for counting, the magnification of microscope used, and the sample matrix [39]. A research team [40] reported that the data assessment done on the sample extract by three different people resulted in inconsistent data quality, with a discrepancy of 1–4 MPs. The color of microplastics is also another important factor in visual inspection for microplastic detection. This shows the data inconsistency that could arise due to the distribution of microplastics of different colors, shapes, and sizes. For better data quality, other techniques such as thermal methods, FTIR techniques, and Raman microscopy are used after visual counting. However, high variance of data quality and error rate from 20 to 70% have been reported in the literature [17, 31]. More than two-thirds of the studies carried on microplastic analysis are still based on visual identification due to its simplicity and less time requirement for analysis.

9.5.2 Fluorescence

Microplastics have been identified and counted by using fluorescence produced by microplastics due to dye tagging on the microplastic surface. The process involves placing the samples upon a PC track etched filter membrane and addition of Nile red solution or other fluorescent dyes to the filter for sample staining. The macromolecular chain of polymers is loosened at high temperature due to thermal expansion and dyes can enter into microplastics and trapping takes place at contraction during cooling. The stained sample kept at 60 °C for 10 minutes allows dyes to get trapped in polymers and to have enhanced stability against strong solvents. The dyed samples when placed under a microscope at a magnification in green excitation of 460 nm and emission of 525 nm allow identification of microplastics due to fluorescence produced. For quantification of automatic particle recognition, fluorescent images are analyzed using imaging software. The fluorescence-based detection provides 100% automatic detection of PE, PP, PS, and nylon-6. Polymers such as PUR, PC, PVC, and PET produced weaker fluorescence. Apart from Nile red, dyes such as Safranin T or fluorescein isophosphate could be used under optimum temperature, staining, and dye combination as a cheaper fluorescent method for microplastic detection [41]. However, this method fails to provide chemical identification of plastic particles, but it is demonstrated as a highly efficient technique for quick, sensitive, and cost-effective quantification of microplastics such as PE, PP, PS, and nylon-6.

9.5.3 Destructive Analysis

The process of isolation, identification, and quantification poses an analytical challenge in microplastic characterization and results in lack of harmonization in data quality. Thermal techniques allow quantitative analysis by monitoring the mass loss of the sample during the heating program in which sample pretreatment is normally not necessary. This makes the technique suitable for the characterization of solid complex samples. Thermal methods are destructive methods in which the mass of the sample against time or temperature is monitored for while under an isothermal or ramped. The methodology is based on the identification of the polymer according to its degradation products. Thermal analysis includes different analytical techniques such as pyrolysis gas chromatography mass spectrometry (py-GC-MS) and thermogravimetry techniques such as TGA-mass spectrometry (TGA-MS), TGA thermal desorption-gas chromatography-mass spectrometry (TGA-TD-GC-MS), TGA-differential scanning calorimetry (DSC), and DSC. Such methods have been useful in microplastic identification as they provide an advantage of simultaneous analysis of both the polymer composition of microplastics along with organic additives present in the sample. These techniques are single-event analysis techniques that depend highly on sample preparation; manual placing of a single particle at a time in the pyrolysis tube making the analysis of microplastics below 500 μm difficult. Pyrolysis products obtained after thermal degradation of polymers also pose a risk of misinterpretation due to similar pyrolysis product from different polymers. Further, such methods can only provide chemical fingerprints while information on microplastic size and its size distribution cannot be obtained.

9.5.3.1 Thermoanalytical Methods

TGA is a thermoanalytical method in which the mass of the sample is monitored for its dependence on time or temperature while the temperature is programmed (isothermal or ramp) under a specific atmosphere (inert or air). This technique allows a quantitative analysis by monitoring the mass loss of the sample during the heating program; sample pretreatment is normally not necessary which makes this technique suitable for the characterization of solid complex samples. A research team [42] developed an online approach in which the gases evolving from the TGA, as the sample is being heated under an inert atmosphere, go directly to the mass spectrometry detector coupled to the TGA instrument. The mass loss measured by the thermo-balance in the TGA system is related to the signal in the MS detector. In this way, the system simultaneously provides a quantitative and qualitative approach using both the MS signal and the mass loss of the sample. Quantification was carried out using the signal characteristic ions on the mass spectrometry detector of the degradation products coming from the PET polymer. The main challenge in this method is the blocking of connection between TGA and detector. Thermogravimetry can also be coupled to differential scanning calorimetry (TGA-DSC). This method has also been shown to be suitable for microplastic characterization, taking advantages of both thermal techniques. DSC is used to know the thermodynamic properties such as enthalpies, heat capacities, and temperatures of the phase transitions. When a sample shows an endothermic or exothermic event, a change in temperature results, and a peak is observed in the DSC system; the area under that peak can be used for quantification purposes [43]. The endothermic phase transition heat flows and peak temperatures of the different plastics are obtained by DSC. The result is then mapped with the library to notice differences. The result showed difference in signatures between PE and PP, while for the rest of materials (PVC, polyamide [PA], polyester [PES], PET, and PU), the temperatures were quite similar and identification in a complex matrix where these polymers may be present would be ambiguous.

In case of Pyrolysis-GC-MS, the sample is thermally decomposed under an inert atmosphere and the resulting pyrolysis fragments of the polymer structure are separated by GC and characterized by MS. The most relevant information obtained in the analytical pyrolysis of polymers is the description of the chemical compounds obtained during and after pyrolysis. Each polymer has characteristic degradation products and indicator ions that can be used for its identification and/or quantification. One of the main limitations of Pyrolysis-GC-MS for MP characterization is the lack of information related to the particle size.

9.5.3.2 High-Performance Liquid Chromatography

Unlike thermogravimetric analysis, high-performance liquid chromatography (HPLC) can be used to identify a very small number of microplastics – PS and PET. In this method, the measurement of molar mass distribution is done by using exclusion chromatography whereby the microplastic samples are dissolved by solvents such as tetrahydrofurane and hexafluoroisopropanol. This method poses limitation in terms of providing information about the size of microplastics and a limited amount of sample that can be assessed at a time [37].

9.5.4 Nondestructive Analysis

Spectroscopic methods are nondestructive methods that allow us to obtain spatial and spectral information from an object under inspection. Infrared spectroscopy and Raman spectroscopy are widely used in identifying the type of microplastics present in a sample. These methods use an excitation source to excite the material which is placed in a suitable filter substrate and the absorbed or scattered light from the object is recorded by the spectrometer and is matched with the reference library to determine the type of microplastic present in the sample. These spectroscopic methods usually require a filter substrate should be water-resistant, mechanically stable, have pores to enable filtration of aqueous samples, and cause minimum spectral interference.

9.5.4.1 Fourier Transform Infrared Spectroscopy

In case of fourier transform (FTIR) infrared spectroscopy, an infrared light source is use to direct the light to the object under interest and the absorption profile is obtained and matched with the reference library to determine the chemical composition of the object to determine the material type. In case of microplastic identification, the sample containing microplastic particle is placed on a filter and is irradiated with IR light for excitation. Specific vibrations that are produced are measured by the spectrometer and is compared with known reference spectra. This means that this method can identify materials that are IR active, mainly molecules with polar functional groups. Transmission, reflection, and attenuated total-reflectance (ATR) are commonly used for microplastic identification in FTIR. In transmission mode, light passed from the source goes through the sample and is collected afterward. Therefore, a transparent filter is required for testing microplastic particles. This limitation can be overcome by using the reflectance mode. In reflectance mode, the incident beam is passed through the sample by reflection on an infrared reflective substrate. The morphology of microplastic particles can often lead to reflection errors due to light scattering. This limitation is overcome by using ATR mode in which placing the sample in an optical contact with a material of high refractive index, usually referred to as an ATR crystal, and surface irradiation with an evanescent wave. The advantage of ATR-FTIR is that it is a quick method which requires minimum sample preparation. Proper care is required while using ATR crystals since the ATR crystal is prone to surface scratching and cracking, the crystal can degrade over time. Another drawback is that this mode is used to analyze particles larger than 500 μm. The resolution can be increased by using micro-FTIR spectroscopy which combines FTIR spectroscopy with microscopy [44]. μFTIR is based on point-by-point mapping approach in which an entire filter needs to be scanned. This process requires a large amount of scanning time. Also, the spectra need to be manually obtained from the several different locations. Focal Plane Array (FPA)-based μFTIR allows acquisition of higher number of spectra from a larger area simultaneously. This technique has been used in microplastic identification of thousands of spectra in few minutes. Thus, microplastic analysis can be done quickly without comprising the spatial resolution [45].

9.5.4.2 Raman Spectroscopy

Raman spectroscopy is also a frequently used and highly reliable technique for polymer identification of microplastics from various environmental matrices. It is a complementary process to Infrared spectroscopy in which identification of microplastics is done by irradiating monochromatic laser beam onto a suspected sample, which results in a different frequency of the backscattered light due to specific molecular structure of the sample. The obtained scattered light is known as Raman shift, which produces a unique spectrum for each polymer. Raman spectroscopy is a nondestructive technique for characterization of microplastics which can characterize particles down to 1 μm. Raman spectroscopy is a high reliability technique which can also be applied to water samples as water produces weak scattering. Therefore, this technique can be realized without actual sample extraction from water samples. When compared to the most frequently used FTIR techniques for microplastics classification, Raman spectroscopy provides higher spatial resolution, wider spectral range, narrower spectral bonds, and lower sensitivity to water interference. A major advantage comes in terms of minimum size that can be identified, i.e., 1 μm while FTIR is only reliable for minimum size of 20 μm [44]. Table 9.1 shows a general comparison of commonly used technique for microplastic identification/classification.

Table 9.1 Techniques used for microplastic identification and/or classification.

Techniques		Size of detection	Advantages	Limitations
Visual sorting		Size down to 500 μm by naked eyes (optical microscope or electron microscopes for lowering detection limits)	Quick counting of cheap, relatively accurate, and fast higher amount of microplastics without requirement of skilled personnel	Large number of microplastics goes uncounted due to smaller size
Fluorescence		20 μm	Fast, cost-effective	Dying of organic debris might lead to overestimation
				No chemical identification
Thermal analysis	py-GC-MS	500 μm	Sensitive and reliable co-polymers in the microplastic samples can be identified	Manual placing of microplastic samples by tweezers
	TED-GC-MS			Destructive analysis
	HPLC	Requires sufficient size enough to chemical extraction for testing	High recovery	Only PS and PET can be analyzed
FTIR	ATR-FTIR	>1 mm	Easy sample preparation	Limitation of size and crystal
			Low cost	
	μFTIR	>20 μm	Lower detection limit	Complex sample preparation
				High cost
	FPA-FTIR	>20 μm	Quick analysis of larger number of samples at the same time	Expensive
Raman spectroscopy		>1 μm	Lower interference from water	High cost
			Low size detection limit	Time-consuming

9.6 Conclusions

The risk of microplastics is therefore very complex to be evaluated since it depends on several aspects including the concentration, the size distribution, shape, surface properties, polymer composition and density of the particles, the duration of exposure, the kinetics of absorption and desorption of contaminants with respect to the plastic, and the organism and the biology of the organism. While a standardized protocol for microplastic assessment is missing, majority of studies done in microplastic prevalence in water bodies is still based on usage of manta nets for sample collection, for which the minimum size of microplastics would be 300 or 100 μm. These are not sufficient to provide correct number of microplastics. Further, identification of microplastics in most of the studies involves counting using naked eyes which is prone to error. With consideration to the sample volume requirement and time requirement, such methods are still in use. However, such unreliable method gives us a wrong estimate and limits the minimum detectable size to the size of mesh in the nets. According to the existing literature, the size of microplastics is critical in the definition of the exposure of human to adverse effect of the contamination, where negative effect in human bodies are typically demonstrated at low particle size, e.g. below 1 μm. In certain cases, such as microplastic ingestion in human beings through tap water in which majority of microplastics below 5 μm are present, Raman spectroscopy which can identify and classify down to 1 μm should be used for microplastic assessment in order to avoid microplastic underestimation and to study exposure of microplastics through drinking water. Although research on microplastic prevalence in the environment, especially in the water bodies and marine animals, has increased exponentially since the last decade, a different way of assessment has led to nonuniformity and non-reliable data interpretation and strategy that has led to collection of data which cannot be fairly compared from one result to another. To summarize, a standardized protocol for microplastic sample collection and same for the classification should be established for the proper assessment of microplastics for uniformity in data collection. Another major challenge that needs to be addressed is high volume of water that needs to be collected. Manual collection of water samples and its transportation to laboratory does not provide real-time monitoring of microplastics at a particular time. Along with that, to avoid underestimation, FTIR and Raman spectroscopy which are used for microplastic classification generally require a huge amount of time. There is a need for new sampling and separation tools that able to increase the efficiency of analysis and broaden the database of contamination distribution in surface waters, marine environment, and food beverages by a more efficient, integrated, and easy set of analytical tools. From the point of view of sample preparation and selection, new filtration tools using microfluidic concept may overcome the current sampling limitation. Current limitations also give us an opportunity to find a better solution for automated sample collection and analysis. For sampling of particles less than 10 μm, techniques such as continuous flow centrifugation or miniaturized hydrocyclones could provide better estimation recommended for efficient microplastic sampling in the aqueous samples [46]. Such systems could be easily integrated with microfluidic system for quicker analysis using spectroscopic or electrical measurements. However, we need to understand the limitation of volume of water required for sampling, the critical dimensions of interest, or range of polymer we wish to characterize. In addition to this, such systems could be integrated to autonomous vehicles having an integrated microplastic separation and identification system that could provide a proper mapping of sample collection and concentration of microplastics at a particular surface level and depth profiles in real time.

In relation to the potential threat of microplastics in human health, risks are often discussed with reference to the effects of microplastics shown in aquatic organisms. Studies done on possible impacts on microplastics on human health often describe that smaller microplastics or nanoplastics could be lethal for human health as such particles could enter our bloodstream or translocate from one organ to another. However, the concentration of microplastics, the additives present in the microplastics, and size that could affect human health are crucial parameters to be studied too. Under such circumstances, analytical systems to detect the particles of nanometer scale will be required. Possible sources which become a route for microplastics to enter human bodies should be further investigated too. It is, therefore, necessary to develop analytical systems to measure the concentration, size, and types of such microplastics. Also, such systems could aid in increasing the amount of research to be done in human health relations with microplastic consumptions before we make incorrect speculations and extrapolate risks without proper scientific findings.

Acknowledgments

The authors would like to thank the European Commission for funding this work through the AQUASENSE (H2020-MSCA-ITN-2018-813680) project.

This research was funded by the H2020-MSCA-ITN-2018-813680 project (AQUASENSE).

References

1 Thompson, R.C., Olsen, Y., Mitchell, R.R. et al. (2004). Lost at sea: where is all the plastic? *Science* 304 (5672): 838. https://doi.org/10.1126/science.1094559.
2 "Plastics - the facts 2019 • plastics Europe," *Plastics Europe*, 22 October 2021. [Online]. https://plasticseurope.org/knowledge-hub/plastics-the-facts-2019/ (accessed 2 April 2023).

3 "Plastics - the facts 2020 • plastics Europe," *Plastics Europe*, 20 January 2022. [Online]. https://plasticseurope.org/knowledge-hub/plastics-the-facts-2020/ (accessed 2 April 2023).

4 Geyer, R., Jambeck, J.R., and Law, K.L. (2017). Production, use, and fate of all plastics ever made. *Sci. Adv.* 3 (7): 25–29. https://doi.org/10.1126/sciadv.1700782.

5 Koelmans, A.A., Mohamed Nor, N.H., Hermsen, E. et al. (2019). Microplastics in freshwaters and drinking water: critical review and assessment of data quality. *Water Res.* 155: 410–422. https://doi.org/10.1016/j.watres.2019.02.054.

6 Wright, S.L. and Kelly, F.J. (2017). Plastic and human health: a micro issue? *Environ. Sci. Technol.* 51 (12): 6634–6647. https://doi.org/10.1021/acs.est.7b00423.

7 Yong, C.Q.Y., Valiyaveetill, S., and Tang, B.L. (2020). Toxicity of microplastics and nanoplastics in mammalian systems. *Int. J. Environ. Res. Public Health* 17 (5): 1509. https://doi.org/10.3390/ijerph17051509.

8 Liu, P., Zhan, X., Wu, X. et al. (2020). Effect of weathering on environmental behavior of microplastics: properties, sorption and potential risks. *Chemosphere* 242: 125193. https://doi.org/10.1016/j.chemosphere.2019.125193.

9 Avio, C.G., Gorbi, S., and Regoli, F. (2017). Plastics and microplastics in the oceans: from emerging pollutants to emerged threat. *Mar. Environ. Res.* 128 (October 2017): 2–11. https://doi.org/10.1016/j.marenvres.2016.05.012.

10 Tibbetts, J., Krause, S., Lynch, I., and Smith, G.H.S. (2018). Abundance, distribution, and drivers of microplastic contamination in urban river environments. *Water* 10 (11): https://doi.org/10.3390/w10111597.

11 "Proceedings of the GESAMP International Workshop on assessing the risks associated with plastics and microplastics in the marine environment," *GESAMP*. [Online]. http://www.gesamp.org/publications/gesamp-international-workshop-on-assessing-the-risks-associated-with-plastics-and-microplastics-in-the-marine-environment (accessed 2 April 2023).

12 Von Moos, N., Burkhardt-Holm, P., and Köhler, A. (2012). Uptake and effects of microplastics on cells and tissue of the blue mussel *Mytilus edulis* L. after an experimental exposure. *Environ. Sci. Technol.* 46 (20): 11327–11335. https://doi.org/10.1021/es302332w.

13 da Costa, J.P., Santos, P.S.M., Duarte, A.C., and Rocha-Santos, T. (2016). (Nano) plastics in the environment – sources, fates and effects. *Sci. Total Environ.* 566–567: 15–26. https://doi.org/10.1016/j.scitotenv.2016.05.041.

14 Khalid, N., Aqeel, M., Noman, A. et al. (2021). Linking effects of microplastics to ecological impacts in marine environments. *Chemosphere* 264: 128541. https://doi.org/10.1016/j.chemosphere.2020.128541.

15 Wang, W. and Wang, J. (2018). Investigation of microplastics in aquatic environments: an overview of the methods used, from field sampling to laboratory analysis. *TrAC Trends Anal. Chem.* 108: 195–202. https://doi.org/10.1016/j.trac.2018.08.026.

16 Guzzetti, E., Sureda, A., Tejada, S., and Faggio, C. (2018). Microplastic in marine organism: environmental and toxicological effects. *Environ. Toxicol. Pharmacol.* 64 (September): 164–171. https://doi.org/10.1016/j.etap.2018.10.009.

17 Peiponen, K.-E., Räty, J., Ishaq, U. et al. (2019). Outlook on optical identification of micro- and nanoplastics in Aquatic Environments. *Chemosphere* 214: 424–429.

18 Cox, K.D., Covernton, G.A., Davies, H.L. et al. (2019). Human consumption of microplastics. *Environ. Sci. Technol.* 53 (12): 7068–7074. https://doi.org/10.1021/acs.est.9b01517.

19 Van Raamsdonk, L.W.D., van der Zande, M., Koelmans, A.A. et al. (2020). Current insights into monitoring, bioaccumulation and potential health effects of microplastics present in the food chain. *Foods* 9 (1): 72.

20 Ziajahromi, S., Neale, P.A., and Leusch, F.D.L. (2016). Wastewater treatment plant effluent as a source of microplastics: review of the fate, chemical interactions and potential risks to aquatic organisms. *Water Sci. Technol.* 74 (10): 2253–2269. https://doi.org/10.2166/wst.2016.414.

21 Schwabl, P., Köppel, S., Königshofer, P. et al. (2019). Detection of various microplastics in human stool: a prospective case series. *Ann. Intern. Med.* 171 (7): 453–457. https://doi.org/10.7326/M19-0618.

22 Ragusa, A., Svelato, A., Santacroce, C. et al. (2021). Plasticenta: first evidence of microplastics in human placenta. *Environ. Int.* 146: 106274. https://doi.org/10.1016/j.envint.2020.106274.

23 Hwang, J., Choi, D., Han, S. et al. (2020). Potential toxicity of polystyrene microplastic particles. *Sci. Rep.* 10 (1): 1–12. https://doi.org/10.1038/s41598-020-64464-9.

24 Horton, A.A., Walton, A., Spurgeon, D.J. et al. (2017). Microplastics in freshwater and terrestrial environments: evaluating the current understanding to identify the knowledge gaps and future research priorities. *Sci. Total Environ.* 586: 127–141. https://doi.org/10.1016/j.scitotenv.2017.01.190.

25 Campanale, C., Massarelli, C., Savino, I. et al. (2020). A detailed review study on potential effects of microplastics and additives of concern on human health. *Int. J. Environ. Res. Public Health* 17 (4): 1212. https://doi.org/10.3390/ijerph17041212.

26 Crawford, C.B. and Quinn, B. (2017). Microplastic collection techniques. *Microplast. Pollut.* 179–202. https://doi.org/10.1016/b978-0-12-809406-8.00008-6.

27 Stock, F., Kochleus, C., Bänsch-Baltruschat, B. et al. (2019). Sampling techniques and preparation methods for microplastic analyses in the aquatic environment – a review. *TrAC Trends Anal. Chem.* 113: 84–92. https://doi.org/10.1016/j.trac.2019.01.014.

28 Campanale, C., Savino, I., Pojar, I. et al. (2020). A practical overview of methodologies for sampling and analysis of microplastics in Riverine Environments. *Sustainability* 12 (17): 6755.

29 Nakajima, R., Lindsay, D.J., Tsuchiya, M. et al. (2019). MethodsX a small, stainless-steel sieve optimized for laboratory beaker-based extraction of microplastics from environmental samples. *MethodsX* 6: 1677–1682. https://doi.org/10.1016/j.mex.2019.07.012.

30 Cutroneo, L., Reboa, A., Besio, G. et al. (2020). Correction to: microplastics in seawater: sampling strategies, laboratory methodologies, and identification techniques applied to port environment. *Environ. Sci. Pollut. Res.* 27 (16): 20571. https://doi.org/10.1007/s11356-020-08704-5.

31 Hidalgo-Ruz, V., Gutow, L., Thompson, R.C., and Thiel, M. (2012). Microplastics in the marine environment: a review of the methods used for identification and quantification. *Environ. Sci. Technol.* 46 (6): 3060–3075. https://doi.org/10.1021/es2031505.

32 Bergmann, M., Gutow, L., and Klages, M. (2015). Marine anthropogenic litter. *Mar. Anthropog. Litter* 1–447. https://doi.org/10.1007/978-3-319-16510-3.

33 Nguyen, B., Claveau-Mallet, D., Hernandez, L.M. et al. (2019). Separation and analysis of microplastics and nanoplastics in complex environmental samples. *Acc. Chem. Res.* 52 (4): 858–866. https://doi.org/10.1021/acs.accounts.8b00602.

34 Prata, J.C., da Costa, J.P., Duarte, A.C., and Rocha-Santos, T. (2019). Methods for sampling and detection of microplastics in water and sediment: a critical review. *TrAC Trends Anal. Chem.* 110: 150–159. https://doi.org/10.1016/j.trac.2018.10.029.

35 Hurley, R.R., Lusher, A.L., Olsen, M., and Nizzetto, L. (2018). Validation of a method for extracting microplastics from complex, organic-rich, environmental matrices. *Environ. Sci. Technol.* 52 (13): 7409–7417. https://doi.org/10.1021/acs.est.8b01517.

36 Frias, J.P.G.L. and Nash, R. (2019). Microplastics: finding a consensus on the definition. *Mar. Pollut. Bull.* 138 (September 2018): 145–147. https://doi.org/10.1016/j.marpolbul.2018.11.022.

37 Li, J., Liu, H., and Paul Chen, J. (2018). Microplastics in freshwater systems: a review on occurrence, environmental effects, and methods for microplastics detection. *Water Res.* 137: 362–374. https://doi.org/10.1016/j.watres.2017.12.056.

38 Miller, M.E., Kroon, F.J., and Motti, C.A. (2017). Recovering microplastics from marine samples: a review of current practices. *Mar. Pollut. Bull.* 123 (1–2): 6–18. https://doi.org/10.1016/j.marpolbul.2017.08.058.

39 Crawford, C.B. and Quinn, B. (2017). Microplastic identification techniques. In: *Microplastic Pollutants*, 219–267. Elsevier.

40 Dekiff, J.H., Remy, D., Klasmeier, J., and Fries, E. (2014). Occurrence and spatial distribution of microplastics in sediments from Norderney. *Environ. Pollut.* 186: 248–256. https://doi.org/10.1016/j.envpol.2013.11.019.

41 Lv, L. et al. (2019). A simple method for detecting and quantifying microplastics utilizing fluorescent dyes – Safranine T, fluorescein isophosphate, Nile red based on thermal expansion and contraction property. *Environ. Pollut.* 255: 113283. https://doi.org/10.1016/j.envpol.2019.113283.

42 David, J., Steinmetz, Z., Kučerík, J., and Schaumann, G.E. (2018). Quantitative analysis of poly (ethylene terephthalate) microplastics in soil via thermogravimetry – mass spectrometry. *Anal. Chem.* 90 (15): 8793–8799. https://doi.org/10.1021/acs.analchem.8b00355.

43 Majewsky, M., Bitter, H., Eiche, E., and Horn, H. (2016). Determination of microplastic polyethylene (PE) and polypropylene (PP) in environmental samples using thermal analysis (TGA-DSC). *Sci. Total Environ.* 568: 507–511. https://doi.org/10.1016/j.scitotenv.2016.06.017.

44 Xu, J.L., Thomas, K.V., Luo, Z., and Gowen, A.A. (2019). FTIR and Raman imaging for microplastics analysis: state of the art, challenges and prospects. *TrAC – Trends Anal. Chem.* 119: 115629. https://doi.org/10.1016/j.trac.2019.115629.

45 Tagg, A.S., Sapp, M., Harrison, J.P., and Ojeda, J.J. (2015). Identification and quantification of microplastics in wastewater using focal plane array-based reflectance micro-FT-IR imaging. *Anal. Chem.* 87 (12): 6032–6040. https://doi.org/10.1021/acs.analchem.5b00495.

46 Hildebrandt, L., Zimmermann, T., Primpke, S. et al. (2021). Comparison and uncertainty evaluation of two centrifugal separators for microplastic sampling. *J. Hazard. Mater.* 414: 125482.

10

Advancements in Drone Applications for Water Quality Monitoring and the Need for Multispectral and Multi-Sensor Approaches

Joao L. E. Simon[1], Robert J. W. Brewin[1], Peter E. Land[2], and Jamie D. Shutler[1]

[1] Centre for Geography and Environmental Science, University of Exeter, Penryn, UK
[2] Plymouth Marine Laboratory, Plymouth, UK

10.1 Introduction

Water is a vital resource for all life on Earth. This resource, however, is often threatened by pollution from anthropogenic sources. It has been estimated that more than half of the world's freshwater lakes and reservoirs show some level of anthropogenic pollution [1, 2]. Water quality must be preserved in order to ensure the health of current and future populations. Water monitoring programs typically focus on physical, biological, and chemical water properties. Dependent upon the application, multivariable approaches are often needed, as the different properties are frequently interconnected. For example, the absence of light within the water column can be caused by intense turbidity due to high sediment concentrations, which can increase anaerobic respiration in bacteria, generating more CO_2 and reducing the pH of the water [3]. Indicators of water quality may be measured *in situ* with portable sensors, but greater accuracy and precision can often be obtained within a laboratory setting [4]. However, collecting biologically or chemically active samples for analysis in a laboratory is challenged by the ability of the sample to change post-sampling [5]. Alternatively, some physical or chemical properties can be quickly analyzed *in situ*, providing valuable, almost instantaneous, information on the water quality, such as electrical conductivity (to determine salinity), temperature, and dissolved oxygen (DO) [6].

Sensing Technologies for Real Time Monitoring of Water Quality, First Edition.
Libu Manjakkal, Leandro Lorenzelli, and Magnus Willander.

Long-term monitoring is imperative for effective management of water resources. It can allow large spatial scale changes and long-term temporal trends to be identified which may not be detectable by irregular sampling [7]. For example, pollutant inputs may be spread over large areas and take time to accumulate to hazardous levels. A priori knowledge of the frequency of phenomena is necessary when deciding the monitoring strategy, since some may be seasonal (e.g. annual ice formation within a lake) while others are sporadic and ephemeral (e.g. algal blooms) and consequently harder to predict. The large datasets produced through long-term sampling can be used to identify trends and develop and parameterize models capable of predicting future ecological change [8].

Harmful algal blooms are natural phenomena commonly triggered by an increase in the nutrient concentration, which can be caused naturally by rain runoff, for example, or by external inputs such as anthropogenic discharges [9]. These phenomena can accumulate toxins produced by the algae in water bodies and impact human health directly or indirectly [10]. Dense algal blooms may cause discoloration of water, drastic changes in chemical concentrations, and fluctuations in DO, causing fish mortality, affecting drinking water processing and negatively impacting multiple levels of the trophic chain, with resulting socioeconomic impacts [11]. Among algal blooms, blue-green algae (e.g. *Microcystis*) have the genetic potential to produce toxins [12] that are harmful to humans and animals, and may cause death to livestock [13].

The ability to monitor water quality effectively by observing and measuring water properties from lakes and reservoirs, including suspended sediment, chemical discharges, and algal bloom occurrences, is a necessity for managing drinking water resources, monitoring ecosystem health, and restoring contaminated land [14].

However, monitoring water bodies *in situ* can be expensive and time-consuming. Drinking water reservoirs are usually located away from urban centers to avoid local pollution, and these water bodies may have restricted access to the water or are located in remote regions with no facilities, increasing operational costs of monitoring [15]. Observing spatial and temporal changes in water quality using discrete stations and *in situ* sampling methods may restrict the sampling to a small number of locations, when the properties being studied are spatially and temporally heterogeneous [16]. The costs increase further in large lakes or lake systems, due to high operational costs of sampling boats and laboratory analysis of collected samples. These methods are also unlikely to capture fully any spatial heterogeneity in events such as algal blooms (Figure 10.1), where concentration levels can drastically change within a few meters or hours [17, 18]. As a result, manual *in situ* sampling of inland waters is increasingly complemented by remote sensing techniques [19].

Remote sensing approaches to water quality monitoring have predominantly focused on optical satellite sensors due to their global coverage and synoptic

Figure 10.1 Photograph of algal bloom on southern UK water reservoir, July 2014. The blooms appear seasonally increasing local maintenance. *Source:* Jamie D. Shutler (Coauthor).

sampling frequency, which can provide spectral observations over large water bodies on a near-daily basis [17, 20–25]. As more advanced satellite sensors are deployed and their numbers grow, the spatial coverage and revisit frequency increase, allowing more frequent monitoring [26]. Satellite data have been successfully applied to estimate a diverse range of water quality variables, such as clarity [27, 28], suspended sediments [29, 30], chlorophyll-a [31], and surface temperature [32, 33].

Airborne remote sensing platforms can fly below the tropospheric cloud base most of the time, and provide observations of higher spatial resolution than satellite approaches (e.g. frequently 2–5 m pixel resolution). This spatial resolution makes it possible to evaluate conditions at a local scale and to calibrate satellite data on a regional scale [34]. However, airborne remote sensing is inherently expensive and is not well suited to regular monitoring or to capturing data on episodic events.

10.2 Airborne Drones for Environmental Remote Sensing

Unmanned aerial vehicles, or drones, carrying remote sensing instruments are powerful tools that are beginning to be used to support water quality monitoring, enabling high spatial and temporal resolution assessment of aquatic systems, particularly in regions that lack extensive *in situ* monitoring [35, 36]. Drone-mounted optical remote-sensing sensors can be used to collect observations of surface conditions and identify surface variability of water quality. For example, such data can be used by hydrologists to visually understand circulation in the waterbody [37], or to systematically map coastal sea grass beds [38].

Drones can provide rapid and responsive surveys with minimal environmental [39], and one key advantage is the ability to collect data from hard-to-reach areas where access to the study site is logistically difficult or unsafe for humans [40, 41]. Drones are generally categorized based on several related attributes, including total weight, maximum flying altitude, payload, endurance, and flight range [42, 43], whether they are powered, or have fixed wings or rotors. Typical classifications consider the total weight of the aircraft and group drones into large (~200 kg), medium (~50 kg), small and mini (>30 kg), and micro and nano (<5 kg) platforms [44].

The collection of detailed and reliable environment observations is essential for the better understanding of the phenomena occurring in that environment. With this in mind, drone-collected data have so far been used for 3D modeling of terrain and buildings [45–49] and reconstruction of topography on eroded environments [50, 51].

Digital cameras, operating in the visible spectrum, have been attached to drones and when combined with the structure-from-motion [52] and multi-view stereopsis [53] techniques, can produce very high-resolution orthoimages and Digital Elevation Models (DEMs), as well as very dense point clouds of the terrain [54, 55]. These have been applied in environmental monitoring studies including coral reef mapping, hydrodynamic models, and soil contamination detection [48, 56]. As expected, these consumer-grade digital cameras excel at remote sensing applications where the features of interest include sharp color gradients, such as in coastal line mapping and burnt vegetation mapping [57]. Drones can be equipped with small versions of satellite LIDARs (Light Detection and Ranging) to produce high-resolution point clouds capturing the position of solid objects within the field of view to construct 3D digital models. Similar results can be achieved using optical cameras and structure-from-motion approaches [38]. An example specific to water systems uses DEMs to identify rain runoff and investigate sediment transport during flood events [58]. The geomorphic impacts of such episodic events include bank and riverbed erosion [59–61], floodplain scour [62], sediment deposition over the bank [63], and widening of the river channel [64].

One of the issues most drone applications face is the lack of precision and accuracy of their internal GPS (Global Positioning System). When in hovering mode, positioning errors can be several meters depending on the number of satellites detected by the GPS [65]. Terrestrial drone remote sensing applications may use visual points of reference (ground control points) and carefully chosen fixed

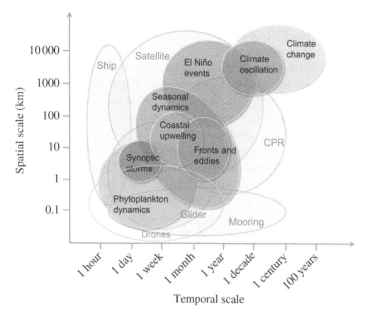

Figure 10.2 Spatial and temporal scale characteristics of climate systems and the coverage offered by most common sampling methods. *Source:* From Ref. [69]/Oxford University Press.

landmarks for image georeferencing [66]. However, more accurate and precise GPS approaches such as drone-capable RTK-GPS (Real Time Kinematics) are now commercially available, and are capable of positioning precision close to 0.3 m when hovering [67].

To date, not many aquatic studies have harnessed the spectral data or investigated the spectral signatures of the targets [68]. With the continuous emergence of smaller and smaller sensors, lighter in weight, and with higher spectral accuracies, drones are now becoming an increasingly effective tool for scientific remote sensing and assist in filling gaps left by other platforms (Figure 10.2) [70].

10.3 Drone Multispectral Remote Sensing

The scientific literature is replete with examples of terrestrial multispectral drone technology being used for scientific applications, including conservation and illegal deforestation mapping [71], wildlife monitoring and evaluation of disturbance impact on the natural habitats [72], field spectrometry precision agriculture [73], biomass estimation and quantification of dryland vegetation structures [74], disaster response in post-earthquake scenarios [75], as well as water-focused applications such as plant inventory mapping of non-submerged aquatic vegetation at very high resolution [76], coastal morphological mapping of dynamic tidal inlets [51], and coral reef monitoring [48].

With the advantage of their low operational requirements, drones are also used widely in civil fields such as urban management, agricultural monitoring, environmental protection, disaster relief, and film and television photography [77–80]. Drones also play an important role in the fields of surveillance, monitoring, electronic confrontation, and damage assessment [81].

Drone-based surveys are increasingly being used in fluvial research [82–85], providing enhanced high-resolution mapping and monitoring of river basins. Drones are also now being used for spatial mapping of water quality [53, 54] and, with the advent of miniaturized thermal cameras, for identifying groundwater discharge and wind-driven turbulent mixing [86].

At present, the majority of available lightweight hyperspectral sensors are linear-array cameras capable of capturing data in narrow spectral bands with a bandwidth of 5–10 nm [73, 87]. These hyperspectral sensors record high volumes of data and have proved useful for estimating wetland biomass, exploring species distribution, verifying water stress, and for geology surveys [88–92]. Their high spectral resolution is typically at the expense of lower spatial resolution ($\sim 50 \times 50$ pixels), compared to equivalent multispectral RGB cameras ($\sim 1024 \times 768$ pixels). These lightweight hyperspectral cameras are normally equipped with about half the spectral range of the larger cameras deployed on aircraft (400–1100 nm or 1100–2500 nm), owing to the weight constraints of drone platforms [93].

Over water, hyperspectral cameras can be employed to measure the optical signal that has been reflected from the water surface and has interacted with its Optically Active Constituents (OACs) through scattering and absorption [94]. The Water Leaving Radiance (WLR) is an apparent optical property that is related to the presence and abundance of OACs in the water column [95], as described by their inherent optical properties (IOPS), and the angular distribution of the light field [96]. The main types of OACs in fresh water are Colored Dissolved Organic Matter (CDOM), phytoplankton, and suspended sediments (organic and nonorganic solids). Each OAC has its own IOPs that help shape the WLR light spectrum [97]. Some IOPs, like phytoplankton absorption, can even contain information on different phytoplankton types [26, 98–101].

Optical sensors capable of detecting light beyond the visible wavelengths are particularly valuable in water quality studies, since some environmental variables can have strong influence in the energy absorption of specific wavelengths, e.g. temperature for mid-infrared light [102]. The near infra-red (NIR) bands (>765 nm) have been used in combination with consumer-grade image data to identify ratios of total phosphorus and Chlorophyll-a [103], and to calculate the Blue Normalized Difference Vegetation Index (BNDVI), in order to investigate surface cyanobacteria (*Microcystis*) concentrations [12]. Larson et al. [104] have also used multispectral imagery to assess turbidity levels in highly turbid fresh water.

Empirical and semi-analytical algorithms are typically used to investigate the relationship between the WLR spectral shape and the abundance of OACs (e.g. [105]). Empirical algorithms typically train statistical regression models using matched *in situ* OAC data and remotely sensed WLR data [106]. Whereas, semi-analytical algorithms combine known optical characteristics of substances found in water bodies with water physics to estimate the variable of choice [107].

However, to date very few studies have used drone-based multispectral cameras to determine WLR and from this infer OAC [108, 109]. In part, this is due to the challenge of carrying multiple sensors (downward and upward viewing sensors) on the same drone, and the need for improved geolocation accuracy. As drone-based technologies advance, these challenges are now being tackled.

10.4   Integrating Multiple Complementary Sensor Strategies with a Single Drone

The miniaturization of sensors is providing new opportunities for drone-based multisource data synergy (Figure 10.3). These sensors bring new opportunities to drone-based remote sensing, including the collection of discrete *in situ* measurements, especially when combined with an optical camera and an atmospheric sampler on the same drone has been demonstrated [111, 112].

Figure 10.3   Quadcopter drone with a hyperspectral imaging system mounted underneath. *Source:* From Ref. [110]/ISPRS, CC BY 4.0.

Drones are now capable of collecting discrete water samples, essential for calibrating and ground-truthing remote sensing algorithms. Such an approach is unlikely to be feasible for all water systems, but shows promise for inland or sheltered reservoirs or rivers where the environmental conditions are often more stable than in the open ocean, for example. Koparan et al. [113] used a drone-mounted sensor array to measure electric conductivity, pH, DO, and temperature within the water by mounting a sensor node to a cable attached to a drone. This also included an adapted water sampler. Ore et al. [114] developed a multi-rotor drone equipped with a water pump sampling tube, which was able to collect three 20 ml samples per flight. While drones have been used to collect water samples, early approaches have identified many operational and technical difficulties, for example, operation in unstable environments (e.g. strong winds and rough seas or lakes) and operation in the presence of waves [37, 114–116].

Drones have also been used for data communication and data download, when access difficulties or hazards prohibit conventional retrieval of data from *in situ* sensors. For example, Trasviña-Moreno et al. [117] constructed a sensor network to monitor a marine-coastal environment. Sensors were placed upon buoys to collect environmental variables including water temperature, and the drone acted as part of the communication network, allowing sensor data to be downloaded from the buoy without the need to physically access each buoy.

10.5 Conclusion

Most natural water resources are under pressure from anthropogenic influence. Novel monitoring approaches are required to ensure future water security and quality. Long-term and responsive water-quality sampling using drones would aid the monitoring of episodic harmful algal blooms in lakes and reservoirs. Monitoring of such reservoirs can be logistically difficult, expensive, and time-consuming.

Drone-based remote sensing could be used to complement discrete sampling, offering a responsive and cost-efficient monitoring tool. However, to date most work has focused on developing their use for terrestrial applications.

One of the weaknesses of using existing drones in aquatic applications is the difficulty of acquiring fixed points of reference over water, reducing the accuracy and precision of onboard positioning instruments, and limiting the use of drones for monitoring water quality. However, higher accuracy drone-enabled navigation systems, such as RTK-GPS, are now commercially available and have the necessary precision and accuracy to solve this issue.

The development of small hyperspectral sensors gives added potential for measurements of WLR that can be used to infer important OAC. Drones are now carrying multiple sensors for sampling the physical and chemical environment. The integration of multiple sensors will enable more in-depth monitoring and move us toward a more complete understanding of environmental phenomena.

Acknowledgment

This project has received funding from the European Union's Horizon 2020 research and innovation programme under the Marie Skłodowska-Curie grant agreement No: H2020-MSCA-ITN-2018-813 680.

References

1 He, B., Oki, K., Wang, Y. et al. (2012). Analyse de la qualité de l'eau et estimation des nutriments en rivière à l'aide de l'imagerie de télédétection Quick Bird. *Hydrol. Sci. J.* 57 (5): 850–860.
2 Dai, X., Zhou, Y., Ma, W., and Zhou, L. (2017). Influence of spatial variation in land-use patterns and topography on water quality of the rivers inflowing to Fuxian Lake, a large deep lake in the plateau of southwestern China. *Ecol. Eng.* 99: 417–428.
3 Li, D. and Liu, S. (2019). Detection of river water quality. In: *Water Quality Monitoring and Management* (ed. D. Li and S. Liu), 211–220. Academic Press.
4 Xu, Z. and Boyd, C.E. (2016). Reducing the monitoring parameters of fish pond water quality. *Aquaculture* 465: 359–366.
5 Stauber, C., Miller, C., Cantrell, B., and Kroell, K. (2014). Evaluation of the compartment bag test for the detection of *Escherichia coli* in water. *J. Microbiol. Methods* 99 (1): 66–70.
6 Chung, W.Y. and Yoo, J.H. (2015). Remote water quality monitoring in wide area. *Sensors Actuators B Chem.* 217: 51–57.
7 Thomas, K.V., Hurst, M.R., Matthiessen, P. et al. (2001). Toxicity characterisation of organic contaminants in stormwaters from an agricultural headwater stream in South East England. *Water Res.* 35 (10): 2411–2416.
8 Malone, T.C., DiGiacomo, P.M., Gonçalves, E. et al. (2014). A global ocean observing system framework for sustainable development. *Mar. Policy* 43 (2014): 262–272.
9 Yankova, Y., Neuenschwander, S., Köster, O., and Posch, T. (2017). Abrupt stop of deep water turnover with lake warming: drastic consequences for algal primary producers. *Sci. Rep.* 7 (1): 1–9.
10 Figgatt, M., Hyde, J., Dziewulski, D. et al. (2017). Harmful algal bloom – associated illnesses in humans and dogs identified through a pilot surveillance system – New York, 2015. *MMWR Morb. Mortal. Wkly Rep.* 66 (43): 1182–1184.

11 Brown, A.R., Lilley, M.K.S., Shutler, J. et al. (2019). Assessing risks and mitigating impacts of harmful algal blooms on mariculture and marine fisheries. *Rev. Aquac.* 1663–1688.

12 Van Der Merwe, D. and Price, K.P. (2015). Harmful algal bloom characterization at ultra-high spatial and temporal resolution using small unmanned aircraft systems. *Toxins* 7: 1065–1078.

13 McGowan, S. (2016). Algal blooms. In: *Biological and Environmental Hazards, Risks, and Disasters* (ed. J.F. Shroder and R. Sivanpillai), 5–43. Boston: Academic Press.

14 Graham, J., Loftin, K., and Kamman, N. (2009). Monitoring recreational freshwaters. *Lakeline* 29 (January 2009): 18–24.

15 Wang, L., Pu, H., and Sun, D.W. (2016). Estimation of chlorophyll-a concentration of different seasons in outdoor ponds using hyperspectral imaging. *Talanta* 147: 422–429.

16 Wawrzyniak, V., Piégay, H., Allemand, P. et al. (2013). Prediction of water temperature heterogeneity of braided rivers using very high resolution thermal infrared (TIR) images. *Int. J. Remote Sens.* 34 (13): 4812–4831.

17 Becker, R.H., Sultan, M.I., Boyer, G.L. et al. (2009). Mapping cyanobacterial blooms in the Great Lakes using MODIS. *J. Great Lakes Res.* 35 (3): 447–453.

18 Shuchman, R.A., Leshkevich, G., Sayers, M.J. et al. (2013). An algorithm to retrieve chlorophyll, dissolved organic carbon, and suspended minerals from Great Lakes satellite data. *J. Great Lakes Res.* 39: 14–33.

19 Binding, C.E., Jerome, J.H., Bukata, R.P., and Booty, W.G. (2010). Suspended particulate matter in Lake Erie derived from MODIS aquatic colour imagery. *Int. J. Remote Sens.* 31 (19): 5239–5255.

20 Binding, C.E., Greenberg, T.A., McCullough, G. et al. (2018). An analysis of satellite-derived chlorophyll and algal bloom indices on Lake Winnipeg. *J. Great Lakes Res.* 44 (3): 436–446.

21 Clark, J.M., Schaeffer, B.A., Darling, J.A. et al. (2017). Satellite monitoring of cyanobacterial harmful algal bloom frequency in recreational waters and drinking water sources. *Ecol. Indic.* 80 (April): 84–95.

22 Kutser, T., Metsamaa, L., Strömbeck, N., and Vahtmäe, E. (2006). Monitoring cyanobacterial blooms by satellite remote sensing. *Estuar. Coast. Shelf Sci.* 67 (1–2): 303–312.

23 Sayers, M., Fahnenstiel, G.L., Shuchman, R.A., and Whitley, M. (2016). Cyanobacteria blooms in three eutrophic basins of the Great Lakes: a comparative analysis using satellite remote sensing. *Int. J. Remote Sens.* 37 (17): 4148–4171.

24 Stumpf, R.P., Wynne, T.T., Baker, D.B., and Fahnenstiel, G.L. (2012). Interannual variability of cyanobacterial blooms in Lake Erie. *PLoS One* 7 (8).

25 Shutler, J.D., Warren, M.A., Miller, P.I. et al. (2015). Computers and geosciences operational monitoring and forecasting of bathing water quality through exploiting satellite Earth observation and models: the AlgaRisk demonstration service. *Comput. Geosci.* 77: 87–96.

26 Sathyendranath, S., Brewin, R.J.W., Brockmann, C. et al. (2019). An ocean-colour time series for use in climate studies: the experience of the ocean-colour climate change initiative (OC-CCI). *Sensors* 19 (19): 4285.

27 Dekker, A.G. and Hoogenboom, H.J. (2009). Remote sensing, ecological water quality modelling and *in situ* measurements: a case study in shallow lakes. *Hydrol. Sci. J.* 41: 531.

28 Olmanson, L.G., Brezonik, P.L., Finlay, J.C., and Bauer, M.E. (2016). Remote sensing of environment comparison of Landsat 8 and Landsat 7 for regional measurements of CDOM and water clarity in lakes. *Remote Sens. Environ.* 185: 119–128.

29 Giardino, C., Oggioni, A., Bresciani, M., and Yan, H. (2010). Remote sensing of suspended particulate matter in Himalayan Lakes. *Mt. Res. Dev.* 30 (2): 157–168.

30 Kaba, E., Philpot, W., and Steenhuis, T. (2014). Evaluating suitability of MODIS-Terra images for reproducing historic sediment concentrations in water bodies: Lake Tana, Ethiopia. *Int. J. Appl. Earth Obs. Geoinf.* 26: 286–297.

31 Hu, C., Lee, Z., Ma, R. et al. (2010). Moderate resolution imaging spectroradiometer (MODIS) observations of cyanobacteria blooms in Taihu Lake, China. *J. Geophys. Res. Oceans* 115: 1–20.

32 Bresciani, M., Giardino, C., and Boschetti, L. (2011). Multi-temporal assessment of bio-physical parameters in lakes Garda and Trasimeno from MODIS and MERIS. *Eur. J. Remote Sens.* 43: 49–62.

33 Lamaro, A.A., Mariñelarena, A., Torrusio, S.E., and Sala, S.E. (2013). Water surface temperature estimation from Landsat 7 ETM+ thermal infrared data using the generalized single-channel method: case study of Embalse del Río Tercero (Córdoba, Argentina). *Adv. Space Res.* 51 (3): 492–500.

34 Fraser, R.H., van der Sluijs, J., and Hall, R.J. (2017). Calibrating satellite-based indices of burn severity from UAV-derived metrics of a burned boreal forest in NWT, Canada. *Remote Sens.* 9 (3): 279.

35 Matthews, M.W. (2011). A current review of empirical procedures of remote sensing in inland and near-coastal transitional waters. *Int. J. Remote Sens.* 32 (21): 6855–6899.

36 Dörnhöfer, K. and Oppelt, N. (2016). Remote sensing for lake research and monitoring – recent advances. *Ecol. Indic.* 64: 105–122.

37 Kaizu, Y., Iio, M., Yamada, H., and Noguchi, N. (2011). Development of unmanned airboat for water-quality mapping. *Biosyst. Eng.* 109 (4): 338–347.

38 Duffy, J., Shutler, J., Witt, M. et al. (2018). Tracking fine-scale structural changes in coastal dune morphology using kite aerial photography and uncertainty-assessed structure-from-motion photogrammetry. *Remote Sens.* 10 (9): 1494.

39 Pirotta, V., Smith, A., Ostrowski, M., and Russell, D. (2017). An economical custom-built drone for assessing whale health. *Front. Mar. Sci.* 4 (DEC): 1–12.

40 Martinez-De Dios, J.R., Lferd, K., De San Bernabé, A. et al. (2013). Cooperation between UAS and wireless sensor networks for efficient data collection in large environments. *J. Intell. Robot. Syst. Theory Appl.* 70 (1–4): 491–508.

41 Sujit, P.B., Lucani, D.E., and Sousa, J.B. (2013). Joint route planning for UAV and sensor network for data retrieval. *Proceedings of the 2013 IEEE International Systems Conference (SysCon)*, Orlando, USA (15–18 April 2013), 688–692. IEEE.

42 Korchenko, A.G. and Illyash, O.S. (2013). The generalized classification of unmanned air vehicles. *Proceedings of the 2013 IEEE 2nd International Conference Actual Problems of Unmanned Air Vehicles Developments Proceedings (APUAVD)*, Kiev, Ukraine (15–17 October 2013), 28–34. IEEE.

43 Dalamagkidis, K. (2015). Classification of UAVs. In: *Handbook of Unmanned Aerial Vehicles* (ed. K.P. Valavanis and G.J. Vachtsevanos), 83–91. Dordrecht: Springer Netherlands.

44 Anderson, K. and Gaston, K.J. (2013). Lightweight unmanned aerial vehicles will revolutionize spatial ecology. *Front. Ecol. Environ.* 11 (3): 138–146.

45 Ventura, D., Bruno, M., Jona Lasinio, G. et al. (2016). A low-cost drone based application for identifying and mapping of coastal fish nursery grounds. *Estuar. Coast. Shelf Sci.* 171: 85–98.

46 Hodgson, A., Kelly, N., and Peel, D. (2013). Unmanned aerial vehicles (UAVs) for surveying marine fauna: a dugong case study. *PLoS One* 8 (11): e79556.

47 Burns, J.H.R., Delparte, D., Gates, R.D., and Takabayashi, M. (2015). Integrating structure-from-motion photogrammetry with geospatial software as a novel technique for quantifying 3D ecological characteristics of coral reefs. *PeerJ* 2015 (7): e1077.

48 Casella, E., Collin, A., Harris, D. et al. (2017). Mapping coral reefs using consumer-grade drones and structure from motion photogrammetry techniques. *Coral Reefs* 36 (1): 269–275.

49 Turner, I.L., Harley, M.D., and Drummond, C.D. (2016). UAVs for coastal surveying. *Coast. Eng.* 114: 19–24.

50 Mancini, F., Dubbini, M., Gattelli, M. et al. (2013). Using unmanned aerial vehicles (UAV) for high-resolution reconstruction of topography: the structure from motion approach on coastal environments. *Remote Sens.* 5 (12): 6880–6898.

51 Long, N., Millescamps, B., Guillot, B. et al. (2016). Monitoring the topography of a dynamic tidal inlet using UAV imagery. *Remote Sens.* 8 (5): 1–18.

52 Fonstad, M.A., Dietrich, J.T., Courville, B.C. et al. (2013). Topographic structure from motion: a new development in photogrammetric measurement. *Earth Surf. Process. Landf.* 38 (4): 421–430.

53 Furukawa, Y. and Ponce, J. (2007). Accurate, dense, and robust multi-view stereopsis. *Proceedings of the IEEE Computer Society Conference on Computer Vision and Pattern Recognition,* Minneapolis, USA (17–22 June 2007). IEEE.

54 Agüera-Vega, F., Carvajal-Ramírez, F., and Martínez-Carricondo, P. (2017). Accuracy of digital surface models and orthophotos derived from unmanned aerial vehicle photogrammetry. *J. Surv. Eng.* 143 (2): 04016025.

55 Martínez-Carricondo, P., Agüera-Vega, F., Carvajal-Ramírez, F. et al. (2018). Assessment of UAV-photogrammetric mapping accuracy based on variation of ground control points. *Int. J. Appl. Earth Obs. Geoinf.* 72 (February): 1–10.

56 Sibanda, M., Mutanga, O., Chimonyo, V.G.P. et al. (2021). Application of drone technologies in surface water resources monitoring and assessment: a systematic review of progress, challenges, and opportunities in the global south. *Drones* 5: 84.

57 Taddia, Y., Stecchi, F., and Pellegrinelli, A. (2020). Coastal mapping using DJI Phantom 4 RTK in post-processing kinematic mode. *Drones* 4 (2): 9.

58 Korup, O. (2012). Earth's portfolio of extreme sediment transport events. *Earth Sci. Rev.* 112 (3–4): 115–125.

59 Prosser, I.P., Hughes, A.O., and Rutherfurd, I.D. (2000). Bank erosion of an incised upland channel by subaerial processes: Tasmania, Australia. *Earth Surf. Process. Landf.* 25 (10): 1085–1101.

60 Milan, D.J. (2012). Geomorphic impact and system recovery following an extreme flood in an upland stream: Thinhope Burn, northern England, UK. *Geomorphology* 138 (1): 319–328.

61 Thompson, C. and Croke, J. (2013). Geomorphic effects, flood power, and channel competence of a catastrophic flood in confined and unconfined reaches of the upper Lockyer Valley, Southeast Queensland, Australia. *Geomorphology* 197: 156–169.

62 Lewis, Q.W., Edmonds, D.A., and Yanites, B.J. (2020). Integrated UAS and LiDAR reveals the importance of land cover and flood magnitude on the formation of incipient chute holes and chute cutoff development. *Earth Surf. Process. Landf.* 45 (6): 1441–1455.

63 Knox, J.C. (2006). Floodplain sedimentation in the Upper Mississippi Valley: natural versus human accelerated. *Geomorphology* 79 (3–4): 286–310.

64 Krapesch, G., Hauer, C., and Habersack, H. (2011). Scale orientated analysis of river width changes due to extreme flood hazards. *Nat. Hazards Earth Syst. Sci.* 2 (8): 2137–2147.

65 Hung, I.K., Unger, D., Kulhavy, D., and Zhang, Y. (2019). Positional precision analysis of orthomosaics derived from drone captured aerial imagery. *Drones* 3 (2): 1–10.

66 James, M.R., Robson, S., and Smith, M.W. (2017). 3-D uncertainty-based topographic change detection with structure-from-motion photogrammetry: precision maps for ground control and directly georeferenced surveys. *Earth Surf. Process. Landf.* 42 (12): 1769–1788.

67 Skoglund, M., Petig, T., Vedder, B. et al. (2016). Static and dynamic performance evaluation of low-cost RTK GPS receivers. *2016 IEEE Intelligent Vehicles Symposium (IV)*, Gotenburg, Sweden (19–22 June 2016), vol. 2016-Augus, no. June, 16–19.

68 Li, D. and Li, M. (2014). Research advance and application prospect of unmanned aerial vehicle remote sensing system. *Wuhan Daxue Xuebao (Xinxi Kexue Ban)/ Geomat. Inf. Sci. Wuhan Univ.* 39: 505–513+540.

69 Racault, M.F., Platt, T., Sathyendranath, S. et al. (2014). Plankton indicators and ocean observing systems: support to the marine ecosystem state assessment. *J. Plankton Res.* 36 (3): 621–629.

70 Zhong, Y., Wang, X., Xu, Y. et al. (2018). Mini-UAV-borne hyperspectral remote sensing: from observation and processing to applications. *IEEE Geosci. Remote Sens. Mag.* 6 (4): 46–62.

71 Koh, L.P. and Wich, S.A. (2012). Dawn of drone ecology: low-cost autonomous aerial vehicles for conservation. *Trop. Conserv. Sci.* 5 (2): 121–132.

72 Christie, K.S., Gilbert, S.L., Brown, C.L. et al. (2016). Unmanned aircraft systems in wildlife research: current and future applications of a transformative technology. *Front. Ecol. Environ.* 14 (5): 241–251.

73 Burkart, A., Cogliati, S., Schickling, A., and Rascher, U. (2014). A novel UAV-based ultra-light weight spectrometer for field spectroscopy. *IEEE Sensors J.* 14 (1): 62–67.

74 Cunliffe, A.M., Brazier, R.E., and Anderson, K. (2016). Ultra-fine grain landscape-scale quantification of dryland vegetation structure with drone-acquired structure-from-motion photogrammetry. *Remote Sens. Environ.* 183: 129–143.

75 Nedjati, A., Vizvari, B., and Izbirak, G. (2016). Post-earthquake response by small UAV helicopters. *Nat. Hazards* 80 (3): 1669–1688.

76 Husson, E., Ecke, F., and Reese, H. (2016). Comparison of manual mapping and automated object-based image analysis of non-submerged aquatic vegetation from very-high-resolution UAS images. *Remote Sens.* 8 (9): 1–18.

77 Karthik Reddy, B.S. and Poondla, A. (2017). Performance analysis of solar powered unmanned aerial vehicle. *Renew. Energy* 104: 20–29.

78 Kumar, G., Sepat, S., and Bansal, S. (2015). Review paper of solar powered UAV. *Int. J. Sci. Eng. Res.* 6 (2): 41–44.

79 Yu, X., Liu, Q., Liu, X. et al. (2017). A physical-based atmospheric correction algorithm of unmanned aerial vehicles images and its utility analysis. *Int. J. Remote Sens.* 38 (8–10): 3101–3112.

80 Lee, J.S. and Yu, K.H. (2017). Optimal path planning of solar-powered UAV using gravitational potential energy. *IEEE Trans. Aerosp. Electron. Syst.* 53 (3): 1442–1451.

81 Guilherme, L., Nanni, M.R., Silva, G.F.C. et al. (2017). Semi professional digital camera calibration techniques for Vis/NIR spectral data acquisition from an unmanned aerial vehicle. *Int. J. Remote Sens.* 38 (8–10): 2717–2736.

82 Entwistle, N., Heritage, G., and Milan, D. (2018). Recent remote sensing applications for hydro and morphodynamic monitoring and modelling. *Earth Surf. Process. Landf.* 43 (10): 2283–2291.

83 Tomsett, C. and Leyland, J. (2019). Remote sensing of river corridors: a review of current trends and future directions. *River Res. Appl.* 35 (7): 779–803.

84 Entwistle, N. and Heritage, G. (2017). An evaluation DEM accuracy acquired using a small unmanned aerial vehicle across a riverine environment. *Int. J. New Technol. Res.* 3 (7): 43–48.

85 Entwistle, N.S. and Heritage, G.L. (2019). Small unmanned aerial model accuracy for photogrammetrical fluvial bathymetric survey. *J. Appl. Remote. Sens.* 13 (01): 014523.

86 Lee, E., Yoon, H., Hyun, S.P. et al. (2016). Unmanned aerial vehicles (UAVs)-based thermal infrared (TIR) mapping, a novel approach to assess groundwater discharge into the coastal zone. *Limnol. Oceanogr. Methods* 14 (11): 725–735.

87 Suomalainen, J., Anders, N., Iqbal, S. et al. (2014). A lightweight hyperspectral mapping system and photogrammetric processing chain for unmanned aerial vehicles. *Remote Sens.* 6 (11): 11013–11030.

88 Adam, E., Mutanga, O., and Rugege, D. (2010). Multispectral and hyperspectral remote sensing for identification and mapping of wetland vegetation: a review. *Wetl. Ecol. Manag.* 18 (3): 281–296.

89 Haboudane, D., Miller, J.R., Pattey, E. et al. (2004). Hyperspectral vegetation indices and novel algorithms for predicting green LAI of crop canopies: Modeling and validation in the context of precision agriculture. *Remote Sens. Environ.* 90 (3): 337–352.

90 Psomas, A., Kneubühler, M., Huber, S. et al. (2011). Hyperspectral remote sensing for estimating aboveground biomass and for exploring species richness patterns of grassland habitats. *Int. J. Remote Sens.* 32 (24): 9007–9031.

91 van der Meer, F.D., van der Werff, H.M.A., van Ruitenbeek, F.J.A. et al. (2012). Multi- and hyperspectral geologic remote sensing: a review. *Int. J. Appl. Earth Obs. Geoinf.* 14 (1): 112–128.

92 Palmason, J.A., Benediktsson, J.A., and Sveinsson, J.R. (2005). Classification of hyperspectral ROSIS data from urban areas. *RAST 2005 – Proceedings of 2nd International Conference on Recent Advances in Space Technologies,* Istanbul, Turkey (9–11 June 2005), vol. 2005, no. 3, 63–69. IEEE.

93 Rufino, G. and Moccia, A. (2005). Integrated VIS-NIR hyperspectral/thermal-IR electro-optical payload system for a mini-UAV. *Collection of Technical Papers – InfoTech at Aerospace: Advancing Contemporary Aerospace Technologies and Their*, Arlington, Virginia (26–29 September 2005), vol. 2, no. September, 915–923.

94 Morel, A. and Prieur, L. (1977). Analysis of variations in ocean color. *Limnol. Oceanogr.* 22 (4): 709–722.

95 Moore, T.S., Dowell, M.D., Bradt, S., and Ruiz Verdu, A. (2014). An optical water type framework for selecting and blending retrievals from bio-optical algorithms in lakes and coastal waters. *Remote Sens. Environ.* 143: 97–111.

96 Preisendorfer, R.W. (1976). *Hydrologic Optics Volume I. Introduction*, 1e. U.S. Department of Commerce, National Oceanic and Atmospheric Administration, Environmental Research Laboratories, Pacific Marine Environmental Laboratory.

97 Werdell, P.J., McKinna, L.I.W., Boss, E. et al. (2018). An overview of approaches and challenges for retrieving marine inherent optical properties from ocean color remote sensing. *Prog. Oceanogr.* 160 (January): 186–212.

98 Bricaud, A., Claustre, H., Ras, J., and Oubelkheir, K. (2004). Natural variability of phytoplanktonic absorption in oceanic waters: influence of the size structure of algal populations. *J. Geophys. Res. Ocean.* 109 (11): 1–12.

99 Brewin, R.J.W., Hardman-Mountford, N.J., Lavender, S.J. et al. (2011). An intercomparison of bio-optical techniques for detecting dominant phytoplankton size class from satellite remote sensing. *Remote Sens. Environ.* 115 (2): 325–339.

100 Ciotti, Á.M., Lewis, M.R., and Cullen, J.J. (2002). Assessment of the relationships between dominant cell size in natural phytoplankton communities and the spectral shape of the absorption coefficient. *Limnol. Oceanogr.* 47 (2): 404–417.

101 Sathyendranath, S., Watts, L., Devred, E. et al. (2004). Discrimination of diatoms from other phytoplankton using ocean-colour data. *Mar. Ecol. Prog. Ser.* 272: 59–68.

102 Yuan, S., Shen, C., Deng, B. et al. (2018). Air-stable room-temperature mid-infrared photodetectors based on hBN/black arsenic phosphorus/hBN heterostructures. *Nano Lett.* 18 (5): 3172–3179.

103 Su, T.C. and Chou, H.T. (2015). Application of multispectral sensors carried on unmanned aerial vehicle (UAV) to trophic state mapping of small reservoirs: a case study of Tain-Pu reservoir in Kinmen, Taiwan. *Remote Sens.* 7 (8): 10078–10097.

104 Larson, M.D., Simic Milas, A., Vincent, R.K., and Evans, J.E. (2018). Multi-depth suspended sediment estimation using high-resolution remote-sensing UAV in Maumee River, Ohio. *Int. J. Remote Sens.* 39 (15–16): 5472–5489.

105 Wei, L., Huang, C., Zhong, Y. et al. (2019). Inland waters suspended solids concentration retrieval based on PSO-LSSVM for UAV-borne hyperspectral remote sensing imagery. *Remote Sens.* 11 (12): 1455.

106 Cheng, K.H., Chan, S.N., and Lee, J.H.W. (2020). Remote sensing of coastal algal blooms using unmanned aerial vehicles (UAVs). *Mar. Pollut. Bull.* 152 (August 2019): 110889.

107 Borges, H.L.F., Branco, L.H.Z., Martins, M.D. et al. (2015). Cyanotoxin production and phylogeny of benthic cyanobacterial strains isolated from the northeast of Brazil. *Harmful Algae* 43 (Complete): 46–57.

108 Novoa, S., Doxaran, D., Ody, A. et al. (2017). Atmospheric corrections and multi-conditional algorithm for multi-sensor remote sensing of suspended particulate matter in low-to-high turbidity levels coastal waters. *Remote Sens.* 9 (1): 61.

109 Shang, S., Lee, Z., Lin, G. et al. (2017). Sensing an intense phytoplankton bloom in the western Taiwan Strait from radiometric measurements on a UAV. *Remote Sens. Environ.* 198 (September): 85–94.

110 Saari, H., Akujärvi, A., Holmlund, C. et al. (2017). Visible, very near IR and short wave IR hyperspectral drone imaging system for agriculture and natural water applications. *Int. Arch. Photogramm. Remote Sens. Spat. Inf. Sci. ISPRS Arch.* 42 (3W3): 165–170.

111 Yang, B. and Chen, C. (2015). Automatic registration of UAV-borne sequent images and LiDAR data. *ISPRS J. Photogramm. Remote Sens.* 101: 262–274.

112 Segl, K., Roessner, S., Heiden, U., and Kaufmann, H. (2003). Fusion of spectral and shape features for identification of urban surface cover types using reflective and thermal hyperspectral data. *ISPRS J. Photogramm. Remote Sens.* 58 (1–2): 99–112.

113 Koparan, C., Koc, A.B., Privette, C.V. et al. (2018). Evaluation of a UAV-assisted autonomous water sampling. *Water* 10 (5): 655.

114 Ore, J.-P., Elbaum, S., Burgin, A. et al. (2015). Autonomous aerial water sampling. In: *Field and Service Robotics: Results of the 9th International Conference* (ed. L. Mejias, P. Corke, and J. Roberts), 137–151. Cham: Springer International Publishing.

115 Eichhorn, M., Ament, C., Jacobi, M. et al. (2018). Modular AUV system with integrated real-time water quality analysis. *Sensors* 18 (6): 1–17.

116 Liu, Y., Wang, T., Ma, L., and Wang, N. (2014). Spectral calibration of hyperspectral data observed from a hyperspectrometer loaded on an unmanned aerial vehicle platform. *IEEE J. Sel. Top. Appl. Earth Obs. Remote Sens.* 7 (6): 2630–2638.

117 Trasviña-Moreno, C.A., Blasco, R., Marco, Á. et al. (2017). Unmanned aerial vehicle based wireless sensor network for marine-coastal environment monitoring. *Sensors* 17 (3): 1–22.

11

Sensors for Water Quality Assessment in Extreme Environmental Conditions

Priyanka Ganguly

Chemical and Pharmaceutical Sciences, School of Human Sciences, London Metropolitan University, London, UK

11.1   Introduction

The rise in population and the ever-increasing human needs have escalated the industrial growth over the past century [1, 2]. This also directly contributed toward building pressure on our water resources. Therefore, water quality assessment has in turn become an important portfolio under the water resource management programs [3, 4]. Different countries in the western world monitor the water quality based on their nation legislation and having monitoring stations built at multiple destinations. Some of the basic parameters collected are the pH, flow rate, turbidity, conductivity, etc. Table 11.1 summarizes the parameters used by different water companies in the United States, the United Kingdom, Australia, Belgium, and Netherlands [5].

In turn, the need for accurate data and long-lasting data acquisition is critical to design any effective remediation or preservation programs. Thus, in terms of data acquisition and storage, the need for labor intensive on-site sampling was prevalent for last several decades [6, 7]. This process involves higher operational cost of transport to remotest locations and the toll on manual labor. Moreover, the absence of standard protocol for data gathering and the missing standards universally has not helped the case. This eventually realizes in compromising the standard and the quality of the data gathered. Thus, to eradicate these challenges, as well as to facilitate high temporal and spatial resolution monitoring, the development of instrumentation system available at proximity and providing continuous analysis has emerged. Thus, to engage in this critical process of data collection,

Sensing Technologies for Real Time Monitoring of Water Quality, First Edition.
Libu Manjakkal, Leandro Lorenzelli, and Magnus Willander.

Table 11.1 Top 10 parameters monitored online by drinking water companies in the United States, Belgium and the Netherlands, the United Kingdom and Australia.

Rate	Parameter USA (%) $n = 52$	Parameter B and N (%) $n = 52$	Parameter UK (%) $n = 52$	Parameter Australia (%) $n = 52$
1	Flow rate 100	Flow rate 100	Flow rate 100	Flow rate 100
2	Turbidity 89	Turbidity 100	Turbidity 100	Turbidity 100
3	pH 79	pH 90	pH 100	pH 100
4	Water temperature 77	Water temperature 80	Water temperature 86	Water temperature 100
5	Conductivity 39	Conductivity 60	Conductivity 72	Conductivity 83
6	Particle count 37	Particle count 30	Nitrate 57	Particle count 83
7	Oxygen 17	Ca/Mg/hardness 50	Pressure 72	Pressure 83
8	Chlorine 14	Biomonitors 50	Chlorine 100	Total Chlorine 50
9	TOC 14	Spectral absorption 30	Oil in water 57	Free Chlorine 100
10	Fluoride 21		Iron 72	Fluoride 83

n, number of water utilities interviewed.

the need for smart sensors to achieve the remote data gathering, monitoring, and storage are paramount [8, 9].

The rising water pollution levels have resulted in looking for various chemical, physical, and biological variables to determine the compliance parameters. Traditional sensing mechanism relied heavily on single probe element, while hosting more than one sensing mechanism in a system would drastically enhance the price of the sensing unit [10]. Printing technologies is revolutionizing the sensing industry by printing sensing materials into miniature scale along with promoting sustainable electronics and low cost of production [9]. Thus, greater public awareness and stricter public health regulations have encouraged the development of portable, low cost, reliable small devices. This essentially opens the need for sensors/detectors to collect data, monitoring for longer duration, and storing the gathered data, so that it could be later collected remotely [5]. These data gathering based on various sensors is critical in assessing several indicators and trends. Significant deviation from the control parameters can raise early warning to the decision makers and provide them with enough time to respond appropriately. This is also critical for remote and smaller communities, where trained personnel are not available and regular upkeep will be challenging [11].

In this chapter, key parameters contributing toward water quality assessment are reviewed and the sensors developed in their monitoring are also discussed.

It also discusses several examples of multisensory modules incorporated with various, physical, chemical, and biological sensors hosted on boats, unmanned aerial vehicles (UAVs), underwater vehicles, and buoys. Thus, aiding in water monitoring even at the remotest environmental conditions. The chapter has been further categorized based on the physical, chemical, and biological parameters.

11.2 Physical Parameters

11.2.1 Electrical Conductivity

The ability of water to transfer electricity (transfer of electrons) defines the electrical conductivity value [12]. The presence of ions aids in the electron transfer process. Therefore, higher the number of ions in the water sample, greater is the electrical conductivity value [13, 14]. The presence of alkali metals, chlorides, sulphides, and even low level of heavy metals contribute toward the conductivity value. While the deionized water or even the distilled water acts as an insulator at many times because of the absence of any ions. The splitting of the molecules into cations and anions increases the ionic content and enhances the conductivity value but overall, the water remains electrically neutral. The presence of higher content of ions in the water would result in higher levels of toxicity and affect the crop growth, when used for farming activities [13, 14]. There are multiple factors affecting the conductivity value, such as temperature, flow and the level of the water, the presence of organic or inorganic content, etc. The increase in temperature results in the increase in the conductivity. Increase in the temperature results in the increase in the ionic mobility and that would result in increase in the conductivity values. Thus, conductivity is measured at standard temperature of 25 °C. While the water flow is one of the most common factors associated with changes in the conductivity values, apart from the increase of pollution levels [13, 14]. Surge in freshwater content, such as from lake or melted ice, etc., would ideally have low ionic content and thus decline in the conductivity value is more evident. On the other hand, the presence of ground water increases the overall ionic content of the water source. Similarly, the water level also plays an important role in defining the overall conductivity. Increase in evaporation results in decline in the water level and therefore increasing the ionic content in the water and therefore increasing the conductivity. While when rain falls, the water level increases and the conductivity declines [15].

There are two major kinds of conductivity sensors: (i) electrode-based sensors and (ii) toroidal or inductive sensors. Figure 11.1 displays a schematic representation of these two different types of conductivity sensors. The sensors based on electrode consist of two, three, or even four electrodes. The conductivity

(a)

(b)

Analyzer measures
Analyzer generates   the current induced
AC drive voltage   in the sense coil

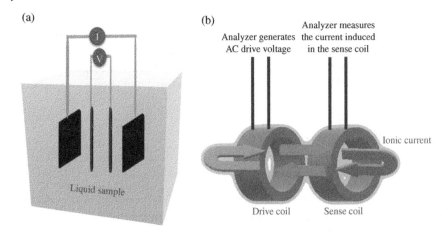

Ionic current

Liquid sample

Drive coil   Sense coil

Figure 11.1 Schematic representation of (a) four electrode-based and (b) toroidal-type conductivity sensors.

measured (σ) is directly proportional to the conductance ($1/R$). The proportionality coefficient (K) depends on the geometry of the sensor which need to be devised corresponding to the proposed conductivity range.

$$\sigma \propto \frac{1}{R} \tag{11.1}$$

$$\sigma = K \frac{1}{R} \tag{11.2}$$

These sensors are ideal for lab setup; however, the metal plates corrode in the presence of corrosive electrolyte and thus not recommended to be used in industrial setup for long time. The inductive or toroidal sensors usually comprise of two coils, sealed within a nonconductive housing. The first coil induces an electrical current in the water, while the second coil detects the magnitude of the induced current, which is proportional to the conductivity of the solution.

There are two kinds of inductive-based conductivity sensor: eddy current type and the transformer type. The magnetic permeability of the solution is governed by the concentration of the electrolyte. This is the principle behind the eddy current-type sensor.

The conductivity of underground water was measured by Julius et al. using an eddy current-type conductivity sensor. The sensor consists of two solenoid coils employed at a frequency above 100 kHz to measure the conductivity. The operational range of these sensors was observed to be between 0.585 and 73.8 mS/cm [16].

Apart from them, printed sensors for conductivity measurements for WQM are not available in common. In one of the works, the authors used gold etched coated glass slide to achieve interdigitated gold electrodes. Copper wires were used for the connection pads. Later, the entire structure with the electrodes and the connection is covered with polydimethylsiloxane (PDMS) to prevent water coming directly in contact with the electrodes. The sensor displayed high repeatability and robustness for conductivity measurements. They were tried at static condition (i.e. no water flow) and dynamic condition. The sensor displayed no significant variation in the conductivity measured at varying flow rate of water (3 and 30 ml/min). The sensor was placed in various solutions with varying degree of conductivities for 30 days. No change in response of conductivity measurement was observed [17]. A low-cost, miniaturized multichannel conductivity sensor was fabricated using micro-USB as the sensing element with gold coating on the connector pins. The probe demonstrated conductivity within the given range of 0.1–15 S/m, with an accuracy of 1%, a resolution observed to be 0.1%, and sampling within 11 ms. However, the authors do not comment anything about the power consumption of the device [18]. Another miniaturized sensor chip was fabricated using Micro-Electro-Mechanical System (MEMS) technique. The device is proposed to perform synchronous measurement of multiple parameters such as pH, conductivity, and temperature of water. Iridium oxide is used as the pH sensing material and a four-terminal conductivity cell with sputtered Pt was used to measure the temperature and the conductivity measurement. The pH sensing displayed effective response within the range of pH 2–11. The conductivity range was observed to be in between a linear determination range from 21.43 μS/cm to 1.99 mS/cm, and the sensitivity of the temperature sensor was measured to be 5.46 Ω/°C. The MEMS sensing device is quite promising, but the power consumption and the accuracy of the measurement platform do not satisfy the desired conditions [19]. There are few reports demonstrating real-time water monitoring using multisensory modules assembled through commercially available sensors [6, 7]. A wireless sensing network based on ISO/IEC/IEEE 21451 sea water probe for monitoring multiple parameters intended for sea WQM such as temperature, salinity/conductivity, turbidity, and chlorophyll-a concentration which was used as biological indicators of water eutrophication. The standard helps in following a definite transparent protocol available for free to be compared, evaluated, and understand the sensor architecture and assembly. The sensory module was used to capture possible severe events and collect extended periods of data [20].

The are several drawbacks of the conductivity sensor. Among them, the value of electrical conductivity is only relevant to access the increase in the pollution levels. It does not provide any insight over the increase of any individual ionic concentration.

11.2.2 Temperature

The temperature is another physical parameter essential to be monitored. Temperature of water is critical for crop growth and can impact the irrigation pattern at a large scale [21]. There are in general four types of temperature sensors [22]. At first is a Negative Temperature Coefficient (NTC) thermistor, as the name suggests a thermistor displayed change in resistance with the change in the temperature. Thus, being called a negative temperature thermistor indicated that as the temperature decreased the resistance increased [23]. The change observed is usually exponential and thus limit its usage to a certain operating temperature. Polymeric and ceramic materials are some of the commonly used thermistor materials. The second type is a Thermocouples, which consists of two different wires joined together. The change in the temperature arises a potential to develop between the metals [24]. This phenomenon is related to *Seebeck effect*, and the voltage generated is proportional to the change in temperature observed. This type of temperature sensor operates at a widest range possible from −200 to 1750 °C. However, the accuracy reported for such sensors are low and often require a lookup table for analysis. The third type is a resistance temperature detector, it is like a thermistor. However, these sensors are basically made up of platinum and consist of two, three, and four wire configurations. The response for these sensors is slow compared to thermocouples. However, they show linear change in resistance measure with the change in temperature, displaying accurate, stable, and repeatable response. The fourth and the final type are the most used temperature sensor made up of semiconductor. The semiconductor-based temperature sensor is commonly incorporated with integrated circuits (ICs) [8]. These sensors employ two same diodes with temperature-sensitive voltage vs current characteristics that are used to observe changes in temperature. A linear response is observed in these sensors but display the lowest accuracy among all the other sensor types. They also display response across a narrowest temperature range (−70 to 150 °C).

There are several multisensory modular set up displaying temperature sensor used for water quality assessments. These different types of sensors are usually observed to be coupled with other sensors such as pH, conductivity, humidity, etc. Poly(3,4-ethylenedioxythiophene) poly(styrene sulfonate) PEDOT:PSS in one of the most commonly used conducting polymer solution for various applications. The change in the resistance observed with the increase in temperature, their impressible mechanical properties make them promising candidate for repeatable and large-scale production. In one of the works, the authors inkjet printed electrodes with PEDOT:PSS ink that is electrically contacted by an interdigitated electrode (IDE) structure using silver nanoparticle (AgNP) ink on a flexible polyethylene terephthalate (PET) substrate (Figure 11.2). The PEDOT:PSS acts as the

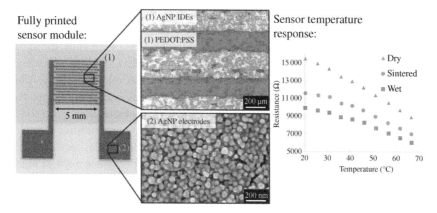

Figure 11.2 A fully inkjet printed sensor with magnified images displaying the microscopic image and the SEM image of the finger structure and the change in resistance observed for various post treatment of the sensor surface with deposition of Ag nanoparticle. *Source:* From Ref. [25]/MDPI, CC BY 4.0.

sensing layer for temperature and humidity sensing. The authors altered the finger dimensions (150–200 μm) and introduced the post deposition treatment (dry/wet/sintered) with Ag nanoparticles to study the influence in the efficiency of their performance. Greater the distance between the fingers resulted in decreased thermal sensitivity. While the deposited layers showed interesting attributes. The dryer samples showed better response compared to the wet and sintered samples [25].

The resistance temperature detector-based ocean monitoring system was developed. Platinum-based electrodes were fabricated using microfabrication technique to create a conductivity/temperature sensor. The sensor had an operational sensitivity of ±0.03 mS/cm and ±0.01 °C for conductivity and temperature, respectively. The multisensory unit is powered by a 9 V/500 mAh PP3 battery and it expends 9.2 mA current. While being inactive, current of 200 uA was consumed. The battery life of the system was up to one month at a sampling time of 10 seconds [26]. In an effort to find solutions for underwater temperature monitoring, 3D printed structures embedded onto the surface of the robot can be easy in terms of integration. The use of polylactic acid (PLA) as the polymer to print electrode structures using fuse deposition modeling (FDM) is reasonable. PLA is insoluble and insensitive to water, and thus making it functional for both terrestrial and aquatic environments. The addition of about 32% of graphene nanorods as fillers in the filament results in effective display of the temperature sensing behavior. The increase in temperature results in the increase of the microscopic distance between the PLA and the graphene, thus resulting in the increase of the resistance. While when the temperature drops down, the resistance drops down as well. The sensor remains functional between the temperature range of 0–70 °C; beyond this value, the structural deformation remains permanent. Thus, the ideal temperature range where the sensor shows repeatable, linear, and stable response is between 0 and 70 °C [27]. The use of graphene has been exploited effectively in designing a flexible graphene-skin with multisensory module for sensing applications in harsh environment conditions. Flexible polymer substrates such as PDMS and polyimide (PI) were used to fabricate sensors using transfer-free flexible single-layered graphene on the polymer and laser scribing-based laser-induced graphene. The temperature sensor displays effective performance even at high temperature up to 650 °C. However, this is only limited by the choice of substrate, altering it with high temperature stable material such as silicon carbide and gallium nitride can effectively increase the potential use if these sensors are at a higher temperature. The single-layer graphene and the laser-induced graphene display unique behavior at different temperature zones. In temperatures up to <210 °C, they display a linear behavior, while at higher temperature, the sensors display thermistor-like property. The sensitivity observed for the linear region was found to be about 260% higher than standard Pt-based resistance temperature detector [28].

11.2.3 Pressure

The need for pressure sensor for water monitoring has expanded across various domains [29]. Conventionally, pressure sensors across pipelines or water storage units were used to monitor the water level. However, as the technology has expanded, the use of pressure sensors has also seen its usage in navigation. Incorporating various kinds of lateral soft sensors across fish-like under water vehicles (UWV) are some of the latest advances observed with regards to pressure sensor. There are three most observed methods to measure the pressure: (i) Absolute water pressure, it is measured against zero and it operates similar to a gas pressure sensor which measures gas pressure against vacuum. (ii) Gauge pressure measurement technique, where the water pressure is measured against the atmospheric pressure across the sensor. Being completely drowned inside water, the pressure sensor has a vent line that allows the flow of air from the water surface and thus providing a reading against the atmospheric pressure. The vent line is usually connected along with the power cable connecting the sensor. (iii) Differential pressure, where the pressure is monitored across two different water bodies such as water tanks or pipes. Thus, change in the water levels across the two bodies enable in measuring the pressure as well as the flow rate in case of water pipelines.

Biomimicking has been the key to resolve various human challenges. The use of printed configuration in the form of skin-like structure embedded with multi-sensory module onto the surface of UWV has been discussed in previous sections. UWV mimicking the translational motion of a fish has been the goal to achieve for deep water research as well as designing robotic units for water monitoring purposes in extreme aquatic conditions. A MEMS-based array of pressure sensor was fabricated using liquid crystal polymer. The liquid polymeric material bestows flexibility and shows strong resistance against harsh sea water environment and moisture. The sensors can be utilized to sense underwater objects by sensing the pressure difference created underwater by the movement of objects. The sensor displays strong ability to detect the velocity of underwater objects pulled beside the system with high accuracy, and an average error of only 2.5%. The sensors demonstrate a pressure sensitivity of 14.3 µV/Pa and a high resolution of 25 mm/s is attained for water flow sensing [30]. In a similar attempt to improve the yaw control and the ability of an UWV, the angle of attack was studied. In this case, five off-the-shelf piezoresistive sensor packaged in a thermoplastic were used as the pressure sensor in the UWV. The pressure range could be monitored within a range of −7 to 7 kPa, at a temperature of 0–85 °C with an error range of maximum 5%. Since the sensing unit is not waterproof, the unit is inbuilt inside the vehicle and the pressure value is monitored from the surface ports via thin tubes filled with air [31]. The use of laser-induced graphene for various electronic devices and sensors has been observed widely in the past decade. The ability of 3D porous graphene to fabricate piezoresistive large range, mechanically flexible, lightweight, and robust pressure sensors is unmatchable (Figure 11.3a and c). Optimizing the geometry through laser patterning results in tuning the pressure-transducing abilities. The sensitivity of the sensor is measured to be in the range of 1.23×10^{-3} kPa and shows a high resolution of 10 Pa. PMMA coating on the sensor is useful in studying the sensor for underwater pressure monitoring at a depth of even 2 km. The increase in pressure brings the graphene layers closer to each other, thereby increasing the conductive pathway (Figure 11.3b). Therefore, the increase in pressure results in decrease of resistance value. The converse is observed when the pressure is decreased or released back to its previous state. The low young modulus ($\simeq$40 kPa) value of the graphene sheets signifies the change in resistance observed is due to the number of turns and dimensions of the meander-shaped LIG electrodes. Therefore, enabling the tailoring process of the pressure sensor geometry controls the pressure sensing applications. Figure 11.3d displays the underwater simulation of the pressure sensor; the results indicate no apparent changes in the morphology of the pressure sensor even when immersed at a depth of about 2 km [32].

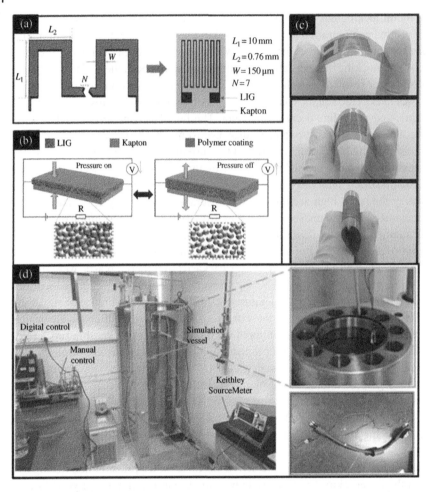

Figure 11.3 Schematic representation of (a) the design of the laser-induced graphene-based pressure sensor, (b) the operating principle of the pressure sensor, (c) photographs of the pressure sensor displaying the flexibility, and (d) photograph of the experimental condition for the high-pressure simulation and real-time testing in the harsh seawater environment. *Source:* From Ref. [32]/John Wiley & Sons, CC BY 4.0.

11.3 Chemical Parameters

11.3.1 pH

The measure of hydrogen cation concentration is defined as pH. It is one of the important parameters to be monitored in order to access the water quality. Change in pH values in our water bodies can cause serious implications on aquaculture, agriculture, crop selection and rotation, food processing, commercial usage, and so forth. Therefore, monitoring the change in pH value is critical and there are several types of pH sensors available in the market for varied range of applications. Simplest pH sensor known would be a litmus paper, which provides a colorimetric response within a small duration. While, to measure pH values accurately and at different conditions, there are several sensors designed. There are several factors crucial for such commercial pH sensors such as the range, sensitivity, portability, cost effectiveness, operational lifetime, etc. The sensing mechanisms could be broadly classified as (i) optical and (ii) electrical pH sensors. Optical pH sensors determine the change in pH through colorimetric observation conducted through various means, for example, pH image sensor, optical fiber pH sensor, etc. On the other hand, the electrical pH sensing mechanism relies on change in potential, resistance, etc., which is observed through various potentiometric, chemi-resistive, ion sensitive field effect transistors (ISFET)-based pH sensor, etc. There are multiple review articles dedicated to the discussion of the basic sensing mechanism used by these various types of pH sensors [33–36]. Thus, in this section, some of the recent sensory modules using these mechanisms for water monitoring is discussed.

The use of multisensory module has been already cited previously in various discussions. Gotor et al. reported an optical pH sensor based on boron–dipyrromethene core as a fluorescent probe for measurement of the entire pH range of 0–14 [37]. Boron–dipyrromethene or BODIPY dyes along with the reference dyes were converted to a hydrogel spot on a test strip. The plastic test strip on its flip side indicated the pH with high precision (uncertainty of less than 0.1 pH). A 3D printed case embedded with the test strip is attached along with the smartphone. The USB port provides the excitation required for the integrated LED which supports the autonomous operation of the sensors in remotest location. An android-based app was designed, which further aids in operation of the sensor even by untrained users. Figure 11.4a displays a schematic illustration of the pH sensor developed by the authors [37].

In one of the recent reports, the authors display a multisensory structure used to monitor pH, temperature, free chlorine, and even harmful pharmaceutical and heavy metal components. All sensors were fabricated on two glass substrates (Figure 11.4b). The pH sensor used palladium/palladium oxide (Pd/PdO) ink as the sensing electrode and Ag/AgCl ink was hand drawn as the reference electrode. A free chlorine sensor was fabricated on the same glass substrate by electrochemically modifying a hand-drawn carbon electrode and an Ag/AgCl reference electrode. The pharmaceuticals, acetaminophen (APAP) and 17β-estradiol (E2), and heavy metal, lead (Pb), sensing were achieved by gluing a single carbon electrode on the same glass substrate. This carbon electrode was further modified by functionalized multiwalled carbon nanotubes and β-cyclodextrin (MWCNT-βCD).

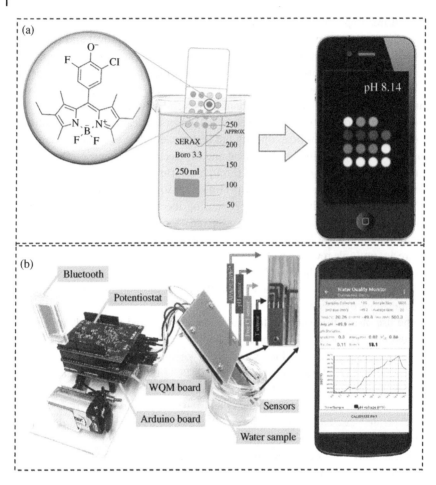

Figure 11.4 (a) Schematic illustration of fluorescent pH probe using boron–dipyrromethene core as a fluorescent probe for measurement of the entire pH range of 0–14. *Source:* From Ref. [37], Reproduced with permission from American Chemical Society. (b) A schematic illustration of a fully integrated electrochemical sensor array for in situ WQM. *Source:* From Ref. [38], Reproduced with permission from American Chemical Society.

The second glass substrate was used to develop a Wheatstone-bridge-based temperature sensor. A silicon wafer was glued on the surface of the glass substrate and PEDOT:PSS ink was used to hand draw to obtain an electrode. Later, the two glass substrates were connected to an Arduino-controlled printed circuit board which wirelessly (Bluetooth low energy) provided data to an android-based app for real-time monitoring of the water samples. While free chlorine sensitivity was observed to be 186 nA/ppm, and the temperature sensor displayed sensitivity of 16.9 mV/°C. While the detection limit for on-demand monitoring of acetaminophen and 17β-estradiol was <10 nM and heavy metal was <10 ppb. The simultaneous monitoring of multiple parameters in water sample makes this system interesting and paves way for several such low-cost system to be developed [38].

Report of portable multisensory module has been gaining attention widely. Multiwalled carbon nanotube (MWCNT)-based ink was used to 3D print IDEs on PDMS substrate to develop a low power, cheap, nitrate, sulphate, and pH sensor. The sensor resistance value changed across different frequency sweep when exposed to different analytes. The fabricated sensor was reported to distinguish between nitrate and phosphate ions even at a lower concentration (from 0.1 to 30 ppm). The sensor displayed substantial variations toward nitrate solutions for the whole range of frequency 10 Hz to 100 kHz, while for phosphate, the response is observed at a range of 10 Hz to 10 kHz. While, pH changes displayed the significant changes between 10 Hz and 300 kHz. The pH range was monitored between 1.71 and 12.59 and the reported measurements were measured across temperature range of 0–45 °C. The sensitivity for temperature, nitrate, phosphate, and pH level are measured to be 1.1974 Ω/°C, 1.9396 Ω/ppm, 0.8839 Ω/ppm, and 1.0295 Ω, respectively. The portable sensor equipped with an Arduino-based system was used for real water sample analysis and displayed precise monitoring with about 5% error to the actual measured value [39].

Thus, in case of pH sensor, the response at wide range of pH at low cost and measurement at extreme conditions is extremely difficult to attain. However, there are reports of multisensory module employing printed electrodes for pH measurements at a wide range of detection [6, 7, 20].

11.3.2 Dissolved Oxygen and Chemical Oxygen Demand

Dissolved oxygen (DO) is one of the important water monitoring indicators along with the chemical oxygen demand (COD). Oxygen gets dissolved in the water source through the sway of wind on the surface of the water or even gets introduced as a by-product of aquatic photosynthesis. The value of DO and COD helps in measuring the quality of the water, as the presence of oxygen in water is critical for the survival of aquatic ecosystem and several microorganisms. The value of DO and COD gets affected when untreated water gets discharged into the water bodies, leaving the level of organic matter high. Higher the value of COD and DO, serious is the pollution levels of the water body. It refers to the amount of oxygen required for the decomposition of organic species and needed by the biological species in water. Traditionally, COD values are measured using two ways, COD_{Cr} and COD_{Mn}, where COD is determined using $K_2Cr_2O_7$ and $KMnO_4$, respectively. Sewage and wastewater bodies are monitored using COD_{Cr} while comparatively cleaner water bodies are assessed using COD_{Mn}. The drawbacks of these traditional techniques are the use of toxic reagents and longer testing time. However, many alternative detection methods such as microwave-assisted digestion methods [40], ultrasound-assisted digestion methods [41], spectrophotometry [42], chemiluminescence methods [43], and flow injection analysis [44] have significantly shortened the detection time. They cannot limit the use of toxic reagents such as sulfuric acid, silver sulphate, etc. Thus, alternative techniques such as photocatalysis, photoelectrocatalysis, ozonation, etc., have been utilized for low cost, quick, and accurate results using radicals (hydroxyl OH^{-*}) to measure the content of organic species [45–47]. In this section, few examples of COD sensors are highlighted using a combination of several of the abovementioned techniques.

Integrating flow injection analysis with multiple other techniques such as photoelectrocatalysis [10], thermal sensing [48], and even optical sensing [49] has enabled in improving the sensing of COD values at a much quicker rate and evading the use of traditional toxic reagents. However, these units are bulky and require numerous components to achieve a single probe to be completed. This is being improved by integrating printing methodology, as explained in previous sections by citing several examples of using printed structures with electrodes being used to probe multiple parameters simultaneously. The commercial advantage of using such multisensory modules equipped with printed array of electrodes enable to perform synchronous detection is huge. The easy production and ultralow cost ($0.2/sensor) makes them extremely attractive venture [50]. Figure 11.5 displays an example of a micro electrode array of device used for temperature, conductivity, DO and pH measurements. These electrodes were fabricated using inkjet printing technique. The use of Kapton tape as the substrate provides unique advantage of flexibility, robustness, and amenability to be integrated across various curved/uncurved surfaces.

The pH and the DO sensors have a three electrode configuration with gold ink used as the working and the counter electrode, while the silver ink is used as the reference electrode. In case of pH sensor, the working electrode is modified by iridium oxide (IrO_2). The conductivity sensor is a four-electrode array of gold printed sensor which measures the resistance of the water sample, which indirectly enables in calculating the dissolved salt in the sample. The temperature sensor is also an array of printed gold electrodes which changes its resistance with increase/decrease of the temperature. The sensing unit was even trialed for more than four weeks in wastewater and waste sludge. The surface of the unit remained intact and displayed excellent robustness and high stability for long-term applications [50]. Autonomous sensing has been the key element for sensors to be deployed at hostile and remote locations. The combination of energy harvesters along with storage solutions has been challenging in integration process. However, in several new advancements, the energy harvesters are also used as sensors. Such applications have gained extensive attention recently for wearable applications.

(a)

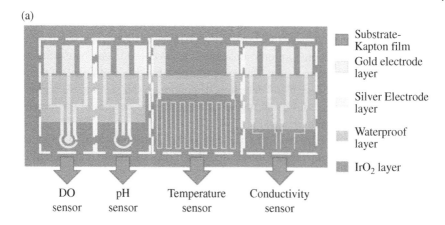

Substrate-
Kapton film

Gold electrode
layer

Silver Electrode
layer

Waterproof
layer

IrO_2 layer

DO	pH	Temperature	Conductivity
sensor	sensor	sensor	sensor

(b)

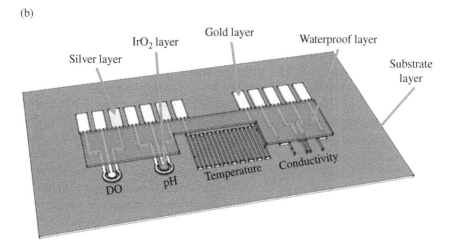

Figure 11.5 Schematic illustration of (a) DO, pH, temperature, and conductivity sensors and (b) the layers on each sensor. *Source:* Reproduced with permission of Ref. [50]. © 2016, Elsevier.

In a similar attempt, a microbial fuel cell is used to produce energy as well utilized as a sensor to measure the COD values [51]. A microbial fuel cell utilizes the oxidation of the organic content present in the wastewater medium at the anode through bacterial respiration, along with the reduction of oxygen at the cathode. Thus, generating electricity as well as contributing to amperometric sensing application, the amount of electricity generated could be directly related to the organic content (COD value) in the water sample [51]. Thus, this provides a complete example of an autonomous nature of a multifunctional system.

11.3.3 Inorganic Content

Apart from heavy metal detection, there are wide range of inorganic content such as pharmaceutical effluents, anionic molecules such as chloride, sulphate nitrates, etc., which are present in water. There are multiple reviews and chapters dedicated exclusively for heavy metal detection. Therefore, in this section, we explore the commonly detected inorganic species and the sensors developed for remote and extreme environment detection.

Nitrate/nitrite sensing is crucial to understand the nitrogen cycle in the aquatic system. Spectroscopic method of nitrate detection relies on nitrate reduction into nitrite. Several reduction techniques involve by reduction by hydrazine, zinc, and even irradiation of UV light. Nitrate detection using Griess assay as miniaturized sensors developed through microfluidic technology was reported [52]. Microfluidic technology overcomes several challenges observed for macroscale systems such as low power supply required for the functioning of the system. Reduced size and low need of reagents makes them cheaper to build. These kinds of fundamental advantages allow the sensors developed to be deployed at remote locations. The system developed show high sensitivity (0.025–350 μM) requires low power requirement (1.5 W) [52]. The need for autonomous system and its need for underwater water monitoring have been explained previously. The autonomous system designed through SALMON (Sea Water Quality Monitoring and Management) research project provides an example of such system [53]. The autonomous underwater vehicle used to mount the sensor system and the other essential unit such as power supply, navigation, propulsion, etc., was from Fraunhofer IOSB-AST. The vehicular design is as shown in Figure 11.6a. The miniaturized sensory system as the payload for the vehicle was equipped with nitrate (43–2000 μg/l), oxygen (0–500 μmol/l), conductivity (0–75 ms/cm), temperature (−5 to 40 °C). The computer aided design (CAD) diagram of the sensor system is shown in Figure 11.6b. The nitrate measurement is done through a spectrometric technique (Figure 11.6c), by irradiating water passed into a cuvette through a hose system. A deuterium lamp irradiates UV radiation, and the absorption measurement is recorded to detect the nitrate levels. Similar to this, in another instance a sensory unit boarded on boat was used to measure multiple parameters such as temperature, pH, nitrate, and conductivity (Figure 11.6d).

The nitrate measurement is done through a similar UV-based spectroscopic technique. The boat is used to collect data in every 15 seconds from Iowa and Cedar Rivers [54]. Fixing such multisensory nodes on traveling and fisheries boats could be used to provide complete information across different parts of the water

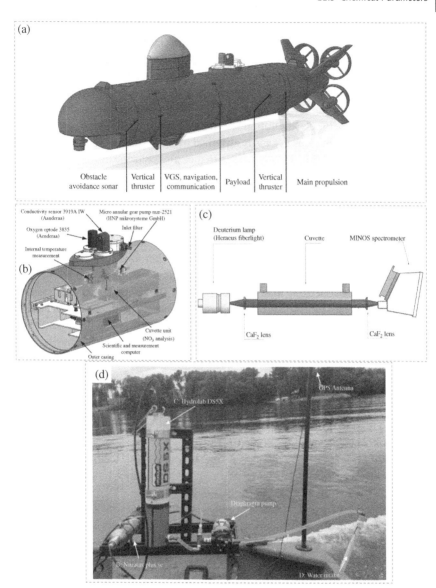

Figure 11.6 Schematic illustration of (a) underwater vehicle CWolf from Fraunhofer IOSB-AST, (b) the computer aided design (CAD) diagram of the sensor system, (c) nitrate determination unit, and (d) picture of a boat with nitrate monitoring sensors added for real-time monitoring in river. *Source:* From Refs. [53, 54]/MDPI, CC BY 4.0.

body. However, the deployment of such sensory units and their operability is often expensive, and time and again results in loss of several nodes.

Free chlorine is another one of the most prevalent anionic species found in water flowing out of water treatment center. It acts as a disinfectant for inactivation of several microbial species such as *Escherichia coli* and *Cryptosporidium*. Therefore, the amount of free chlorine to be discharged out of water should be below 5 ppm. Smaller concentration is essential for crop growth, but they can be toxic at higher levels. They get easily absorbed by plants and could result in dried tissues of the leaves. The older technique to measure chlorine in water samples is using argentometric or Mohr's method. It is one of the titration techniques, which is prone to greater number of errors and depends on calorimetric end points. However, recent technological advances have paved the way for portable, robust, and low-cost sensors. The common problems encountered by these sensors are their limit of detection and hysteresis of the detection pattern on repeated experiments. Apart from them, interferences caused by other ions such as nitrate, sulphate, etc., tend to cause issues with the detection system. There are several reports of expensive sensing materials such as gold, boron-doped diamond, glassy carbon used as free chlorine sensor. Similarly, there are other hazardous materials such as aniline and oligomers used as sensing materials. The chemical tendency of free chlorine reacting with amine groups on cyclic molecules has arose the use of carbon-based materials such as carbon nanotube and graphene as sensing materials. In one of the work, carbon nanotubes were used as the sensing material. The single wall carbon nanotube (SWCNT) is functionalized using phenyl capped aniline tetramer (PCAT-SWCNT). The device is built of two parallel gold electrodes deposited by photolithography and sputtering. The electrodes are further connected through PCAT-SWCNT methanol suspension. Finally, a microchannel of PDMS was bonded via soft lithographical technique and the PCAT-SWCNT suspension is flowed through the microchannel. The device displayed detection within the range of 60–0.06 mg/l [55].

In another report, a commercially available 2B pencil lead which is just a simple graphite was further electrochemically aminated to be used as a sensing material to detect free chlorine. In this work, chronoamperometric technique was used for the determination process. The sensitivity measured was in the range of $0.302\,\mu A/$ ppm/cm^2 and the presence of other interfering ions such as chloride, nitrate, and sulphate do not show any change in the response. Thus, displaying the selectivity of such low-cost sensor, however, the present system faces issue related to miniaturization and remote monitoring [56].

The present pencil lead-based graphite sensor could be integrated with other sensors as multinodular system to detect free chlorine, as discussed in the previous section. Printed graphite ink and further aminating could be integrated with other printed structures.

11.4 Biological Parameters

Biological safety and stability of water is extremely crucial for both drinking water and industrial water requirements. Uncontrolled growth of bacteria and other pathogens could be extremely dangerous hygienic problems. This leads to issues in fouling, change in the taste and smell of the water. Water with high proportions of bacteria, fungi, etc., could cause serious health-related challenges in the community. Therefore, the need to understand the number of biological species of various forms is critical for WQM. The water used in industries need to be of pristine standards, as use of them to develop sensitive products could tamper the quality of the product developed. There are varied ranges of pathogens trying to grow in the limited scale of nutrient available. The present biosensing techniques rely on detecting definite proteins or enzymes released by the microorganisms. The selective release of these enzymes and the protein interaction by the sensing material enables in qualitatively detecting their presence in the water sample [57]. However, integrating electronics along with the sensor results in precise identification as well as quantification. In one of the report, a multisensory chip has been used for simultaneous detection of enzymes, pH, cell density, etc., for biotechnological process, such as in bioreactors. A similar design could be utilized for water monitoring aspect. The chip has been fabricated using Complementary Metal Oxide Semiconductor (CMOS) technique. Figure 11.7a displays the fabricated chip with four individual sensors. pH measurements were done by a top layer of Ta_2O_5 and the electrodes developed of gold and platinum. The presence of microbial cells could be detected by simply measuring the impedance using the IDEs. The live cells could be polarized in electric field while dead cells show no polarization because they devoid the presence of unbroken cellular membrane. This technique is known as dielectrophoresis (DEP). The time-averaged DEP force is represented as [60]

$$\langle \text{FDEP} \rangle = 2\pi \varepsilon_m \varepsilon_0 \varepsilon a^3 \, \text{Re}\{K\} + \nabla \left| E_{\text{mrms}} \right|^2 \tag{11.3}$$

where ε_m is the relative permittivity of the suspending medium, ε_0 is the vacuum permittivity, a is the radius of the cell, Re$\{K\}$ is the real part of the Clausius–Mossotti (CM) factor K, and $|E_{\text{mrms}}|$ is the magnitude of the electric field. The effective polarizability between the cells and suspending medium is

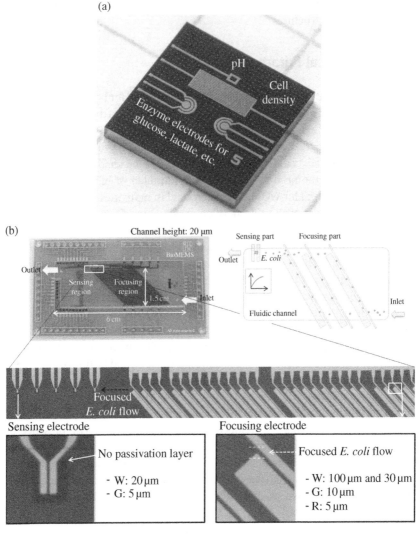

Figure 11.7 (a) Photograph of multi-sensor chip, chip size 7.16 mm × 7.16 mm and (b) the schematic illustration of the microorganism sensor. The sensor consists of a region that utilizes positive dielectrophoresis (pDEP) for *E. coli* focusing, and an *E. coli* sensing region that employs dielectrophoretic impedance measurements. *Source:* From Refs. [58, 59], Reproduced with permission from Elsevier.

governed by the CM value. Greater the polarizability of the cells than the medium, the greater they travel toward the higher electric field zone, also known as positive DEP force. While cells that display low levels of polarizability than their medium would end up traveling to the lower electric field region, also known as negative DEP (nDEP) force.

Combining impedance measurement with DEP has been used extensively to detect microbial contamination in water. However, there remains a catch, the polarization occurs till it reaches a certain frequency, which is characteristic to the cell type. Once exceeding the characteristic frequency, the polarization of the cells no more aligns. Thus, measuring the impedance below the characteristic frequency enables in estimating the cell density. There are several other crucial factors to be taken into account for using impedance spectroscopy for measuring the cell density with a defined electrode [58].

Using the dielectrophoretic impedance measurement technique, another work was reported to develop a biosensor to detect the presence of *E. coli*. The device uses DEP force-mediated *E. coli* to focus and then further uses the dielectrophoretic impedance measurement to sense the *E. coli*. The focusing electrodes consist of IDE pairs aligned at an angle of 45° with the fluidic channel. The widths of the IDE were measured to be 100 and 30 µm and a gap of 10 µm (Figure 11.7b). While the sensing electrode had a width of 20 µm and a gap of 5 µm. The channel length, width, and height were 51 mm, 15 mm, and 20 µm, respectively. The designed system displayed low limit of detection, high-throughput, and label-free operation. The authors observed a detection limit of 300 CFU/ml and the sensitivity of the instrument could be improved by altering the geometry of the electrodes [59]. Park et al. in a similar instance report a smartphone-based *E. coli* detection system using microfluidic paper. A three-channel paper chip is used to detect *E. coli* at low and high concentrations. One of the channels is loaded with bovine serum albumin (BSA)-conjugated beads and other two detection channels are loaded with anti-*E. coli*-conjugated beads. The sample water is introduced either by dipping or pipetting and the antigens from the *E. coli* travel via capillary action. The paper fibers enable in filtering the soil/dust and algae. The detection channel with antibody-conjugated beads immunoagglutinate in the presence of *E. coli* antigens, while BSA-conjugated beads do not. The magnitude of immunoagglutination is measured by assessing Mie scatter intensity from the digital pictures. These images could be taken at an optimized angle and distance using a smartphone. The results observed show similar observation with the MacConkey plate results, i.e. the count of viable *E. coli*. Such systems provide portable solution, however, the detection limit is single-cell-level and the total assay time is almost 90 seconds [61].

11.5 Sensing in Extreme Water Environments

Synchronous detection of various physical and chemical parameters was discussed through various examples in the previous sections. In this section, we discuss the challenges and the vision for water monitoring systems for extreme environment.

The paramount observation to be made is the choice of the sensing technique opted for any application. Small, portable, multisensory array-based device are now launched in small boats, buoys, ships, and even on UWV for applications which requires minimal human interactions. Monitoring the same parameters at a water treatment center would require more sturdy, reliable, and long-lasting sensors or detection units. Moreover, hosting several sensing mechanisms comes with its own challenges. The primary challenge resides on integration of the sensory unit to a steady source of power. The key concern lays over the production of high current and a steady voltage to power the device and later store the excess produced energy into an energy storage unit such as supercapacitors. This topic opens its own challenges of integrating synchronous detection, energy harvesters, and energy storage unit into one singular module. There are several examples highlighting such modules for wearable applications, but less has culminated into their usage in water monitoring applications.

Not everything appears ominous in this context as newer technologies are paving way to resolve such crucial gaps in the existing technologies. Among them, use of UAVs or drones has been a great choice for water sampling at various depths of remotest part of water bodies. Figure 11.8 illustrates one such example of water sampling, monitoring, and visualization using an unmanned aerial vehicle [62]. Off-the-shelf components were picked up and assembled with commercially obtained UAV. The water sampling unit was 3D printed and later attached to the UAV. This system demonstrated successful evidence of water sampling at remote location. However, the present system could be improved by integrating

Figure 11.8 Real-time WQM, sampling, and visualization platform using unmanned aircraft system. *Source:* From Ref. [62]/Elsevier, CC BY 4.0.

various optical sensors, collision avoidance sensor, and/or light detection and ranging (Lidar) sensor as highlighted in the work by the authors. Moreover, systems integrated with water monitoring sensors have also been reported in one of the recent works published in the form of a thesis structure. The authors use an off-the-shelf components and sensors to measure conductivity, temperature, pH, and even DO of the water using custom-built hexacopter. The required parts and accessories are obtained through 3D printing technique. Such systems are bulky in nature and integration of individual commercial sensors increases the overall weight of the designed hexacopter [63]. This could be improved by integrating printed structure-based micro sensory modules to analyze multiple physical and chemical parameters.

Moreover, the UAVs suffer some basic challenges of poor GPS reception while in operation at remote location. In addition to that, weather has also been a challenge in operation of UAVs. Wind flow at higher rate and even heavy rain and severe weather events such as hurricanes disrupt the operations and make it challenging for steady usage. Thus, these aspects serve as the challenges to be overcome for future technologies to be built.

Similar to sensing operation above water, underwater sensing applications is extremely complex and challenging. Incomprehensible underwater environment with multiple variable and numerous numbers of sensing parameters makes it an extremely difficult aspect. Therefore, the modalities available to use in air cannot be used as it is for underwater sensing applications. The interaction of the sensing material to the water sample will have a completely different dynamic. Exposed to strong pressure and inconsistent turbidity and ever-changing temperature are some of the basic factors to be considered. The overall communication strategy is required to be altered as electromagnetic waves propagation is tampered in sea water. Even fundamental issues such as packaging of the sensory module become a concern. As much attention need to be rerouted to ensure the steady monitoring as well as safety of the module. Deployment techniques such as under water mobile robotic assistance aid in broadening the scope of deploying multisensory modules. Artificial intelligence (AI) algorithms can enable in effective communication of the data as well as help in forecasting. Combining existing technology such as satellite imagery and time-averaged spatial analysis tool can enable in providing effective forecasting as well as support in precise decision-making [64].

Figure 11.9 displays a futuristic scenario of combining various forms of sensing mechanisms along with various data gathered through different vehicular strategies. Several examples as highlighted in the previous section discuss concepts of UWV and robotic modules in the form of fish. Thus, exploring the combined effort of multiple avenues of sensing various physical, chemical, and biological parameters.

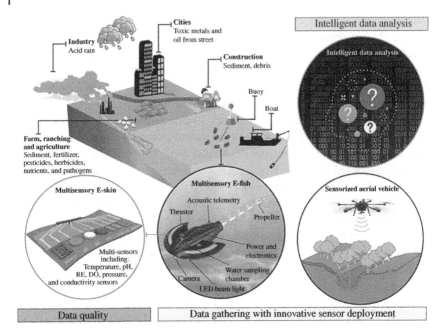

Figure 11.9 Schematic illustration depicting various human activities contributing to the deterioration of water quality and different monitoring, including using multiparametric sensor patches or electronic skin (e-Skin), traditional methods of sensor deployment such as using sensorized buoys, and advanced deployment using underwater robots or multisensory e-Fish. *Source:* From Ref. [11]/IEEE, CC BY 4.0.

11.6 Discussion and Outlook

The above sections provided in-depth understanding of the various available sensors used to measure physical, chemical, and biological parameters. The new improved sensory materials along with unique deployment techniques which are equipped with data gathering and forecasting equipment provide excellent opportunities for environmental monitoring. The network of sensors and the data collected enables various stakeholders to effectively improve in water quality, distribution, and consumption.

Even though there are several technological advances observed in robotics, sensor-based technologies, AI-based advancements, and our improving ability in other technical know-how, but less has translated into a complete system. There still exists serious disconnect between the development of an advance sensory unit to function autonomously in gathering data, communicating the same, and supporting the end user to make sound decision using its forecasting abilities. This enables us to think toward potential solutions starting from developing proper quality assurance procedures for sensor networks. Identifying common parameters for monitoring purposes and henceforth determining the sensors required for the module are some of the initial steps. The choice of the fabrication technique opted for assembling or even fabricating a sensory module would be critical in determining the cost of the overall product. This leads to the next important factor to be borne in mind is the choice of material used in the fabrication process. Sustainable and biocompatible materials as sensing component have been discussed widely in several works previously. Among them, polymers and carbon-based materials have attracted much of the attention. PVC and PET are some of the commonly used substrates for sensors and they require prolong time to degrade and end up as another source of microplastics. Thus, paper or cloth-based substrate has opened up new discussion of sustainable sensing [35, 65]. Similarly, the use of expensive metals such as Au/Ag/Pt and Pd as electrodes provides precise results. However, the cost of the overall unit becomes a challenge for commercialization. In such cases, polymers such as PEDOT:PSS has emerged as best alternative [66].

In addition, operating of such systems in an autonomous nature could be only feasible in the presence of an autonomous power system. As discussed previously, energy independence is critical in operation of such systems in desolated locations. Integrating with solar cells, triboelectric nanogenerators (TENG) as power generators and equipped with battery and supercapacitors to store the energy produced could be the way forward for such systems. Apart from all these factors, the reliability of these sensors is also crucial while in operating at remote location. Therefore, robustness of the sensors produced is crucial. Another factor is the packaging of the overall device structure, as less has translated into effective options available. These all contribute toward the final cost of the assembled device. In low-income countries, manual sampling is the common method and transporting the sample to laboratory setup accounts for the big chunk of the cost incurred. The marginal cost of such testing in the labs is about ~$7.25 per test [67]. Thus is a major influence that is impeding the examining at large scale such as in urban settings. Hence, price of the unit is crucial for commercialization and its wide applicability, as use of robotic system for deployment would increase the price of the system. The use of 3D/4D printing techniques are paving the route to modular structure and also for low-cost deployment and packaging alternatives [68]. With this arises the question of data gathering and data handling. The acquisition rate would determine the amount of data gathered and safely enabling data observations through satellite-based assessment techniques. The advent of AI-based modeling and data acquisition provides newer scope of water quality assessment.

11.7 Conclusion

The present chapter provided a glimpse of the new developing technologies for the water quality assessments. The diverse parameters and the environmental conditions make the sensors developed quite unique and niche in their detection abilities. The robustness, sensitivity, biocompatibility, and the sustainability of the overall system are one of the biggest challenges to overcome. Moreover, the WQM system can thus be summarized into following steps: the forecasting and modeling of the parameters to be assessed, fabricating of the sensors, integration of the sensory module and the power system, devising the deployment module, and integrating with communication unit for data acquisition and transmission. Thus, summarizing the scope of the work is to be done ahead.

References

1 Suyana, P., Ganguly, P., Nair, B.N. et al. (2021). Structural and compositional tuning in g-C_3N_4 based systems for photocatalytic antibiotic degradation. *Chem. Eng. J. Adv.* 8: 100148.

2 Meadows, D. and Randers, J. (2012). *The Limits to Growth: The 30-Year Update.* Routledge.

3 Mathew, S., Ganguly, P., Kumaravel, V. et al. (2020). Solar light-induced photocatalytic degradation of pharmaceuticals in wastewater treatment. In: *Nano-Materials as Photocatalysts for Degradation of Environmental Pollutants* (ed. P. Singh, A. Borthakur, P.K. Mishra, and D. Tiwary), 65–78. Elsevier.

4 Miller, K.A. and Belton, V. (2014). Water resource management and climate change adaptation: a holistic and multiple criteria perspective. *Mitig. Adapt. Strateg. Glob. Chang.* 19 (3): 289–308.

5 Raich, J. (2013). Review of sensors to monitor water quality. *European Reference Network for Critical Infrastructure Protection (ERNCIP) Project, 2013.*

6 Geetha, S. and Gouthami, S. (2016). Internet of things enabled real time water quality monitoring system. *Smart Water* 2 (1): 1–19.

7 Cloete, N.A., Malekian, R., and Nair, L. (2016). Design of smart sensors for real-time water quality monitoring. *IEEE Access* 4: 3975–3990.

8 Huang, Y., Zeng, X., Wang, W. et al. (2018). High-resolution flexible temperature sensor based graphite-filled polyethylene oxide and polyvinylidene fluoride composites for body temperature monitoring. *Sensors Actuator A Phys.* 278: 1–10.

9 Dervin, S., Ganguly, P., and Dahiya, R. (2021). Disposable electrochemical sensor using graphene oxide–chitosan modified carbon-based electrodes for the detection of tyrosine. *EEE Sens. J.* .

10 Chen, J., Zhang, J., Xian, Y. et al. (2005). Preparation and application of TiO_2 photocatalytic sensor for chemical oxygen demand determination in water research. *Water Res.* 39 (7): 1340–1346.

11 Manjakkal, L., Mitra, S., Petillot, Y.R. et al. (2021). Connected sensors, innovative sensor deployment and intelligent data analysis for online water quality monitoring. *IEEE Internet Things J.* 8 (18): 13805–13824.

12 Ramos, P.M., Pereira, J.D., Ramos, H.M.G., and Ribeiro, A.L. (2008). A four-terminal water-quality-monitoring conductivity sensor. *IEEE Trans. Instrum. Meas.* 57 (3): 577–583.

13 Hui, S.K., Jang, H., Gum, C.K. et al. (2020). A new design of inductive conductivity sensor for measuring electrolyte concentration in industrial field. *Sensors Actuator A Phys.* 301: 111761.

14 Glasgow, H.B., Burkholder, J.M., Reed, R.E. et al. (2004). Real-time remote monitoring of water quality: a review of current applications, and advancements in sensor, telemetry, and computing technologies. *J. Exp. Mar. Biol. Ecol.* 300 (1–2): 409–448.

15 Paepae, T., Bokoro, P.N., and Kyamakya, K. (2021). From fully physical to virtual sensing for water quality assessment: a comprehensive review of the relevant state-of-the-art. *Sensors* 21 (21): 6971.

16 Parra, L., Sendra, S., Lloret, J., and Bosch, I. (2015). Development of a conductivity sensor for monitoring groundwater resources to optimize water management in smart city environments. *Sensors* 15 (9): 20990–21015.

17 Banna, M.H., Najjaran, H., Sadiq, R. et al. (2014). Miniaturized water quality monitoring pH and conductivity sensors. *Sensors Actuators B Chem.* 193: 434–441.

18 Carminati, M. and Luzzatto-Fegiz, P. (2017). Conduino: affordable and high-resolution multichannel water conductivity sensor using micro USB connectors. *Sensors Actuators B Chem.* 251: 1034–1041.

19 Zhou, B., Bian, C., Tong, J., and Xia, S. (2017). Fabrication of a miniature multi-parameter sensor chip for water quality assessment. *Sensors* 17 (1): 157.

20 Adamo, F., Attivissimo, F., Carducci, C.G.C., and Lanzolla, A.M.L. (2014). A smart sensor network for sea water quality monitoring. *EEE Sens. J.* 15 (5): 2514–2522.

21 Liu, J. (2009). A GIS-based tool for modelling large-scale crop-water relations. *Environ. Model Softw.* 24 (3): 411–422.

22 Potter, D. (1996). Measuring temperature with thermocouples – a tutorial. *Natl. Instrum.* 340904B-01.

23 Turkani, V.S., Maddipatla, D., Narakathu, B.B. et al. (2018). A carbon nanotube based NTC thermistor using additive print manufacturing processes. *Sensors Actuators A Phys.* 279: 1–9.

24 Bentley, J. (1984). Temperature sensor characteristics and measurement system design. *J. Phys. E Sci. Instrum.* 17 (6): 430.

25 Rivadeneyra, A., Bobinger, M., Albrecht, A. et al. (2019). Cost-effective PEDOT: PSS temperature sensors inkjetted on a bendable substrate by a consumer printer. *Polymers* 11 (5): 824.

26 Huang, X., Pascal, R.W., Chamberlain, K. et al. (2011). A miniature, high precision conductivity and temperature sensor system for ocean monitoring. *EEE Sens. J.* 11 (12): 3246–3252.

27 Sajid, M., Gul, J.Z., Kim, S.W. et al. (2018). Development of 3D-printed embedded temperature sensor for both terrestrial and aquatic environmental monitoring robots. *3D Print. Addit. Manuf.* 5 (2): 160–169.

28 Shaikh, S.F. and Hussain, M.M. (2020). Multisensory graphene-skin for harsh-environment applications. *Appl. Phys. Lett.* 117 (7): 074101.

29 Liu, H., Wang, W., Xiang, H. et al. (2022). Based flexible strain and pressure sensor with enhanced mechanical strength and super-hydrophobicity that can work under water. *J. Mater. Chem. C* 10 (10): 3908–3918.

30 Kottapalli, A.G., Asadnia, M., Miao, J. et al. (2012). A flexible liquid crystal polymer MEMS pressure sensor array for fish-like underwater sensing. *Smart Mater. Struct.* 21 (11): 115030.

31 Gao, A. and Triantafyllou, M. (2012). *Bio-Inspired Pressure Sensing for Active Yaw Control of Underwater Vehicles*. IEEE.

32 Kaidarova, A., Alsharif, N., Oliveira, B.N.M. et al. (2020). Laser-printed, flexible graphene pressure sensors. *Global Chall.* 4 (4): 2000001.

33 Manjakkal, L., Szwagierczak, D., and Dahiya, R. (2020). Metal oxides based electrochemical pH sensors: current progress and future perspectives. *Prog. Mater. Sci.* 109: 100635.

34 Khan, M.I., Mukherjee, K., Shoukat, R., and Dong, H. (2017). A review on pH sensitive materials for sensors and detection methods. *Microsyst. Technol.* 23 (10): 4391–4404.

35 Manjakkal, L., Dervin, S., and Dahiya, R. (2020). Flexible potentiometric pH sensors for wearable systems. *RSC Adv.* 10 (15): 8594–8617.

36 Vivaldi, F., Salvo, P., Poma, N. et al. (2021). Recent advances in optical, electrochemical, and field effect pH sensors. *Chemosensors* 9 (2): 33.

37 Gotor, R., Ashokkumar, P., Hecht, M. et al. (2017). Optical pH sensor covering the range from pH 0–14 compatible with mobile-device readout and based on a set of rationally designed indicator dyes. *Anal. Chem.* 89 (16): 8437–8444.

38 Alam, A.U., Clyne, D., Jin, H. et al. (2020). Fully integrated, simple, and low-cost electrochemical sensor array for in situ water quality monitoring. *ACS Sens.* 5 (2): 412–422.

39 Akhter, F., Siddiquei, H.R., Alahi, M.E.E. et al. (2021). An IoT-enabled portable water quality monitoring system with MWCNT/PDMS multifunctional sensor for agricultural applications. *IEEE Internet Things J.* 9 (16): 14307–14316.

40 Heng, W., Zhang, W., Zhang, Q. et al. (2016). Photoelectrocatalytic microfluidic reactors utilizing hierarchical TiO_2 nanotubes for determination of chemical oxygen demand. *RSC Adv.* 6 (55): 49824–49830.

41 Canals, A., Cuesta, A., Gras, L., and Hernández, M.R. (2002). New ultrasound assisted chemical oxygen demand determination. *Ultrason. Sonochem.* 9 (3): 143–149.

42 Li, J., Tao, T., Li, X.-b. et al. (2009). A spectrophotometric method for determination of chemical oxygen demand using home-made reagents. *Desalination* 239 (1–3): 139–145.

43 Yao, H., Wu, B., Qu, H., and Cheng, Y. (2009). A high throughput chemiluminescence method for determination of chemical oxygen demand in waters. *Anal. Chim. Acta* 633 (1): 76–80.

44 Dan, D., Sandford, R.C., and Worsfold, P.J. (2005). Determination of chemical oxygen demand in fresh waters using flow injection with on-line UV-photocatalytic oxidation and spectrophotometric detection. *Analyst* 130 (2): 227–232.

45 Zheng, Q., Zhou, B., Bai, J. et al. (2008). Self-organized TiO_2 nanotube array sensor for the determination of chemical oxygen demand. *Adv. Mater.* 20 (5): 1044–1049.

46 Ganguly, P., Panneri, S., Hareesh, U. et al. (2019). Recent advances in photocatalytic detoxification of water. In: *Nanoscale Materials in Water Purification* (ed. S. Thomas, D. Pasquini, S.-Y. Leu, and D.A. Gopakumar), 653–688. Elsevier.

47 Nair, K.M., Ganguly, P., Mathew, S. et al. (2021). TiO_2 based Z-scheme photocatalysts for energy and environmental applications. In: *Heterostructured Photocatalysts for Solar Energy Conversion* (ed. S. Ghosh), 257–282. Elsevier.

48 Yao, N., Liu, Z., Chen, Y. et al. (2015). A novel thermal sensor for the sensitive measurement of chemical oxygen demand. *Sensors* 15 (8): 20501–20510.

49 Chen, J., Liu, S., Qi, X. et al. (2018). Study and design on chemical oxygen demand measurement based on ultraviolet absorption. *Sensors Actuators B Chem.* 254: 778–784.

50 Xu, Z., Dong, Q., Otieno, B. et al. (2016). Real-time in situ sensing of multiple water quality related parameters using micro-electrode array (MEA) fabricated by inkjet-printing technology (IPT). *Sensors Actuators B Chem.* 237: 1108–1119.

51 Shabani, F., Philamore, H., and Matsuno, F. (2021). An energy-autonomous chemical oxygen demand sensor using a microbial fuel cell and embedded machine learning. *IEEE Access* 9: 108689–108701.

52 Beaton, A.D., Cardwell, C.L., Thomas, R.S. et al. (2012). Lab-on-chip measurement of nitrate and nitrite for in situ analysis of natural waters. *Environ. Sci. Technol.* 46 (17): 9548–9556.

53 Eichhorn, M., Ament, C., Jacobi, M. et al. (2018). Modular AUV system with integrated real-time water quality analysis. *Sensors* 18 (6): 1837.

54 Meulemans, M.J., Jones, C.S., Schilling, K.E. et al. (2020). Assessment of spatial nitrate patterns in an eastern Iowa watershed using boat-deployed sensors. *Water* 12 (1): 146.

55 Hsu, L.H., Hoque, E., Kruse, P., and Ravi Selvaganapathy, P. (2015). A carbon nanotube based resettable sensor for measuring free chlorine in drinking water. *Appl. Phys. Lett.* 106 (6): 063102.

56 Pan, S., Deen, M.J., and Ghosh, R. (2015). Low-cost graphite-based free chlorine sensor. *Anal. Chem.* 87 (21): 10734–10737.

57 Kirsanov, D., Korepanov, A., Dorovenko, D. et al. (2017). Indirect monitoring of protein a biosynthesis in *E. coli* using potentiometric multisensor system. *Sensors Actuators B Chem.* 238: 1159–1164.

58 Mross, S., Zimmermann, T., Winkin, N. et al. (2016). Integrated multi-sensor system for parallel in-situ monitoring of cell nutrients, metabolites, cell density and pH in biotechnological processes. *Sensors Actuators B Chem.* 236: 937–946.

59 Kim, M., Jung, T., Kim, Y. et al. (2015). A microfluidic device for label-free detection of *Escherichia coli* in drinking water using positive dielectrophoretic focusing, capturing, and impedance measurement. *Biosens. Bioelectron.* 74: 1011–1015.

60 Jones, T. (1995). *Electromechanics of Particles*. Cambridge University Press.

61 San Park, T. and Yoon, J.-Y. (2014). Smartphone detection of *Escherichia coli* from field water samples on paper microfluidics. *EEE Sens. J.* 15 (3): 1902–1907.

62 Ryu, J.H. (2022). UAS-based real-time water quality monitoring, sampling, and visualization platform (UASWQP). *HardwareX* 11: e00277.

63 Koparan, C. (2020). *UAV-Assisted Water Quality Monitoring*. Clemson University.

64 González-Márquez, L.C., Torres-Bejarano, F.M., Torregroza-Espinosa, A.C. et al. (2018). Use of LANDSAT 8 images for depth and water quality assessment of El Guájaro reservoir, Colombia. *J. S. Am. Earth Sci.* 82: 231–238.

65 Manjakkal, L., Franco, F.F., Pullanchiyodan, A. et al. (2021). Natural jute fibre-based supercapacitors and sensors for eco-friendly energy autonomous systems. *Adv. Sustain. Syst.* 5 (3): 2000286.

66 Manjakkal, L., Dang, W., Yogeswaran, N., and Dahiya, R. (2019). Textile-based potentiometric electrochemical pH sensor for wearable applications. *Biosensors* 9 (1): 14.

67 Crocker, J. and Bartram, J. (2014). Comparison and cost analysis of drinking water quality monitoring requirements *versus* practice in seven developing countries. *Int. J. Environ. Res. Public Health* 11 (7): 7333–7346.

68 Banna, M., Bera, K., Sochol, R. et al. (2017). 3D printing-based integrated water quality sensing system. *Sensors* 17 (6): 1336.

Section IV

Sensing Data Analysis and Internet of Things with a Case Study

12

Toward Real-Time Water Quality Monitoring Using Wireless Sensor Networks

Sohail Sarang, Goran M. Stojanović, and Stevan Stankovski

Faculty of Technical Sciences, University of Novi Sad, Novi Sad, Serbia

12.1 Introduction

Water is a key natural resource that is indispensable for the well-being of humankind [1]. The earth's surface is covered with around 71% of water and only 2.5% is considered freshwater. Around 20% of the world's population lacks access to safe drinking water [2]. The fast-growing urbanization, slum settlements in cities, inadequate sanitation, and the waste of mining companies have devastating impacts on the water environment [3, 4]. Moreover, a study in Europe detected the presence of SARS-CoV-2 genetic material in sewage and river water, which is a serious concern [5]. Because almost all capital cities of Europe have at least one major river (Danube, Thames, Tiber, Liffey, Spree, and Vistula) and are facing contamination issues [6, 7]. The Danube river that passes through the largest cities like Vienna, Bratislava, Budapest, and Belgrade [8, 9] is seeing a rise in anthropogenic contamination [10]. Water pollution leads to negative impacts on public health. Approximately 200 million cases are reported due to water-related diseases annually, which cause 3.4 million deaths [11]. Particularly in the Europe region, every day 14 diarrhea deaths are reported due to inadequate water and sanitation, and hygiene [12]. Also, the excessive level of chlorinated and *Escherichia coli* (*E. coli*) bacteria in the water leads to cancer and serious poisoning [13]. Therefore, water quality monitoring (WQM) is highly important to measure parameters like temperature, oxidation–reduction potential (ORP), power of hydrogen (pH), dissolved oxygen (DO), conductivity, and turbidity from water resources such as rivers, streams, lakes, and others [14]. Water quality is defined

Sensing Technologies for Real Time Monitoring of Water Quality, First Edition.
Libu Manjakkal, Leandro Lorenzelli, and Magnus Willander.
Published 2023 by John Wiley & Sons, Inc.

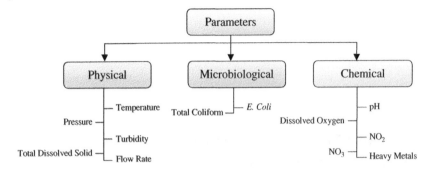

Figure 12.1 Water quality parameters. *Source:* Adapted from Ref. [1].

as the process of checking the physical, biological, and chemical characteristics of water, and the method of sampling and analyzing these characteristics periodically is called WQM [1, 15]. The classification of water quality parameters is given in Figure 12.1.

12.2 Water Quality Monitoring Systems

In the past, several WQM systems [16–34] have been developed to monitor different water sources for providing safe water. These can be categorized into laboratory and wireless sensor networks (WSNs)-based WQM [15].

12.2.1 Laboratory-Based WQM (LB-WQM)

The laboratory-based WQM approach consists of three basic phases: water sampling, laboratory testing, and analysis of results [2, 15, 35]. Figure 12.2 shows a laboratory-based WQM system. First, water sampling is crucial which involves the collection of water samples manually, followed by the transportation of collected water samples to the laboratory for testing. Second, the laboratory testing process uses standard methods to detect microbes and chemical contaminations in water. The final phase is decision-making based on laboratory results that are performed

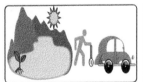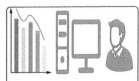

Figure 12.2 Laboratory-based WQM system [2, 15].

by the laboratory specialist. The laboratory-based approach is widely used for many years; however, it has the following limitations. The water sampling phase comprises special apparatuses which require trained workers for sample collections. Thus, human interactions during the sample collection and on transportation to the laboratory waste a lot of time and also increase cost [36]. Additionally, data can be lost during manual processes. Moreover, it is not feasible to collect water samples frequently and as a result, it is also difficult to detect the change in conditions and conduct trend analysis. Also, the laboratory analysis does not guarantee accurate results as they can be based on outdated apparatuses and can also be affected by the transportation process. In many cases, the laboratory specialist does not perform testing immediately as samples are stored first for a plan which increases further delay in making the decision [37]. Finally, the design and implementation of the method have high costs in terms of time, difficulty in deployment, large size, maintenance, and investment for fixed deployment of data collection points and hardware for testing [36]. These shortcomings motivate researchers to use modern approaches such as WSNs for WQM which can support real-time monitoring on-site with minimal human interactions, provide more accurate results, and a longer network operation [38].

12.2.2   Wireless Sensor Networks-Based WQM (WSNs-WQM)

Recently, WSNs have gained significant popularity in both industry and academia and also considered a core part of the Internet of Things (IoT) [39–41]. These networks consist of multiple tiny nodes which are low power, compact dimensions, and low cost. They can sense, store, and process the gathered data such as temperature, conductivity, DO, and pH from a region of interest such as WQM [42, 43]. Each sensor node can communicate with each other wirelessly and can forward sensed data to multiple sinks (base station) using relay nodes. The sink node collects the data from the sender nodes and provides a gateway service to manage the data remotely. A WSN node is composed of four components that include sensor unit, power source, microcontroller, and transceiver. The sensor node can also use external sources such as solar and thermal to harvest energy from the environment, as shown in Figure 12.3. It is also necessary to mention that soiling issues in the solar panel such as snow, dust, and pollen degrade the performance of solar panels. Thus, superhydrophobic coatings can be utilized to address these issues [44]. In the node, sensor unit is responsible for sensing the physical phenomenon, e.g. pH, DO, and temperature parameters of water. Table 12.1 shows commercially available sensors for WQM. The microcontroller and transceiver are used to process and transmit data, respectively. The node uses a small battery, i.e. AA battery as an energy source to power up its components. The NiMH and Alkaline batteries are widely used; however, the battery level decreases with time

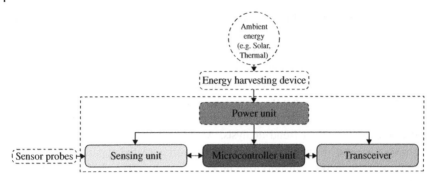

Figure 12.3 Structure of energy harvesting WSN node. *Source:* Adapted from Ref. [2].

Table 12.1 Suggested water quality parameters for monitoring [2, 16].

Parameter	Quality range	Commercially available sensors	Measurement cost
Temperature	15 °C (drinking water)	Dascore Six-Sense Sonde, Smart Water (Libelium), 600XL, WQMS, YSI 6600 Sonde, 6820 V2, Hydrolab Data Sonde 4a PT100	Low
pH	6.5–8.5	WQ101, GLI PHD, WQ201, Hydrolab Data Sonde 4a	Low
Oxidation–reduction potential (ORP)	650–800 mV	WQ600, Dascore Six-Sense Sonde, Hydrolab Data Sonde 4a	Low
Electrical conductivity	500–1000 μS/cm	WQ-Cond, GLI 3422, Dascore Six-Sense Sonde, Hydrolab Data Sonde 4a	Low
Turbidity	0–5 NTU	Hach 1720 D, WQ720, WQ730, Hydrolab Data Sonde 4a	Medium
Free residual chlorine	0.2–2 mg/l	Hach A-15 Cl-17, ATI, Dascore Six-Sense Sonde, Hydrolab Data Sonde 4a	High
Dissolved oxygen	5–6 mg/l	WQ40, Dascore Six-Sense Sonde, Hydrolab Data Sonde 4a,	Medium
Nitrates	<10 mg/l	Smart Water (Libelium)	High

and the water quality sensor can operate until its battery energy reaches an unusual level. The applications of WSNs are not limited to water monitoring but also widely used in smart homes, agriculture, industrial monitoring, healthcare, and food monitoring. In each application, sensors generate different types of data and

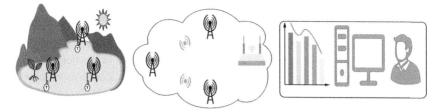

Figure 12.4   Water monitoring using WSNs [2, 15].

offer unique features that require distinctive solutions. For example, in underwater WSNs (UWSNs), nodes are employed to perform different tasks such as oceanic data collection, marine surveillance, and pollution monitoring. However, nodes which are deployed underwater are more prone to failures and they also face challenges like limited bandwidth, efficient data communication and resource utilization, and routing [45]. Therefore, WQM using traditional WSNs provides efficient real-time monitoring and ensures a timely response to water contamination. In addition, the application also provides a longer network lifetime, low deployment, and maintenance costs, and supports data transfer promptly. Therefore, researchers have widely used WSN as a promising alternative to laboratory-based WQM [2, 46–48]. Figure 12.4 describes WQM using WSNs. Moreover, these methods can be further categorized into solar-powered and fixed battery-powered.

12.2.2.1   Solar-Powered Water Quality Monitoring

1) A low-cost WQM system [17] is proposed to monitor the large aquatic area. The sensor, sink node, base station, and web-based portal are the main components of the system. Commercially available sensors developed by Atlas Scientific Company are used to measure DO, pH, and temperature after every 30 minutes. The Zigbee and GSM modules are used to transmit the sensed data to the sink and base station, respectively. The sink node uses a GSM module for remote communication, which consumes a significant amount of energy. A solar cell is used to harvest energy from the surrounding environment. Furthermore, the gathered data is delivered to the database located in the base station. A user-friendly interface through a web-based portal is designed to update data in graphical and tabular formats for analysis. Also, to send parameter values to predefined users via SMS. The developed system provides long-term operation, high flexibility, and reproducibility.

2) WQM system [18] incorporates a data averaging scheme to reduce data traffic in the network. In this scheme, the node collects several data samples in fixed time duration. Then, it calculates the average value every five minutes and

sends it to the neighbor receiver node for further processing. As a result, the data traffic decreases, which helps to improve data reliability and energy efficiency. Moreover, the flooding routing protocol has been implemented that helps neighbor nodes to transfer collected data efficiently to the field servers which are placed at 10 minutes from each other. The network consists of TelosB nodes that use YSI 600XL sensor module to collect water samples every one minute such as DO, pH, conductivity, turbidity, temperature, and depth of water. Each node is equipped with a 12 V battery and solar panel to power up its components. However, the calculation of the average value may increase packet delay. Also, the routing protocol uses additional overhead which increases energy consumption. The developed system can also be used to detect water containment in streams, coasts, and rivers.

3) A low-cost and self-powered WQM [19] is implemented on Lake Victoria Basin that provides continuous water monitoring. The system consists of sensor nodes, gateway, and user applications. Each node consists of water quality sensors, Arduino MCU, power source, RF transceiver, and GPS modules. A solar cell of 10 W is also used to charge the battery during the daytime. The sensor measures temperature, DO, pH, and conductivity every 20 minutes and passes to the gateway wirelessly. Then, it goes to sleep to conserve energy. The GPS module provides location and time stamping value. The gateway node utilizes 1 GB compact memory to store data and uses the GPRS module to connect with the cellular network. The WaGoSy system database is also used to store data and allow users to analyze data in graphical forms. The developed system achieves acceptable results and supports long-term outdoor monitoring. However, it does not provide local data analysis.

4) A WQM system for UWSNs is developed to measure pH, turbidity, and oxygen level [20]. A prototype is proposed where each node is equipped with a 12 V rechargeable battery and solar cell. The nodes collect samples from the water body and forward them to the base station wirelessly, which follows the synchronization approach to collect data. Then, data is gathered at the computer using serial communication connection for further analysis using a time division approach. The gathered data can be displayed and analyzed using visual basic 6.0 graphical user interface (GUI) and MATLAB tools for graphical and numerical results. The developed system shows high flexibility and provides low energy consumption in the network. However, the system does not consider long-range communication and does not offer remote data management services.

5) WQM using solar-powered WSNs is presented [21] that measures the pH, redox, and turbidity in the network. The developed system provides feedback in real-time that can be seen on the computer. The network consists of several nodes powered by solar energy that perform multi-sensing and a base station,

which are distributed randomly. The base station is responsible for collecting data from all sender nodes wirelessly and then forwards the data to the user. The SunSPOT transceiver and regulator are used for wireless data transmission and stabilize the output voltage received from the solar panel.

12.2.2.2 Battery-Powered Water Quality Monitoring

6) The authors deployed a low-cost sensors network to monitor in-pipe water in real-time [16]. The network structure consists of PIC32 MCU as the central node for sensing, Zigbee RF transceiver, ARM MCU for local storage, and PIC MCU-based board for sending the notification to the user. The central node consists of several low-cost, lightweight electrochemical and optical sensors to collect temperature, turbidity, pH, electrical conductivity, and ORP samples every five seconds. The MCU-based board is used to receive the notification from the central node using Zigbee. The system provides good detection accuracy with low deployment and operational cost. In addition, it also enables large-scale deployment for water monitoring. However, in case of any malfunction, the node that performs multi-sensing may lose all water quality data.

7) The real-time freshwater monitoring system is deployed to measure temperature range, pH scale, turbidity, and DO [22]. The nodes are distributed at different locations that enable the multi-sensors interface to collect different water quality samples. The gathered data is sent to the coordinator node through the multi-hop scenario. Each node uses a 1000 mAH power supply and employs PIC16F886 MCU for data processing. A remote monitoring station stores data for further analysis. The developed system uses the XBEE module which supports a maximum range of up to 30 minutes.

8) The river monitoring system [23] is presented based on the Cyberwater platform that supports the user in decision-making. The system measures pH, temperature, pressure, conductivity, and DO data using sensors that are distributed on the river site. The gathered data is sent to the gateway using the Zigbee protocol. The sensor nodes can communicate with each other wirelessly. The gateway transmits the information using the GSM module to the base station computer for storage and analysis, which can be accessed using the Hypertext Transfer Protocol (HTTP). This method is cost-effective and can be used for other monitoring applications.

9) The lake WQM system [24] is developed that provides online monitoring within a range of up to 30 minutes. The central node is connected with four sensors in the network. These nodes measure temperature, water level, turbidity, and salinity, and then forward to the coordinator node. Furthermore, the system displays parameter readings in graph forms on the computer. An energy-efficient WQM system [25] is designed and implemented to measure pH levels in real-time. The proposed system uses a scheduling mechanism to

increase the network lifetime. In a scheduling mechanism, a node remains active for monitoring in the specific area while other nodes go to sleep and save energy. The network architecture consists of Squid bee nodes, pH sensor, base station, and data center. Furthermore, the base station is connected to the data center via the internet. The information center is accessible to the public, registered users, and administrators. The proposed system supports dynamic distributions of nodes on water resources and can easily reconfigure and reuse for other water monitoring applications such as a lake and pond.

10) The industrial sewerage monitoring system is presented using WSNs [26]. The monitoring system consists of three tiers (infrastructure, route, and user). The network consists of seven nodes that collect pH and temperature from the industrial wastewater and forward to the upper tire using the CC2530 wireless module. The nodes support a long communication range of up to 1.5 km and provide an output power of +25.99 dBm. The dynamic routing protocol selects the route for data transmission-based received signal strength indication (RSSI) value. However, the use of ARM-7LPC2138 introduces computational complexity and runs application code in addition to the operating system, which causes higher energy consumption. Moreover, the routing mechanism also uses overheads which lead to further energy consumption in the network.

11) WQM system is designed that consists of low-power sensors and a remote data center [27]. The hardware includes ZigBee modules to measure pH, temperature, and turbidity in the network. These nodes collect data and forward it to the remote data center wirelessly. Then, send over the internet. The system provides good network life of more than six months and can also be widely used in industrial and agriculture applications. However, the developed system does not have any local database for storage and it also requires an internet connection to access the information.

12) A real-time WQM is implemented in Yixing Jiangsu province of China to collect pH, temperature, and DO samples from the pond water [28]. The system architecture comprises three layers: data acquisition, transport, and application. The data acquisition layer consists of tiny sensor nodes which are responsible for sample collection on user demand and send to the router node using a wireless transmission module. The router nodes select suitable routes and forward to the base station for further processing. The base station transports data to the monitoring system using GPRS. The application layer helps users to monitor, analyze, and make control decisions based on expert knowledge. The developed system uses different energy-efficient techniques to avoid any disruption in monitoring while providing better water quality.

13) A remote WQM system is developed that provides long-range monitoring of water [29]. The system architecture is divided into three layers: data

acquisition, management, and sharing. The data acquisition layer consists of low-power sensors that collect temperature, pH, DO, and conductivity samples from water and send data to the coordinator using CC2430 transceiver in the multi-hop scenario. The coordinator forwards data to the management layer using GPRS for display and storage. Furthermore, the management layer is connected to the sharing layer via the internet that shows the status of water quality which helps to observe the trend remotely.

14) A new design of WQM is proposed that supports long-range monitoring, flexible configuration, and uses low energy [30]. The system consists of sensing nodes, base stations, and remote monitoring centers. The network structure follows a clustering approach where several nodes connect with a base station. Each node requires an input voltage of 5–9 V from the battery. In the experiment, a total of five nodes are deployed on an artificial lake to collect temperature and pH level samples every one hour in real-time and forward them to the base station using Zigbee communication. The base station uses low-power MCU MSP430F1611 and CC2430 to control and transmit data, respectively. Furthermore, the base station transfers data to the remote monitoring center using GPRS. The developed system can be widely used in other monitoring applications. However, they have considered only two parameters which are not sufficient.

15) An energy-efficient WQM [31] is implemented in Malawi that addresses the internetworking issue and reduces energy consumption in the network. The system proposes low power and lightweight Linux-based gateway that enables internetworking between different protocols and provides good flexibility in different scenarios. Also, it saves the monthly data that can be accessed using a web application. The network consists of sensor nodes, SunSPOT base station, and gateway. The 90-FLT sensors collect the sample of temperature, turbidity, pH, DO, total dissolved solids, and conductivity every one hour and send data to SunSPOT base station using Zigbee protocol. Furthermore, the base station is connected to the gateway. The system prolongs the network lifetime by reducing energy consumption by turning off the devices and sensors when they are not in use. However, the developed system does not consider long-term outdoor environments.

16) SmartCoast [32] is proposed to monitor DO, pH, conductivity, turbidity, water level, temperature, and phosphate in the water. The SmartCoast uses novel sensors and supports plug and play interface with the system. Additionally, a phosphate sensor has been developed and integrated with the system to measure phosphate concentration. Furthermore, the network consists of Tyndall motes that use low-power multi-sensor and employee Zigbee for low-power wireless communication. The summary of WSNs-based WQM is given in Table 12.2.

Table 12.2 Summary of measured water quality parameters in different WQM systems.

Reference and year	Measured parameters
[16], 2014	Temperature, turbidity, pH, electrical conductivity, and ORP
[17], 2019	DO, pH, and temperature
[18], 2015	Dissolved oxygen, pH, conductivity, turbidity, temperature, and depth of water
[19], 2014	Temperature, DO, pH, and conductivity
[20], 2013	pH, turbidity, and oxygen level
[21], 2012	pH, redox, and turbidity
[22], 2011	Temperature, pH, turbidity, and DO
[23], 2015	pH, temperature, pressure, conductivity, and DO
[24], 2013	Temperature, water level, turbidity, and salinity
[25], 2013	pH
[26], 2012	pH and temperature
[27], 2012	pH, temperature, and turbidity
[28], 2011	pH, temperature, and DO
[29], 2009	Temperature, pH, DO, and conductivity
[30], 2009	Temperature and pH
[31], 2009	Temperature, turbidity, pH, DO, total dissolved solids, and conductivity
[32], 2007	DO, pH, conductivity, turbidity, water level, temperature, and phosphate
[33], 2020	DO, pH, conductivity, calcium ion (Ca^{2+}), NO_3, ORP, temperature, and fluoride ion (F^-)
[34], 2019	Total dissolved solids, pH, and turbidity

17) A smart river monitoring system, reported in [33], measures physical and chemical parameters using the Libelium Waspmote sensor nodes. The node architecture consists of four units: sensing, processing, power, and communication. Each sensor node is equipped with different probes capable of measuring parameters such as DO, pH, conductivity, calcium ion (Ca^{2+}), NO_3, ORP, temperature, and fluoride ion (F^-) and forward directly to the gateway. Moreover, a rechargeable battery is used to power up the node. Additionally, a solar panel is also attached to the node to harvest solar energy from the surrounding environment and recharge the battery. The Libelium Waspmote node supports communication technologies such as 3G/4G, WIFI, GPRS,

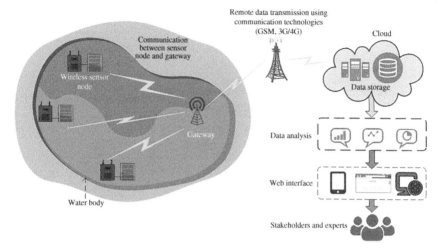

Figure 12.5 General architecture of smart river monitoring system. *Source:* Adapted from Ref. [33].

802.15.4/ZigBee, and others to transmit the data to the cloud for real-time monitoring and analysis. Furthermore, the web portal is designed to track the node position, visualize, and share the sensor data with the stakeholders using web portals and SMS alerts. The general architecture of the system is given in Figure 12.5.

18) A wide-area WQM system, developed in [34], consists of multiple unmanned surface vehicles (USVs) and an online analysis platform for real-time monitoring. Each USV consists of low-power sensors to measure different water parameters (total dissolved solids, pH, and turbidity) and a lifting device that enables data gathering at different water depths. The lifting device contains a DC motor and lifting gear shaft. Furthermore, a monocrystalline silicon solar cell is also attached to the USV to harvest energy from the surrounding environment. In addition, an automatic positioning system and communication modules are also embedded in the USV. The architecture of the developed system is shown in Figure 12.6. The MCU is responsible for power control and establishing wireless links to the communication module. There are two wireless modules incorporated to connect the USV to WQMS and the cloud. The Wi-Fi module is intended for communication between the USV and WQM, and GPRS is used to establish communication between the USV and the cloud. The cloud platform gathers the data and stores it for further analysis.

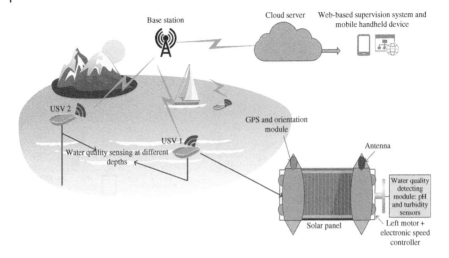

Figure 12.6 Unmanned surface vehicles-based WQM system. *Source:* Adapted from Ref. [34].

12.3 The Use of Industry 4.0 Technologies for Real-Time WQM

Traditional WQM systems lack real-time water monitoring and have implementation issues such as time consumption and high maintenance costs. These shortcomings drive to utilize the latest industry 4.0 technologies such as industrial IoT (IIoT), cloud computing, big data analytics, machine learning, blockchain, and 5G/NB-IoT communication to enable more efficient, cost-effective, secure, and real-time WQM [49–60]. Figure 12.7 shows a typical IoT-based WQM system that comprises different sensors that have the ability to sense, gather, and transmit

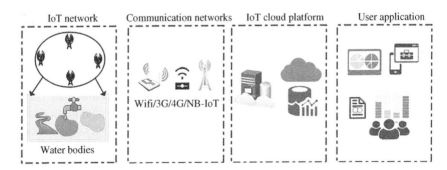

Figure 12.7 A typical IoT-based WQM system.

data to the remote server using communication technologies. These IoT sensors gather a large quantity of data that requires cloud services for data storage to extract useful information. For instance, WQM system developed in [58] consists of IoT sensors that sense data and transmit it to a long distance using narrow band internet of things (NB-IoT) to the cloud platform. Keil tool was used to burn the program into STM32 MCU that contains specific commands such as data acquisition, data sending, and others. Java programming was used to develop the back-end application that accesses on the cloud platform, and processes the data for further analysis and decision-making. The developed WQM system in [61] collects pH, turbidity, water level, temperature, and humidity parameters and sends to the ThingSpeak server for storage using ESP8266 Wi-Fi module. ThingSpeak is an IoT application that enables the authorized users to access real-time data. Likewise, a secure IoT-based WQM was designed in [60] that applied blockchain technology to measure water quality in real-time and maintain the reliability and transparency of the system.

Currently, big data analytics and machine learning are gaining strong attention for data analysis, modeling, and predictions. The data gathered at the cloud is further processed for data analysis to get useful trends and information using data analytics and machine learning algorithms. For instance, authors in [56], used a large and diverse dataset of various water quality parameters gathered from water monitoring devices and utilized machine learning algorithms to classify and predict the water quality index. Likewise, the linear polynomial neural network was employed to predict DO concentration in the Danube river in Serbia [62]. The authors in [54] discussed applications of various ML algorithms in WQM systems that can help to optimize the system performance and support in decision-making. Moreover, big data analytics framework was developed in [63] that finds the correlation between water quality parameters (WQPs) obtained using sensors. Figure 12.8 shows the architecture and components of the developed framework [63].

12.4 Conclusion

In the past, a large number of WQM systems have been designed that include manual water samples collection and laboratory testing for WQM. In addition, these traditional systems need highly skilled workers and experts for water sampling and analysis. Currently, WSNs have gained strong attention for real-time WQM to facilitate more efficient and cost-effective solutions. Therefore, this chapter has focused on WQM systems using WSNs that employ different sensors which are powered by solar and battery energy sources. In battery-powered WQM, sensor nodes are powered up using small and fixed-size batteries which have limited

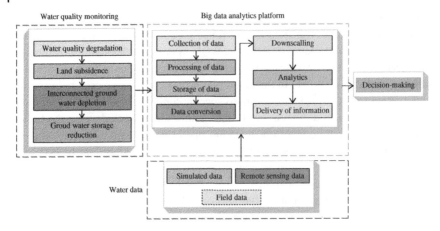

Figure 12.8 Big data analytics framework for WQM system. *Source:* Adapted from Ref. [63].

capacity. More frequent collection of water samples may lead to a faster depletion of battery energy. This drives to development of more sustainable WQMs that can harvest energy using external sources from the surrounding environment, named solar-powered -based WQMs. In these systems, nodes can harvest a sufficient amount of energy using solar cells and can use the harvested energy to recharge their batteries to extend the network lifetime. Finally, an overview of the industry 4.0 technologies for real-time WQM system has been discussed that involves latest IoT tools for massive data gathering, transmission, storage, analysis and predictions using IIoT, cloud computing, wireless communications, big data analytics, and machine learning, respectively.

References

1 Olatinwo, S.O. and Joubert, T.H. (2019). Enabling communication networks for water quality monitoring applications: a survey. *IEEE Access* 7: 100332–100362.

2 Adu-Manu, K.S., Tapparello, C., Heinzelman, W. et al. (2017). Water quality monitoring using wireless sensor networks: current trends and future research directions. *ACM Trans. Sens. Netw.* 13 (1): 1–41.

3 Derbew, Y. and Libsie, M. (2014). A wireless sensor network framework for large-scale industrial water pollution monitoring. *Proceedings of the IST-Africa Conference and Exhibition (IST-Africa '14)*, Mauritius (7–9 May 2014), 1–8.

4 Marinković, V., Obradović, L., Bugarin, M., and Stojanović, G. (2014). The impact of polluted wastewater on water quality of the Bor River and surrounding groundwater. *Min. Metall. Eng. Bor* (3): 33–40.

5 Shutler, J.D., Zaraska, K., Holding, T. et al. (2021). Rapid assessment of SARS-CoV-2 transmission risk for fecally contaminated river water. *ACS ES&T Water* 1 (4): 949–957.

6 Amirat, L., Wildeboer, D., Abuknesha, R.A., and Price, R.G. (2012). *Escherichia coli* contamination of the river Thames in different seasons and weather conditions. *Water Environ. J.* 26 (4): 482–489.

7 Jushi, A., Pegatoquet, A., and Le, T.N. (2016). Wind energy harvesting for autonomous wireless sensor networks. *Proceedings of the Euromicro Conference on Digital System Design (DSD)*, Limassol, Cyprus (31 August–2 September 2016), 301–308.

8 Besemer, K., Moeseneder, M.M., Arrieta, J.M. et al. (2005). Complexity of bacterial communities in a river-floodplain system (Danube, Austria). *Appl. Environ. Microbiol.* 71 (2): 609–620.

9 Páll, E., Niculae, M., Kiss, T. et al. (2013). Human impact on the microbiological water quality of the rivers. *J. Med. Microbiol.* 62 (11): 1635–1640.

10 Kondor, A.C., Jakab, G., Vancsik, A. et al. (2020). Occurrence of pharmaceuticals in the Danube and drinking water wells: efficiency of riverbank filtration. *Environ. Pollut.* 265: 114893.

11 Pule, M., Yahya, A., and Chuma, J. (2017). Wireless sensor networks: a survey on monitoring water quality. *J. Appl. Res. Technol.* 15 (6): 562–570.

12 World Health Organization (2019). Environmental Health Inequalities in Europe: Second Assessment Report. WHO/EURO:2019-3507-43266-60638.

13 DeZuane, J. (1997). *Handbook of Drinking Water Quality*. New York: Wiley.

14 Simić, M., Stojanović, G.M., Manjakkal, L., and Zaraska, K. (2016). Multi-sensor system for remote environmental (air and water) quality monitoring. *Proceedings of the 24th telecommunications forum (TELFOR)*, Belgrade, Serbia (22–23 November 2016), 1–4.

15 Olatinwo, S.O. and Joubert, T.-H. (2018). Energy efficient solutions in wireless sensor systems for water quality monitoring: a review. *IEEE Sens. J.* 19 (5): 1596–1625.

16 Lambrou, T.P., Anastasiou, C.C., Panayiotou, C.G., and Polycarpou, M.M. (2014). A low-cost sensor network for real-time monitoring and contamination detection in drinking water distribution systems. *IEEE Sens. J.* 14 (8): 2765–2772.

17 Demetillo, A.T., Japitana, M.V., and Taboada, E.B. (2019). A system for monitoring water quality in a large aquatic area using wireless sensor network technology. *Sustain. Environ. Res.* 29 (1): 12.

18 Chung, W.-Y. and Yoo, J.H. (2015). Remote water quality monitoring in wide area. *Sensors Actuators B Chem.* 217: 51–57.

19 Faustine, A., Mvuma, A.N., Mongi, H.J. et al. (2014). Wireless sensor networks for water quality monitoring and control within Lake Victoria Basin: prototype development. *Wireless Sens. Netw.* 6 (12): 281–290.

20 Amruta, M.K. and Satish, M.T. (2013). Solar powered water quality monitoring system using wireless sensor network. *International Mutli-Conference on Automation, Computing, Communication, Control and Compressed Sensing (iMac4s)*, Kottayam, India (22–23 March 2013), 281–285.

21 Yue, R. and Ying, T. (2012). A novel water quality monitoring system based on solar power supply & wireless sensor network. *Procedia Environ. Sci.* 12: 265–272.

22 Nasirudin, M.A., Za'bah, U.N., and Sidek, O. (2011). Fresh water real-time monitoring system based on wireless sensor network and GSM. *Proceedings IEEE Conference on Open Systems*, Langkawi, Malaysia (25–28 September 2011), 354–357.

23 Vacariu, L., Cret, O., Hangan, A., and Bacotiu, C. (2015). Water parameters monitoring on a cyberwater platform. *Proceedings IEEE 20th International Conference on Control Systems and Computer Science*, Bucharest, Romania (27–29 May 2015), 797–802.

24 Khetre, A. and Hate, S. (2013). Automatic monitoring & reporting of water quality by using WSN technology and different routing methods. *Int. J. Adv. Res. Comput. Eng. Technol.* 2 (12): 3255–3260.

25 Nasser, N., Ali, A., Karim, L., and Belhaouari, S. (2013). An efficient wireless sensor network-based water quality monitoring system. *Proceedings ACS International Conference on Computer Systems and Applications (AICCSA)*, Ifrane, Morocco (27–30 May 2013), 1–4.

26 Shao, X., Liu, X., and Zhang, H. (2012). Monitoring system of sewerage treatment based on wireless network. *Proceedings IEEE International Conference on Optoelectronics and Microelectronics*, Changchun, Jilin, China (23–25 August 2012), 490–492.

27 He, D. and Zhang, L.-X. (2012). The water quality monitoring system based on WSN. *Proceedings 2nd International Conference on Consumer Electronics, Communications and Networks (CECNet)*, Yichang, China (21–23 April 2012), 3661–3664.

28 Ming-fei, Z. and Lian-zhi, W. (2011). A WSN-based monitor system for water quality combined with expert knowledge. *Proceedings IEEE International Conference on Electronics, Communications and Control (ICECC)*, Ningbo, China (9–11 September 2011), 105–108.

29 Wang, Z., Wang, Q., and Hao, X. (2009). The design of the remote water quality monitoring system based on WSN. *Proceedings 5th International Conference on Wireless Communications, Networking and Mobile Computing*, Beijing, China (24–26 September 2009), 1–4.

30 Jiang, P., Xia, H., He, Z., and Wang, Z. (2009). Design of a water environment monitoring system based on wireless sensor networks. *Sensors* 9 (8): 6411–6434.

31 Zennaro, M., Floros, A., Dogan, G. et al. (2009). On the design of a water quality wireless sensor network (WQWSN): an application to water quality monitoring in

Malawi. *Proceedings International Conference on Parallel Processing Workshops,* Vienna, Austria (22–25 September 2009), 330–336.

32 O'Flynn, B., Martinez-Catala, R., Harte, S. et al. (2007). SmartCoast: a wireless sensor network for water quality monitoring. *Proceedings 32nd IEEE Conference on Local Computer Networks (LCN 2007),* Dublin, Ireland (15–18 October 2007), 815–816.

33 Adu-Manu, K.S., Katsriku, F.A., Abdulai, J.-D., and Engmann, F. (2020). Smart river monitoring using wireless sensor networks. *Wireless Commun. Mobile Comput.* 2020: 1–19.

34 Cao, H., Guo, Z., Wang, S. et al. (2020). Intelligent wide-area water quality monitoring and analysis system exploiting unmanned surface vehicles and ensemble learning. *Water* 12 (3): 1–15.

35 Madrid, Y. and Zayas, Z.P. (2007). Water sampling: traditional methods and new approaches in water sampling strategy. *TrAC Trends Anal. Chem* 26 (4): 293–299.

36 Bhardwaj, J., Gupta, K.K., and Gupta, R. (2015). A review of emerging trends on water quality measurement sensors. *Proceedings International Conference on Technologies for Sustainable Development (ICTSD),* Mumbai, India (4–6 February 2015), 1–6.

37 Paschke, A. (2003). Consideration of the physicochemical properties of sample matrices–an important step in sampling and sample preparation. *Trends Anal. Chem.* 22 (2): 78–89.

38 Olatinwo, S.O. and Joubert, T.-H. (2020). Energy efficiency maximization in a wireless powered IoT sensor network for water quality monitoring. *Comput. Netw.* 176: 107237.

39 Sarang, S., Drieberg, M., Awang, A., and Ahmad, R. (2018). A QoS MAC protocol for prioritized data in energy harvesting wireless sensor networks. *Comput. Netw.* 144: 141–153.

40 Sohail, M., Khan, S., Ahmad, R. et al. (2019). Game theoretic solution for power management in IoT-based wireless sensor networks. *Sensors* 19 (18): 1–20.

41 Sarang, S., Stojanović, G.M., Stankovski, S. et al. (2020). Energy-efficient asynchronous QoS MAC protocol for wireless sensor networks. *Wirel. Commun. Mob. Comput.* 2020: 1–13.

42 Yu, Q., Li, G., Hang, X., and Fu, K. (2017). An energy efficient MAC protocol for wireless passive sensor networks. *Future Internet* 9 (2): 1–12.

43 Sarang, S., Drieberg, M., and Awang, A. (2017). Multi-priority based QoS MAC protocol for wireless sensor networks. *Proceedings 7th IEEE International Conference on System Engineering and Technology (ICSET),* Shah Alam, Malaysia (2–3 October 2017), 54–58.

44 Mishra, A., Bhatt, N., and Bajpai, A. (2019). Nanostructured Superhydrophobic Coatings for Solar Panel Applications. In: *Nanomaterials-Based Coatings* (ed. P.N. Tri, S. Rtimi, and C.M. Ouellet Plamondon), 397–424. Elsevier

45 Zaman, F., Lee, S., Rahim, M.K., and Khan, S. (2019). Smart antennas and intelligent sensors based systems: enabling technologies and applications. *Wirel. Commun. Mob. Comput.* 2019: 1–13.

46 Olatinwo, S.O. and Joubert, T.-H. (2018). Optimizing the energy and throughput of a water-quality monitoring system. *Sensors* 18 (4): 1–21.

47 Olatinwo, S. and Joubert, T. (2018). Maximizing the throughput and fairness of a water quality monitoring wireless sensor network system. *Int. J. Commun. Antenna Propag.* 8 (6): 448–460.

48 Kageyama, T., Miura, M., Maeda, A. et al. (2016). A wireless sensor network platform for water quality monitoring. *IEEE Sensors*, Orlando, FL (30 October–3 November 2016), 1–3.

49 Garrido-Momparler, V. and Peris, M. (2022). Smart sensors in environmental/ water quality monitoring using IoT and cloud services. *Trends Environ. Anal. Chem.* 35: 1–7.

50 Jha, B.K., Sivasankari, G., and Venugopal, K. (2020). Cloud-based smart water quality monitoring system using IoT sensors and machine learning. *Int. J. Adv. Trends Comput. Sci. Appl.* 9 (3): 1–7.

51 Jan, F., Min-Allah, N., and Düştegör, D. (2021). IoT based smart water quality monitoring: recent techniques, trends and challenges for domestic applications. *Water* 13 (13): 1–37.

52 Madhavireddy, V. and Koteswarao, B. (2018). Smart water quality monitoring system using IoT technology. *Int. J. Eng. Technol.* 7 (4.36): 636.

53 Bharani Baanu, B. and Jinesh Babu, K. (2022). Smart water grid: a review and a suggestion for water quality monitoring. *Water Supply* 22 (2): 1434–1444.

54 Zhu, M., Wang, J., Yang, X. et al. (2022). A review of the application of machine learning in water quality evaluation. *Eco-Environ. Health* 1 (2): 1–10.

55 Koditala, N.K. and Pandey, P.S. (2018). Water quality monitoring system using IoT and machine learning. *Proceedings 2018 International Conference on Research in Intelligent and Computing in Engineering (RICE)*, El Salvador (22–24 August 2018), 1–5.

56 Aldhyani, T.H., Al-Yaari, M., Alkahtani, H., and Maashi, M. (2020). Water quality prediction using artificial intelligence algorithms. *Appl. Bionics Biomech.* 2020: 1–12.

57 Jamroen, C., Yonsiri, N., Odthon, T. et al. (2023). A standalone photovoltaic/ battery energy-powered water quality monitoring system based on narrowband internet of things for aquaculture: design and implementation. *Smart Agric. Technol.* 3 (1–14): 100072.

58 Huan, J., Li, H., Wu, F., and Cao, W. (2020). Design of water quality monitoring system for aquaculture ponds based on NB-IoT. *Aquac. Eng.* 90: 1–10.

59 Saravanan, S., Renugadevi, N., Sudha, C.N., and Tripathi, P. Industry 4.0: smart water management system using IoT. In: *Security Issues and Privacy Concerns in*

Industry 4.0 Applications (ed. S. David, R.S. Anand, V. Jeyakrishnan, and M. Niranjanamurthy), 1–14. Wiley.

60 Alharbi, N., Althagafi, A., Alshomrani, O. et al. (2021). A Blockchain based secure IoT solution for water quality management. *Proceedings 2021 International Congress of Advanced Technology and Engineering (ICOTEN)*, Taiz, Yemen (4–5 July 2021), 1–8.

61 Pasika, S. and Gandla, S.T. (2020). Smart water quality monitoring system with cost-effective using IoT. *Heliyon* 6 (7): 1–9.

62 Tomić, A.Š., Antanasijević, D., Ristić, M. et al. (2018). A linear and non-linear polynomial neural network modeling of dissolved oxygen content in surface water: inter-and extrapolation performance with inputs' significance analysis. *Sci. Total Environ.* 610: 1038–1046.

63 Kimothi, S., Thapliyal, A., Akram, S.V. et al. (2022). Big data analysis framework for water quality indicators with assimilation of IoT and ML. *Electronics* 11 (13): 1–35.

13

An Internet of Things-Enabled System for Monitoring Multiple Water Quality Parameters

Fowzia Akhter[1], H. R. Siddiquei[1], Md. E. E. Alahi[2], and S. C. Mukhopadhyay[1]

[1] *Faculty of Science and Engineering, Macquarie University, Sydney, NSW, Australia*
[2] *Shenzhen Institute of Advanced Technology, Chinese Academy of Science, Shenzhen, China*

13.1   Introduction

Water contamination is a major environmental issue worldwide. Managing the proper balance of the water quality parameters is the key to ensure safe, healthy, and productive aquaculture [1]. Water contaminants negatively affect the production, growth, development, and mortality of the fish living on the farm. As a result, the profit of the farm reduces [2]. Marine water and freshwater fisheries contribute significantly to the economics of many countries like Australia, the Philippines, Japan, and Vietnam [3]. Maintaining good water quality is the key to ensure a productive and profitable aquaculture industry. Efficient water quality management is only possible if the water is monitored regularly. Therefore, researchers are seeking interest in developing the Internet of Things (IoT)-enabled water quality monitoring system.

The most significant advantage of including IoT with the physical sensors is monitoring the farm condition from any remote location. Collection of long-term data in IoT cloud server and sharing the data with the experts enables to receive critical feedback from the agronomist whenever necessary [4, 5]. Realizing the importance of IoT, different agricultural countries such as the United States, Australia, New Zealand, and Japan have started including IoT in agriculture [6]. Figure 13.1 shows the schematics of smart fisheries.

Several research works have been carried out in developing intelligent water quality monitoring [7, 8]. However, the systems are too expensive because of

Sensing Technologies for Real Time Monitoring of Water Quality, First Edition.
Libu Manjakkal, Leandro Lorenzelli, and Magnus Willander.
© 2023 The Institute of Electrical and Electronics Engineers, Inc.
Published 2023 by John Wiley & Sons, Inc.

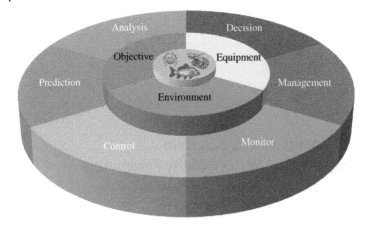

Figure 13.1 A typical diagram of smart fisheries.

using commercial sensors. Moreover, water quality monitoring is limited to very few parameters such as temperature, pH, dissolved oxygen (DO), conductivity, and velocity [9, 10]. Simultaneously, various water quality parameters such as nitrate, phosphate, calcium, and magnesium play vital roles in the development and production of fishes. Numerous researchers have developed low-cost sensors to detect these parameters [11, 12], but it still needs vigorous testing to apply in a real-life scenario.

This research proposes a planar interdigital sensor for detecting and determining temperature, nitrate, phosphate, calcium, and magnesium. PDMS is chosen for developing the substrate because of its biocompatibility, hydrophobicity, and cost-effectiveness [13]. Multi-walled carbon nanotubes (MWCNTs) are used for forming the electrodes as this kind of sensors is required to produce in mass production to deploy for large, distributed networks [14]. MWCNTs are better than SWCNTs as it has certain characteristics such as easy for mass production, functionalize with PDMS, and enhanced stabilities [15]. An IoT-enabled autonomous system is developed for checking water quality parameters from any remote location. A machine learning algorithm is used for predicting the levels of the parameters in real-life samples. The farmers can get experts' suggestions whenever necessary, no matter where the location of the farm is.

13.2 Water Quality Parameters and Related Sensors

Water temperature is one of the most critical factors deciding the metabolism of fishes. The fishes living at higher temperature have a more significant metabolism than the fishes residing at a lower temperature [16]. The temperature requirement varies from fishes to fishes. If the cold-water fishes are put inside the warm water, they become less active and stop consuming food. The optimum temperature requirement for cold-water fishes is less than 20 °C, but for warm-water fishes, it is between 24 and 27 °C [17]. Hence, it is essential to maintain and monitor optimum temperature condition in fisheries for maximum growth and productivity of the fishes. Some researchers have developed fiber Bragg grating (FBG) sensors, whereas some researchers developed an interdigital sensor with lithium niobate (LiNbO$_3$) to measure water temperature [18, 19]. Although these sensors provide a wide detection range, they require complex electronics and vigorous testing to apply in a real-life scenario. The most popular sensor for agricultural application is DS18B20, and many researchers have used it [20–22]. This commercially available sensor detects a wide range of temperature (−55 to +125 °C) and responds very fast but not cost-efficient [23]. Therefore, developing a low-cost, low-power, wide range temperature sensor is essential for fresh water and marine fisheries applications.

pH level is another essential factor that needs to be maintained in any fisheries. Different aquatic animals are adapted to live in specific pH. The optimum pH level for freshwater fishing is 6.5–9.0, and for marine fishes, it is from 7.5 to 8.5 [24]. Fishes' lives are impacted significantly, such as slow growth, becoming unproductive, and reaching a point of acid or alkaline death if the pH level changes rapidly [25]. Therefore, efficient pH management in aquaculture is necessary for both economic and environmental reasons. Some researchers developed pH sensors using ruthenium oxide (RuO$_2$), polysulfone/polyaniline, and Ag/AgCl, respectively. There is also a commercially available sensor (SEN-10972) for pH detection. Most lab-based sensors are low-cost and detect a wide range of pH levels but need improvement to fit in actual scenarios. The commercially available sensors are IoT compatible, respond fast, provide a wide range of pH, but not cost-efficient. Table 13.1 shows a comparative study of the existing sensors for pH determination in water.

Nitrate (NO$_3^-$) is an essential water nutrient required for aquatic plants growth. It acts as a source of nitrogen for them [32]. Although nitrate levels do not directly impact aquatic animals' lives, they are difficult to live when present in excessive amounts [33]. Nitrates are the food source for algae. When a large amount of nitrate exists in water, the growth of algae becomes uncontrollable. The higher the algae, the more significant DO fluctuation [34]. Consequently, fishes suffer from various problems such as disorientation, loss of equilibrium, acting dazed, high respiration rates, rapid gill movement, and loss of appetite [35]. The optimum nitrate levels are dependent on the type of fisheries; for marine water, it is 40 ppm, whereas, for freshwater fishing, it is 110 ppm [36]. Hence, nitrate management in water is the key to ensure sustainable aquaculture. Some researchers developed colorimetric sensor, whereas some researchers developed the graphene-based

Table 13.1 Comparison among existing pH sensors.

Sensing material	Detection range	Detection limit	Response time (s)	References
Ruthenium oxide	1.5–12	1.5	100	[26]
Polysulfone membranes and polyaniline nanofibers coating	4–12	4	<4	[27]
L-carrageenan, *Clitoria* sp., and *Brassica* sp.	1–13	1	30	[28]
Commercially available optical pH sensor	1–14	1	1	[29]
Microfabricated platinum	4–10	4	—	[30]
Silver and silver chloride	4.01–6.86	4.01	0.083	[31]
MWCNTS/PDMS	1.3–12.4	1	1	This work

sensor for nitrate detection in water. Table 13.2 shows a comparative study among the existing nitrate sensors. These sensors provide a wide detection range, but it still needs a lot to improve in terms of low-power, low-cost, response time, high-sensitivity, and easily operable sensors applicable to real-time monitoring in any environmental condition.

Table 13.2 Comparison among the existing research on nitrate detection.

Sensing material	Detection range (ppm)	Detection limit (ppm)	Response time (s)	References
Nickel (Ni) @ platinum (Pt) on graphene sheet	10–1500	10	250	[37]
Sulfonated graphene and Gold nanoparticles	10–3960	0.2	3	[38]
Reduced graphene oxide and benzyltriethylammonium chloride	0.2–200	0.2	2–7	[39]
Poly(decyl methacrylate) and poly(methyl methacrylate) copolymer	0.5–10	0.55	—	[40]
Gold nanoparticles on graphene sheet	0.3–720	0.1	96	[41]
Gold nanoparticles and reduced graphene oxide	0.1–20	0.1	98	[42]
MWCNTs/PDMS	0–25	1	1	This work

Table 13.3 Comparison among phosphate sensors for water bodies.

Sensing material	Detection range (ppm)	Detection limit (ppm)	Response time (s)	References
Copper monoamino-phthalocyanine	0.001–2680.7	0.0002	10	[46]
Copper phthalocyanine–acrylate-polymer	0.00003–2.680	0.0003	—	[47]
Nanocomposite made of reduced graphene oxide and silver	5–6000	1.20	100	[48]
Ferritin and reduced graphene oxide	16.7–500	0.806	—	[49]
Polyurethane and poly vinyl chloride	0.096–960	0.096	5	[50]
Reduced graphene oxide and aluminum oxide	1–10	1	50	[51]
MWCNTs/PDMS	0–25	1	1	This work

Phosphate (PO_4^{3-}) is also an essential mineral that generates energy from sunlight for growing and reproducing the cells [43]. Higher phosphate level in water reduces the DO level in lakes and rivers, which significantly impacts flora and fauna's life inside rivers or lakes [44]. If the phosphate level in lakes or reservoirs exceeds more than 1 ppm, the water bodies' natural balance gets affected [45]. Efficient phosphate management in lakes or rivers is necessary for both economic and environmental reasons. Some researchers developed graphene-based sensors, whereas some researchers fabricated colorimetric sensor for phosphate detection in water samples. Although the existing sensors detect a wide range of phosphate concentration, adequate improvements are necessary for reducing cost, power, increasing sensitivity, and applying in any environmental condition. Table 13.3 shows the comparison among the research works conducted on phosphate detection on water.

Calcium (Ca) and magnesium (Mg) are two essential minerals required for successful aquaculture. Fishes receive these nutrients from both food and water. Calcium is a crucial water nutrient for fish hatchery. If the calcium concentration in any fish hatchery is inadequate, the eggs dehydrate and do not hatch appropriately in that water [52]. Magnesium (Mg) is another vital nutrient for freshwater fish development. The growth of freshwater fishes largely depends on Mg and protein diets. When the magnesium level in water reduces, the fishes' growth reduces, and skin hemorrhages may appear [53]. Various methods are available for determining these nutrients, such as chromatography [54], High-performance liquid chromatography (HPLC) [55], and UV-spectroscopy [56]. Although these techniques' accuracy is high, these techniques are complicated, consume a very long time, and required regular maintenance. Therefore, it is essential to develop simple and easily operable sensors for detecting these nutrients accurately in any environmental condition in different water bodies for both economical and efficient farming.

13.3 Design and Fabrication of the Proposed Sensor

After reviewing the existing sensors, it is decided to use MWCNTs and PDMS for developing the sensor to detect nitrate in water. Mold-based casting method is applied to fabricate the sensor as this method provides reproducible sensors. Additionally, the dimensions can be precisely controlled. The mold is designed and printed using a 3D printer. The PDMS and the curing agent are mixed at a 10 : 1 ratio to use in the fabrication process. Initially, the conductive ink is prepared by mixing PDMS with MWCNTs nanopowder at 10 : 2 ratio. The MWCNTs are adequately mixed with PDMS by ultrasonication first and then mechanical stirring at 400 rpm for 45 minutes. The optimum ratio of PDMS and MWCNTs are found as 10 : 2 by varying the concentration and analyzing the resulting conductivity value. When the amount of MWCNTs increases, the conductivity increases and then reaches a saturation point. Figure 13.2 shows the impact of PDMS and MWCNTs ratio on the conductivity of the composite. It is clear from Figure 13.2 that the optimum ratio is 10 : 2 for developing electrodes.

The MWCNTs/PDMS composite is cast on the mold first and then cured in the oven at 65 °C for an hour to solidify the nanocomposite. After filling the mold with

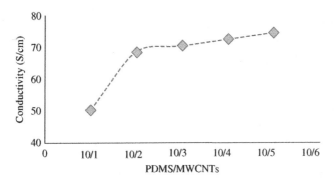

Figure 13.2 Impact of PDMS and MWCNTs ratio on the conductivity of the nanocomposite.

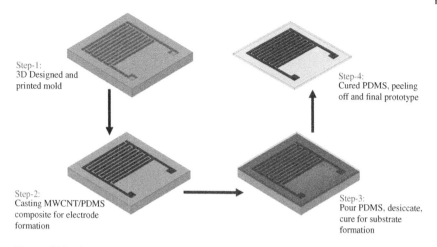

Step-1:
3D Designed and
printed mold

Step-4:
Cured PDMS, peeling
off and final prototype

Step-2:
Casting MWCNT/PDMS
composite for electrode
formation

Step-3:
Pour PDMS, desiccate,
cure for substrate
formation

Figure 13.3 Schematic representation of the sensor fabrication process.

PDMS, the height is adjusted to 1000 μm using a casting knife for substrate formation. The prototype is then desiccated, which is followed by curing at 65 °C for four hours to solidify. Finally, the sensor is peeled off from the mold and used for experimentation after washing several times with ethanol. Figure 13.3 represents the schematics of the sensor fabrication process. The interdigitated electrodes' width, length, and thickness play a vital role in electric field distributions between the electrodes. The width, length, and thickness of the proposed sensor's electrodes are 1.5 mm, 30 mm, and 0.5 mm, respectively. Figure 13.4 shows the top and cross-sectional view of the interdigitated electrode.

(a)   (b)

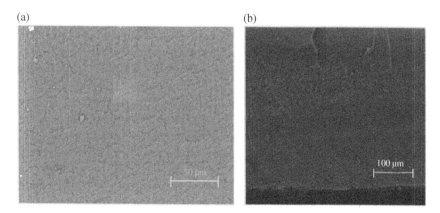

Figure 13.4 Scanning electron microscopic (SEM) images showing the (a) top view and (b) cross-section of the electrodes.

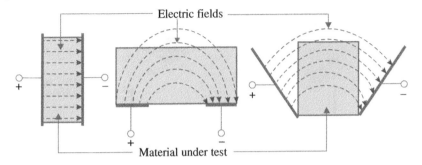

Figure 13.5 Sensor's working principle.

13.3.1 Sensor's Working Principle

The operating mechanism of the proposed sensor is similar to a capacitor. Figure 13.5 shows the transformation of electric field distribution from parallel plate capacitor to an interdigital sensor. In these types of sensors, when a low-amplitude ac voltage is supplied to the positive (excitation) electrode, an electric field is generated between the positive and negative (sensing) electrodes. Whenever a material comes into close contact between them, the electric field distribution changes. As a result, the impedance of the sensor changesrs. The sensor impedance is a function of the tested material. Hence, the characteristic of the tested material can be understood by studying the change in impedance characteristics.

13.4 Experimental Process

At the beginning of the experiment, different solutions are prepared. Basic and acidic pH solutions are prepared by mixing sodium hydroxide (NaOH) and sulfuric acid (H_2SO_4), respectively, with deionized water. Differently, concentrated nitrate solutions are made by mixing potassium nitrate (KNO_3) with deionized water. Initially, 100 ppm stock solutions are prepared. Then, the serial dilution method is applied to prepare the rest of the solutions. Various phosphate, calcium, and magnesium solutions are prepared by mixing sodium phosphate (Na_3PO_4), calcium sulphate ($CaSO_4$), and magnesium sulphate ($MgSO_4$), respectively, with deionized water by applying a similar technique.

Figure 13.6 refers to the schematics of the experimental setup for experimenting with the proposed sensor. It requires an impedance analyzer (HIOKI IM 3536) to obtain impedance response for a wide frequency range, a computer for data collection, and a proposed sensor. The sensing surface is dipped appropriately

Figure 13.6 Schematic of experimental process using MWCNTs/PDMS sensor.

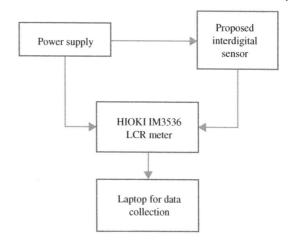

Figure 13.6 Schematic of experimental process using MWCNTs/PDMS sensor.

inside the beaker to read the impedance data correctly. The frequency of the impedance analyzer varies from 10 Hz to 100 kHz. The sensor's response in terms of resistance, reactance, and impedance is obtained by applying the electrochemical impedance spectroscopy (EIS) method.

13.5 Autonomous System Development

This research proposes an autonomous water quality monitoring system. Table 13.4 shows the electronic components used to develop the prototype. Figures 13.7 and 13.8a,b show the connection diagram of the proposed system, inset of the prototype, and final system, respectively. A chamber is allocated for water collection. Two pumps are used. One for water collection from the creek and the other one empty the chamber after finishing the measurement. LM298 motor controller controls all these. The pump goes to ON and OFF state based on NAND gate logic. When both inputs are in the same state, the pump remains OFF, and when both inputs are at different states, the pump goes ON.

AD5933 impedance analyzer is used to read each sensor data and convert it into meaningful water quality parameters applying a data processing algorithm discussed in the subsequent section. Once the sensor's data are read, those can be transferred to the ThingSpeak server. The necessary coding for operating the system is written in Arduino IDE (Integrated Development Environment). A channel is open in ThingSpeak [65], and different fields are allocated to store each data. While writing the server code, the channel number and application programming interface (API) key are given to keep the allocated channel's system data.

Table 13.4 Necessary electronics components for the autonomous system.

Name	Description	References
Arduino Uno Rev3	Development board	[57]
LoRa Shield for Arduino 915 MHz	Long-range transceiver	[58]
12/24V 10A Solar charge controller with USB	Power management block	[59]
12V 10W Solar panel with clips	Solar panel	[60]
12V 12Ah SLA battery	Sealed rechargeable battery	[61]
Seaflo water pressure pump	Water collection and execution	[62]
Stepper motor controller module	Controlling two DC motors	[63]
ANT-916-CW-HWR-SMA	External antenna	[64]

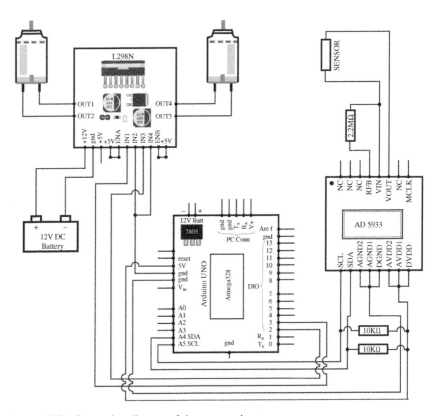

Figure 13.7 Connection diagram of the proposed system.

(a)

Water collection
and sensing unit

Electronics unit

Inlet pump

Outlet pump

(b)

Figure 13.8 (a) Inset of the proposed system and (b) the final system installed in the location.

13.5.1 Algorithm for Data Classification

Actual water samples have many contaminants that influence the sensor's response. Therefore, understanding the sensor's response to the mixed sample is essential. For this purpose, 972 samples are prepared by mixing various solutions of nitrate (0.1, 1, and 10 ppm), phosphate (0.1, 1, and 10 ppm), calcium (10, 50, and 100 ppm), and magnesium (10, 50, and 100 ppm) with different pH (1.8, 6.7, and 10.2) values for various temperatures (5, 15, and 25 °C). The sensor is operated at its operating frequency; sensor response such as impedance (Z) and resistance (R) are obtained. Table 13.5 shows the complete dataset matrix. These data are saved

Table 13.5 Dataset matrix to train the system.

T (°C)	pH	Nitrate (ppm)	Phosphate (ppm)	Calcium (ppm)	Magnesium (ppm)	Z (Ω)	R (Ω)
5	1.8	0.1	0.1	10	10	Z_1	R_1
5	1.8	0.1	0.1	10	50	Z_2	R_2
—	—	—	—	—	—	—	—
—	—	—	—	—	—	—	—
5	1.8	0.1	0.1	50	10	Z_4	R_4
5	1.8	0.1	0.1	50	50	Z_5	R_5
—	—	—	—	—	—	—	—
—	—	—	—	—	—	—	—
5	1.8	0.1	0.1	100	10	Z_7	R_7
5	1.8	0.1	0.1	100	50	Z_8	R_8
—	—	—	—	—	—	—	—
—	—	—	—	—	—	—	—
5	1.8	0.1	1	10	10	Z_{10}	R_{10}
5	1.8	0.1	1	10	50	Z_{11}	R_{11}
—	—	—	—	—	—	—	—
—	—	—	—	—	—	—	—
5	1.8	0.1	10	10	10	Z_{19}	R_{19}
5	1.8	0.1	10	10	50	Z_{20}	R_{20}
—	—	—	—	—	—	—	—
—	—	—	—	—	—	—	—
5	1.8	1	0.1	10	10	Z_{28}	R_{28}
5	1.8	1	0.1	10	50	Z_{29}	R_{29}
—	—	—	—	—	—	—	—
—	—	—	—	—	—	—	—
5	1.8	10	0.1	10	10	Z_{55}	R_{55}
5	1.8	10	0.1	10	50	Z_{56}	R_{56}
—	—	—	—	—	—	—	—
—	—	—	—	—	—	—	—
5	6.7	0.1	0.1	10	10	Z_{82}	R_{82}
5	6.7	0.1	0.1	10	50	Z_{83}	R_{83}
—	—	—	—	—	—	—	—
—	—	—	—	—	—	—	—

Table 13.5 (Continued)

T (°C)	pH	Nitrate (ppm)	Phosphate (ppm)	Calcium (ppm)	Magnesium (ppm)	Z (Ω)	R (Ω)
5	10.2	0.1	0.1	10	10	Z_{163}	R_{163}
5	10.2	0.1	0.1	10	50	Z_{164}	R_{164}
—	—	—	—	—	—	—	—
—	—	—	—	—	—	—	—
15	1.8	0.1	0.1	10	10	Z_{244}	R_{244}
15	1.8	0.1	0.1	10	50	Z_{245}	R_{245}
—	—	—	—	—	—	—	—
—	—	—	—	—	—	—	—
25	1.8	0.1	0.1	10	10	Z_{487}	R_{487}
25	1.8	0.1	0.1	10	50	Z_{488}	R_{488}
—	—	—	—	—	—	—	—
—	—	—	—	—	—	—	—
25	10.2	10	10	100	100	Z_{972}	R_{972} ($n = 972$)

as a matrix in Arduino IDE and used to train the Arduino UNO microcontroller applying the K-nearest neighbor algorithm to predict each parameter's nearest possible value. This algorithm provides a more precise and accurate algorithm for data calibration and classification for smaller dataset and used by many researchers for water quality monitoring [10, 66].

The sensor's response is considered unclassified data. After that, the Euclidean distance's formula is used to determine the nearest distance between the new data and the trainer matrix in the KNN algorithm [67].

$$d(z,\theta) = d(\theta,z) = \sqrt{\sum_{i=1}^{n}\left(z_i - \theta_i\right)^2} \tag{13.1}$$

After that, the algorithm calculates the value of "k" from the training matrix dataset and new entry, verifying the closest distant group from the matrix's rows. Later, the mean deviation of those distances is used to determine the most nearly matched probable value. Figure 13.9 shows the workflow of the system.

All the necessary coding such as client and server codes are written in Arduino IDE to operate the system. As soon as the system is powered, it initialized the AD5933 impedance analyzer. AD5933 reads the impedance and resistance values of the sensor at a specific frequency for the tested solution. After reading the sensor's responses, those are processed to predict the meaningful water quality

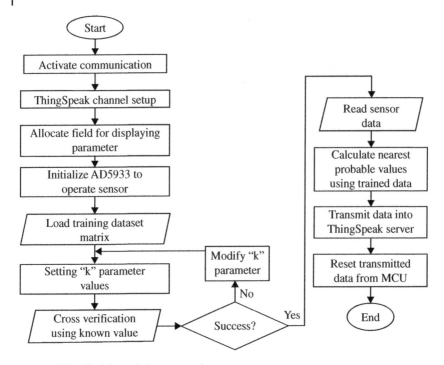

Figure 13.9 Workflow of the proposed system.

parameter values applying the K-NN algorithm. Finally, those data are transferred to the allocated ThingSpeak channel to collect and process further. The channel identification number and API keys are specified in the server code for transferring sensor data into the allocated channel. The spreading factor (SF) value is selected as seven for transferring the data at the fastest speed. This open-access database of the ThingSpeak allows the agronomist to check farms' water condition in real-time and advise the farmers accordingly to prevent the farm from any ecological risk.

13.6 Experimental Results

The multifunctional sensor is characterized to determine various water quality parameters, including temperature, pH, nitrate, phosphate, calcium, and magnesium. Additionally, the calibration curves are also obtained. Repeatability, stability, and validation tests are also carried. All the experimental outcomes are discussed and analyzed in this section.

13.6.1 Sensor Characterization for Temperature, pH, Nitrate, Phosphate, Calcium, and Magnesium Measurement

This research proposes a sensor for determining multiple water quality parameters in indoor and outdoor conditions. Water temperature changes significantly in real life and thus affecting the impedance response of the interdigital sensor. Therefore, the effect of temperature variations on the performance of the MWCNTs/PDMS sensor is studied initially. The water temperature inside a beaker is varied using IEC Hotplate (Cat no. 1920-001). The sensor's impedance responses such as resistance, reactance, and phase angle are obtained for a temperature range from 0 to 50 °C. The actual water temperature is also measured simultaneously using a mercury thermometer.

The sensor's responses are observed for a broad frequency range from 10 Hz to 100 kHz. The variation in resistance for different frequencies is more significant than the change in reactance. Therefore, the frequency versus resistance curve (Figure 13.10a) for different temperatures is included here. Additionally, the calibration curve (Figure 13.10b) to determine water temperature in the unknown water sample is also obtained from resistance values at 1.2 kHz frequency. The calibration equation is specified at 1.2 kHz frequency as it falls in the sensitive area where all the temperatures are separable. The determination coefficient (R^2) obtained after the linear fit is 0.9909 proves that the temperature and sensor's resistance are well-correlated. The higher the temperature, the lower the value of resistance. When the temperature of water increases, the mobility of electron increases. As a result, the sensor's conductivity increases and which in turn reduces the sensor's resistance. The equation to measure water temperature using the proposed sensor is as follows:

$$T(^\circ C) = \frac{3 \times 10^7 - R_{s\_T}}{544705} \tag{13.2}$$

Here, T is the water temperature (°C), R_{s_T} is the resistance of the sensor for a particular temperature, and the temperature coefficient is 544 705 Ω/°C.

The MWCNTs/PDMS sensor is also characterized for various pH level, nitrate, phosphate, calcium, and magnesium concentrations. The resistance and reactance changes for a broad frequency range (10 Hz to 100 kHz) is observed for various nitrate concentrations, phosphate, calcium, magnesium, and different pH values. Frequency responses for all the tested solutions regarding resistance are shown in Figures 13.11a, 13.12a, 13.13a, 13.14a, and 13.15a, respectively. Experimental results show that the sensor can detect and differentiate the various levels of pH and water nutrients. The changes in sensor responses are due to the solution resistance (R_s), adsorption capacitance (C_{ad}), and double-layer capacitance (C_{dl}). The sensitive regions vary from parameters to parameters. A typical

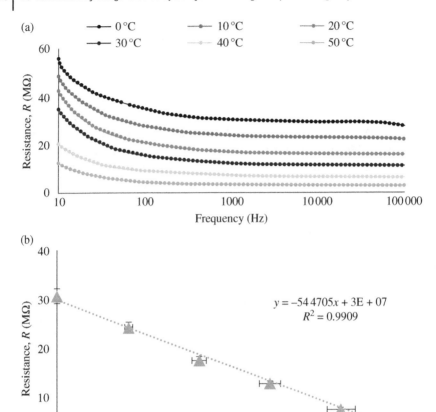

Figure 13.10 (a) Change in sensor's resistances with temperature variations (0–50 °C) and (b) the calibration curve for measuring the water temperature from the sensor's resistances at 1.2 kHz.

frequency of 1.2 kHz is chosen as the operating frequency where the sensor can differentiate various parameters at their different levels due to using a sensor for determining multiple minerals. The calibration curves for determining pH levels, nitrate, phosphate, calcium, and magnesium concentrations are obtained based on the sensor's resistance values at 1.2 kHz frequency and shown in Figures 13.11b, 13.12b, 13.13b, 13.14b, and 13.15b, respectively. The determination coefficient (R^2) for various parameters is more than 0.99 signifies the sensor can detect each

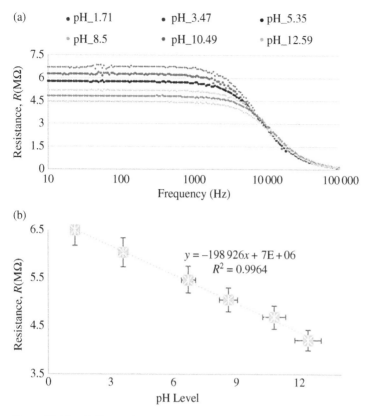

Figure 13.11 (a) Change in sensor's resistances with pH variations (1.3–12.4) and (b) the calibration curve for measuring the water pH from the sensor's resistances at 1.2 kHz.

parameter's values with more than 99% accuracy when only one parameter is present in water. The equations for determining pH, nitrate, phosphate, calcium, and magnesium are as follows:

For pH value,

$$\mathrm{pH} = \frac{7 \times 10^6 - R_{\mathrm{s\_pH}}}{193\,248} \tag{13.3}$$

For nitrate,

$$C_{\mathrm{n}} = \frac{2 \times 10^6 - R_{\mathrm{s\_n}}}{472\,150} \tag{13.4}$$

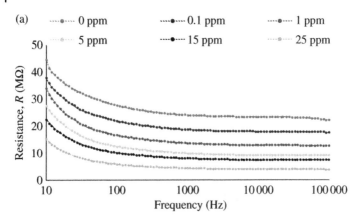

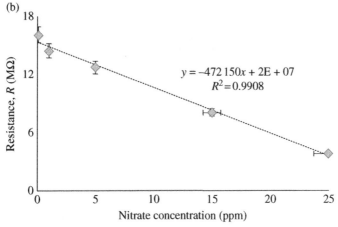

Figure 13.12 (a) Change in sensor's resistances with nitrate concentration variations (0–25 ppm) and (b) the calibration curve for measuring nitrate concentrations from the sensor's resistances in water at 1.2 kHz.

For phosphate,

$$C_p = \frac{2\times10^7 - R_{s\_p}}{626\,825} \tag{13.5}$$

For calcium,

$$C_c = e^{\frac{2\times10^7 - R_{s\_c}}{3\times10^6}} \tag{13.6}$$

For magnesium,

$$C_m = e^{\frac{2\times10^7 - R_{s\_m}}{4\times10^6}} \tag{13.7}$$

(a)

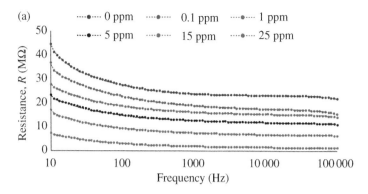

(b)

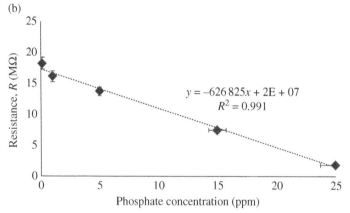

Figure 13.13 (a) Change in sensor's resistances with phosphate concentration variations (0–25 ppm) and (b) the calibration curve for measuring phosphate concentrations from the sensor's resistances in water at 1.2 kHz.

Here, pH refers to the pH level of water, C_n (ppm), C_p (ppm), C_c (ppm), and C_m (ppm) represents the nitrate, phosphate, calcium, and magnesium concentrations, whereas R_{s_pH} (Ω), R_{s_n} (Ω), R_{s_p} (Ω), R_{s_c} (Ω), and R_{s_m} (Ω) are sensor's resistance for a particular amount of water quality parameters.

13.6.2 Repeatability

Consistency in the sensor's performance is an essential aspect of deciding the reliability of the sensor. Therefore, the sensor's resistances for various pH solutions and nitrate, phosphate, calcium, and magnesium concentrations are obtained three times to verify the sensor's reliability. Figure 13.16a–e shows the data collected for different solutions and relative standard deviation (RSD) is calculated to understand the stability of the sensor's response. This illustrates the change in the

(a)

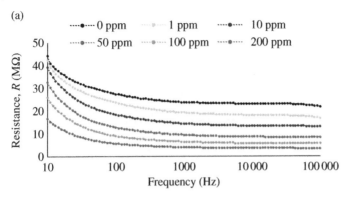

(b)

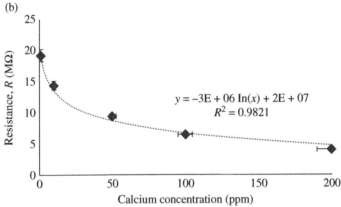

Figure 13.14 (a) Change in sensor's resistances with calcium concentration variations (1–200 ppm) and (b) the calibration curve for measuring calcium concentrations from the sensor's resistances in water at 1.2 kHz.

standard deviation of collected data as compared to the mean. The RSD of the measured data is calculated as follows:

$$\text{RSD} = \frac{S}{X_{\text{Avg}}} \times 100 \tag{13.8}$$

where, X_{Avg} is the average value of a dataset and S is the standard deviation of that dataset. The value of X_{Avg} and S of a dataset, starting from X_1, X_2, and X_3 are calculated as follows:

$$X_{\text{Avg}} = \frac{X_1 + X_2 + X_3}{3} \text{ and } S = \sqrt{\frac{\left(X_1 - X_{\text{Avg}}\right)^2 + \left(X_2 - X_{\text{Avg}}\right)^2 + \left(X_1 - X_{\text{Avg}}\right)^2}{2}}$$

(a)

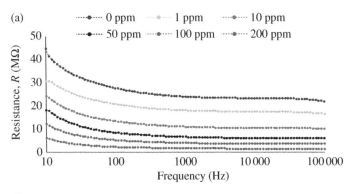

(b)

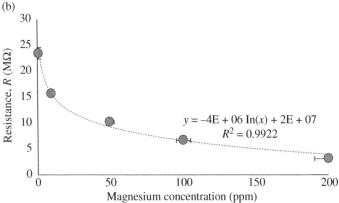

Figure 13.15 (a) Change in sensor's resistances with magnesium concentration variations (1–200 ppm) and (b) the calibration curve for measuring magnesium concentrations from the sensor's resistances in water at 1.2 kHz.

The lower the RSD value, the more consistent the sensor's results are and vice versa [68]. The RSD value found to be below 2% proves the repeatability of the sensor for all the tested solutions.

13.6.3 Reproducibility

A reproducibility test is also carried out, and five sensors are fabricated initially for this purpose. Sensor's performances for various water quality parameters are observed, and the collected data are summarized in Table 13.6. The sensors' performances are almost the same; having the highest RSD of about 4% proves the proposed fabrication method's outstanding reproducibility.

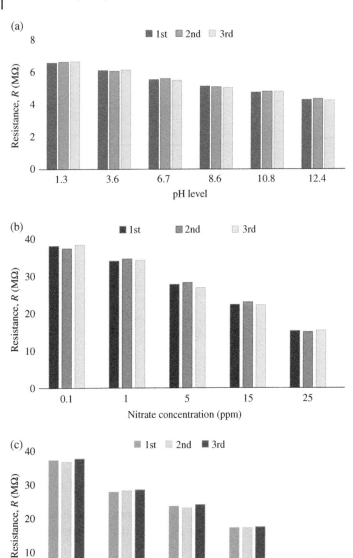

Figure 13.16 Repeatability test of sensor for various solutions of (a) pH (1.3–12.4), (b) nitrate (0.1–25 ppm), (c) phosphate (0.1–25 ppm), (d) calcium (1–200 ppm), and (e) magnesium (1–200 ppm).

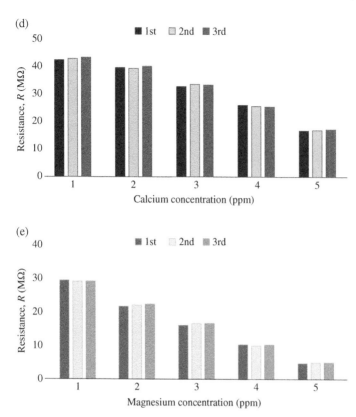

Figure 13.16 (Continued)

13.6.4 Real Sample Measurement and Validation

Once the proposed system is trained using the K-NN method, it is used to measure actual water samples of various streams, creeks, rivers, tap water, and lakes of Sydney, Australia. The water samples are brought to the lab and the sensor is used for measuring pH, nitrate, phosphate, calcium, and magnesium value. The sensor's responses for each sample are recorded, and relevant water quality parameters are predicted by applying the data processing algorithm. The sensor's measurements are validated using the standard lab-based method. Water temperature measurements are validated against mercury thermometer measurements. The pH values measured by the sensor are compared with the pH meter (IC-Starter300) for validation. The nitrate, phosphate, calcium, and magnesium measurements are compared with the UV–vis spectrometry method to validate the

Table 13.6 Testing reproducibility of the MWCNTs/PDMS sensor.

Parameter	Sensor number	Actual value	Sensor's measurement	RSD (%)
	1		6.80	
	2		6.70	
pH	3	6.80	6.60	1.78
	4		6.85	
	5		6.90	
	1		2.00	
	2		2.15	
Nitrate	3	2.00	2.05	2.77
	4		2.03	
	5		2.08	
	1		1.00	
	2		1.08	
Phosphate	3	1.00	1.06	4.18
	4		1.12	
	5		1.09	
	1		50.00	
	2		50.85	
Calcium	3	50.00	50.65	0.67
	4		50.46	
	5		50.22	
	1		30.00	
	2		30.48	
Magnesium	3	30.00	30.37	0.82
	4		30.63	
	5		30.18	

sensor's results. Figure 13.17a–e shows the experimental outcomes of the validation experiments. The proposed sensor can detect water quality parameters in real samples with less than 5% error, which is due to the presence of various nutrients in the water sample. However, the proposed system will provide more accurate results if the system is trained with larger dataset and using a higher memory-based microcontroller board.

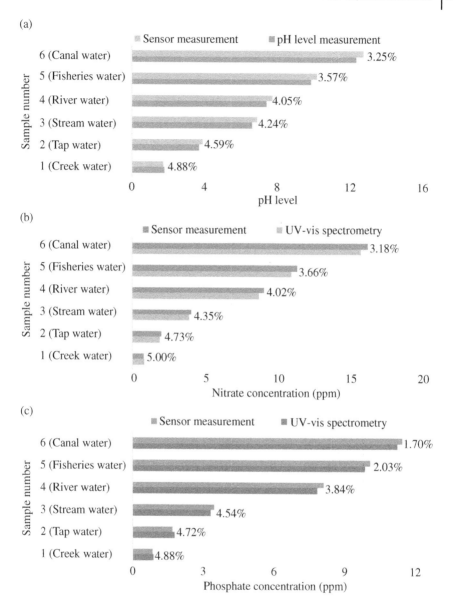

Figure 13.17 Validation of sensor's measurement with standard method for (a) pH, (b) nitrate, (c) phosphate, (d) calcium, and (e) magnesium.

(d)

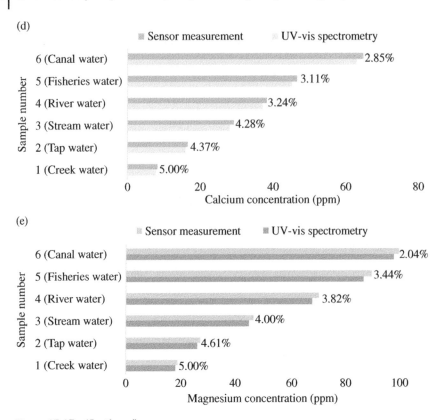

Figure 13.17 (Continued)

13.6.5 Data Collection

After developing the sensor node, it is installed at Macquarie University creek for data collection. Figure 13.18 shows the preliminary results of the collected data into the ThingSpeak cloud server. Sensors' performances deteriorate when used in the long term. Therefore, an auto-calibration algorithm will be developed based on the collected data in the future, enhancing the proposed system's reliability.

13.6.6 Power Consumption

Water is pumped from Macquarie creek to the sensing system every one hour. After that, the sensor data are read and transmitted. Most of time the MSCNTs/PDMS sensor, AD5933 IC, pump controllers, and Lora Shield are kept in idle condition for reducing the power consumption and increasing the battery life. The current drawn by each is calculated by determining voltage across a 1 Ω resistance

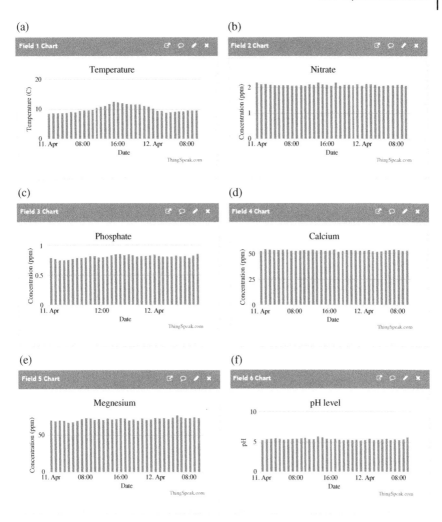

Figure 13.18 Sensor data collected into ThingSpeak server: (a) temperature, (b) nitrate, (c) phosphate, (d) calcium, (e) magnesium, and (f) pH.

connected in series connection with the Arduino UNO. The current drawn by the devices of the sensor node in a cycle (one hour) is shown in Table 13.7.

Total power consumed by the sensing system including transmission is given by

$$P_s = V_s \times I_n + V_s \times I_{td} \times t_{td} \times f_{td}$$

Here,

V_s = Source voltage = 12 V

Table 13.7 Current drawn of the proposed system in one cycle (one hour).

Components	Time	Current drawn (A)	Current consumption in one cycle (ASec)
Arduino Uno	1 h	20×10^{-3}	$20 \times 10^{-3} \times 60 \times 60 = 72$
MWCNTs/PDMS	3 s	1×10^{-3}	$1 \times 10^{-3} \times 3 = 3 \times 10^{-3}$
Two LM298 pump controller including pump (active)	4 min	2.5	$2.5 \times 4 \times 60 = 600$
LM298 pump controller (idle)	56 min	1×10^{-6}	$1 \times 10^{-6} \times 56 \times 60 = 3.36 \times 10^{-3}$
LoRa shield (idle)	59 min 59 s 940 ms	1×10^{-3}	$1 \times 10^{3} \times (59 \times 60 + 59 + 940/1000)$ $= 3.59994$

I_n = Current drawn by the system in one hour = 0.187 mA
I_{td} = Current drawn during data transmission = 0.120 A
t_{td} = Required time to transmit data = 60 msec

$$f_{td} = \text{frequency of data transmission} = \frac{1}{60 \text{ minutes}} = \frac{1}{60 \times 60} = \frac{1}{3600} \text{Hz.}$$

The total current consumed by the sensor node in one hour is

$$I_n = \frac{72 + 3 \times 10^{-3} + 600 + 3.36 \times 10^{-3} + 3.59994}{60 \times 60} = 0.187 \text{A}$$

Power consumption, P_s, is calculated as

$$\Rightarrow P_s = \left(12 \times 0.187 + 12 \times 0.120 \times \frac{60}{10^3} \times \frac{1}{3600} \right) \text{VA}$$
$$\Rightarrow P_s = 2.24 \text{ VA}$$

The minimum mVA required from the battery if the efficiency of the power converter is 85% is

$$P_{bs} = \frac{2.24}{0.85} = 2.65 \text{VA}$$

The required discharge current from the 12 V battery is

$$I_{dis} = \frac{2.65}{12} = 0.22 \text{A}$$

Therefore, the life span of 12 V/12 Ah battery is calculated as

$$B_{\text{Life}} = \frac{12}{0.22} = 54.54 \, \text{hours}$$

If the battery is 100% discharged, the 12 V/12 Ah battery can power the system for 54 hours. The 12 V/10 W solar panel is used during the day to charge the battery. If the battery is not charged due to weather, the node can sustain two days without interruption. Therefore, the voltage across the battery is continuously measured to understand the percentage of charge. When the percentage of charge reaches 25%, an emergency message is sent to user for changing the battery.

13.7 Conclusion

Design, fabrication, and implementation of a novel MWCNTs/PDMS sensor for water quality monitoring are successfully presented in this research. The proposed sensor is characterized to detect various water quality parameters. An autonomous system is also developed for remotely monitoring water quality parameters. An intelligent data processing algorithm is also developed to predict the sensor's response to actual samples' water quality parameters. After applying the KNN algorithm, the system can detect the nitrate with 95% accuracy compared to the actual samples' standard lab-based method. Using a machine learning algorithm rather than selective coating, the proposed sensor can be used for the long term. Additionally, the proposed autonomous system's cost can be reduced significantly because of using a sensor for multiple parameters detection. This type of intelligent system's existence in any fisheries farm will help detect ecological risk early to take an effective countermeasure beforehand.

Acknowledgment

The authors would like to thank Macquarie University for providing necessary facilities to carry out this research.

References

1 Aura, C.M., Musa, S., Yongo, E. et al. (2018). Integration of mapping and socio-economic status of cage culture: towards balancing lake-use and culture fisheries in Lake Victoria, Kenya. *Aquac. Res.* 49: 532–545.

2 Giacomazzo, M., Bertolo, A., Brodeur, P. et al. (2020). Linking fisheries to land use: how anthropogenic inputs from the watershed shape fish habitat quality. *Sci. Total Environ.* 717: 1–11.

3 MacNeil, M.A., Mellin, C., Matthews, S. et al. (2019). Water quality mediates resilience on the Great Barrier Reef. *Nat. Ecol. Evol.* 3: 620–627.

4 Fowzia, A., Sam, K., Hasin, R.S. et al. (2019). IoT enabled intelligent sensor node for smart city: pedestrian counting and ambient monitoring. *Sensors* 5: 1–19.

5 Fowzia, A., Sam, K., Jordan, L. et al. (2019). Design and development of an IoT enabled pedestrian counting and environmental monitoring system for a smart city. *Proceedings of the Thirteenth International Conference on Sensing Technology (ICST)*, Sydney, Australia (2–4 December 2019).

6 Fowzia, A., Hasin, R.S., Alahi, M.E.E., and Subhas, C.M. (2021). Recent advancement of the sensors for monitoring the water quality parameters in smart fisheries farming. *Computers* 10: 1–20.

7 Achim, W., Robert, F., Robert, H., and Nina, B. (2017). Smart farming is key to developing sustainable agriculture. *Proc. Natl. Acad. Sci.* 114: 6148–6150.

8 Jungsu, P., Keug, T.K., and Woo, H.L. (2020). Recent advances in information and communications technology (ICT) and sensor technology for monitoring water quality. *Water* 510: 1–24.

9 Dieisson, P., Paulo, D.W., Edson, T. et al. (2018). Scientific development of smart farming technologies and their application in Brazil. *Inf. Process. Agric.* 5: 21–32.

10 Demi, A. and Nico, S. (2019). Hydroponic nutrient control system based on internet of things. *Proceedings of the International Conference on Computer, Control, Informatics and its Applications (IC3INA)*, Tangerang, Indonesia (23–24 October 2019).

11 Shajulin, B., Nila, G., Deepak, G., and Sreelakshmi, N. (2018). Real time water quality analysis framework using monitoring and prediction mechanisms. *Proceedings of the Conference on Information and Communication Technology (CICT)*, Jabalpur, India (26–28 October 2018).

12 Emerson, N., Nuno, C., and António, P. (2020). A systematic review of IoT solutions for smart farming. *Sensors* 20: 1–29.

13 da Costa, T.H. and Choi, J.W. (2017). A flexible two dimensional force sensor using PDMS nanocomposite. *Microelectron. Eng.* 174: 64–69.

14 Dutta, J.C. and Sharma, P.K. (2018). Fabrication, characterization and electrochemical modeling of CNT based enzyme field effect acetylcholine biosensor. *IEEE Sensors J.* 18: 3090–3097.

15 Ma, L., Dong, X., Chen, M. et al. (2017). Fabrication and water treatment application of carbon nanotubes (CNTs)-based composite membranes: a review. *Membranes* 7: 1–16.

16 Graf, R., Zhu, S., and Sivakumar, B. (2019). Forecasting river water temperature time series using a wavelet–neural network hybrid modelling approach. *J. Hydrol.* 578: 1–12.

17 Bowerman, T., Roumasset, A., Keefer, M.L. et al. (2018). Prespawn mortality of female Chinook salmon increases with water temperature and percent hatchery origin. *Trans. Am. Fish. Soc.* 147: 31–42.

18 Zhaozhao, T., Wenyan, W., and Jinliang, G. (2018). A wireless passive SAW delay line temperature and pressure sensor for monitoring water distribution system. *Proceedings of the IEEE Sensors Conference*, New Delhi, India (28–31 October 2018).

19 Martin, A., Peter, H., Shima, T. et al. (2017). Fibre optic temperature and humidity sensors for harsh wastewater environments. *Proceedings of the Eleventh International Conference on Sensing Technology (ICST)*, Sydney, Australia (4–6 December 2017).

20 Pushkar, S. and Sanghamitra, S. (2016). Arduino-based smart irrigation using water flow sensor, soil moisture sensor, temperature sensor and esp8266 WiFi module. *Proceedings of the IEEE Region 10 Humanitarian Technology Conference (R10-HTC)*, Agra, India (21–23 December 2016).

21 Muhammad, D.K., Endro, A., and Sidik, P. (2018). Design and implementation of smart bath water heater using Arduino. *Proceedings of the 6th International Conference on Information and Communication Technology (ICoICT)*, Bandung, Indonesia (3–5 May 2018).

22 Stiyawan, E.A. (2020). Design of total dissolve solid (TDS) measuring using conductivity sensor and temperature sensor DS18B20. *J. Appl. Electr. Sci. Technol.* 2: 25–29.

23 DFRobot (2021) Waterproof DS18B20 Digital temperature sensor. https://core-electronics.com.au/waterproof-ds18b20-digital-temperature-sensor.html (accessed 28 January 2021).

24 Dong, Y., Son, D.H., Dai, Q. et al. (2018). AlGaN/GaN heterostructure pH sensor with multi-sensing segments. *Sensors Actuators B Chem.* 260: 134–139.

25 Solovyev, M.M., Izvekova, G.I., Kashinskaya, E.N., and Gisbert, E. (2018). Dependence of pH values in the digestive tract of freshwater fishes on some abiotic and biotic factors. *Hydrobiologia* 807: 67–85.

26 Wade, L., Magdalena, W., and Kamal, A. (2017). RuO_2 pH sensor with super-glue-inspired reference electrode. *Sensors* 2036: 1–11.

27 Nedal, A.T., Yunusa, U., Elaref, R. et al. (2016). A flexible optical pH sensor based on polysulfone membranes coated with pH-responsive polyaniline nanofiber. *Sensors* 986: 1–13.

28 Ahmad, N.A., Yook Heng, L., Salam, F. et al. (2019). A colorimetric pH sensor based on *Clitoria* sp. and *Brassica* sp. for monitoring of food spoilage using chromametry. *Sensors* 19 (21): 4813.

29 Wongmeekaew, T., Boonkirdram, S., and Phimphisan, S. (2019). Wireless sensor network for monitoring of water quality for pond tilapia. *2019 Twelfth International Conference on Ubi-Media Computing (Ubi-Media)*, 294–297. IEEE.

30 Lin, W.C., Brondum, K., Monroe, C.W., and Burns, M.A. (2017). Multifunctional water sensors for pH, ORP, and conductivity using only microfabricated platinum electrodes. *Sensors* 17 (7): 1655.

31 Higuchi, S., Okada, H., Takamatsu, S., and Itoh, T. (2020). Valve-actuator-integrated reference electrode for an ultra-long-life rumen pH sensor. *Sensors* 20 (5): 1249.

32 Wongkiew, S., Hu, Z., Chandran, K. et al. (2017). Nitrogen transformations in aquaponic systems: a review. *Aquac. Eng.* 76: 9–19.

33 Alahi, M.E.E., Nag, A., Mukhopadhyay, S., and Burkitt, L. (2018). A temperature-compensated graphene sensor for nitrate monitoring in real-time application. *Sensors Actuators A Phys.* 269: 79–90.

34 Alahi, M.E.E., Mukhopadhyay, S., and Burkitt, L. (2018). Imprinted polymer coated impedimetric nitrate sensor for real-time water quality monitoring. *Sensors Actuators B Chem.* 259: 753–761.

35 Alahi, M.E.E., Xie, L., Mukhopadhyay, S., and Burkitt, L. (2017). Temperature compensated smart nitrate-sensor for agricultural industry. *IEEE Trans. Ind. Electron.* 64: 7333–7341.

36 Alahi, M.E.E., Xie, L., Zia, A.I. et al. (2016). Practical nitrate sensor based on electrochemical impedance measurement. *Proceedings of the International Instrumentation and Measurement Technology Conference*, Taipei, Taiwan (23–26 May 2016).

37 Hameed, R.A. and Medany, S.S. (2019). Construction of core-shell structured nickel@platinum nanoparticles on graphene sheets for electrochemical determination of nitrite in drinking water samples. *Microchem. J.* 145: 354–366.

38 Li, S.J., Zhao, G.Y., Zhang, R.X. et al. (2013). A sensitive and selective nitrite sensor based on a glassy carbon electrode modified with gold nanoparticles and sulfonated graphene. *Microchim. Acta* 180 (9): 821–827.

39 Chen, X., Pu, H., Fu, Z. et al. (2018). Real-time and selective detection of nitrates in water using graphene-based field-effect transistor sensors. *Environ. Sci. Nano* 5 (8): 1990–1999.

40 Choosang, J., Numnuam, A., Thavarungkul, P. et al. (2018). Simultaneous detection of ammonium and nitrate in environmental samples using on ion-selective electrode and comparison with portable colorimetric assays. *Sensors* 18 (10): 3555.

41 Wang, P., Wang, M., Zhou, F. et al. (2017). Development of a paper-based, inexpensive, and disposable electrochemical sensing platform for nitrite detection. *Electrochem. Commun.* 81: 74–78.

42 Amanulla, B., Palanisamy, S., Chen, S.M. et al. (2017). Selective colorimetric detection of nitrite in water using chitosan stabilized gold nanoparticles decorated reduced graphene oxide. *Sci. Rep.* 7 (1): 1–9.

43 Prabhu, A.J., Schrama, J.W., and Kaushik, S.J. (2013). Quantifying dietary phosphorus requirement of fish–a meta-analytic approach. *Aquac. Nutr.* 19: 233–249.

44 Hongwei, C., Linlu, Z., Fabiao, Y., and Qiaoling, D. (2019). Detection of phosphorus species in water: technology and strategies. *Analyst* 144: 7130–7148.

45 Unni, S., Lena, R., Sanu, K.A. et al. (2020). Ultrasensitive electrochemical sensing of phosphate in water mediated by a dipicolylamine-zinc(II) complex. *Sensors Actuators B Chem.* 321: 1–8.

46 Abbas, M.N., Radwan, A.L.A., Nooredeen, N.M., and El-Ghaffar, M.A.A. (2016). Selective phosphate sensing using copper monoamino-phthalocyanine functionalized acrylate polymer-based solid-state electrode for FIA of environmental waters. *J. Solid State Electrochem.* 20 (6): 1599–1612.

47 Barhoumi, L., Baraket, A., Nooredeen, N.M. et al. (2017). Silicon nitride capacitive chemical sensor for phosphate ion detection based on copper phthalocyanine–acrylate-polymer. *Electroanalysis* 29 (6): 1586–1595.

48 Bhat, K.S., Nakate, U.T., Yoo, J.Y. et al. (2019). Nozzle-jet-printed silver/graphene composite-based field-effect transistor sensor for phosphate ion detection. *ACS Omega* 4 (5): 8373–8380.

49 Mao, S., Pu, H., Chang, J. et al. (2017). Ultrasensitive detection of orthophosphate ions with reduced graphene oxide/ferritin field-effect transistor sensors. *Environ. Sci. Nano* 4 (4): 856–863.

50 Kumar, P., Kim, D.M., Hyun, M.H., and Shim, Y.B. (2010). An all-solid-state monohydrogen phosphate sensor based on a macrocyclic ionophore. *Talanta* 82 (4): 1107–1112.

51 Zhou, G., Jin, B., Wang, Y. et al. (2020). Ultrasensitive sensors based on aluminum oxide-protected reduced graphene oxide for phosphate ion detection in real water. *Mol. Syst. Des. Eng.* 5 (5): 936–942.

52 Fowzia, A., Anindya, N., Alahi, M.E.E. et al. (2020). Electrochemical detection of calcium and magnesium in water bodies. *Sensors Actuators A Phys.* 305: 1–10.

53 Global Seafood Alliance (2015). Calcium and magnesium use in aquaculture. https://www.aquaculturealliance.org/advocate/calcium-and-magnesium-use-in-aquaculture (accessed 11 April 2021).

54 Alahi, M.E.E. and Mukhopadhyay, S.C. (2018). Detection methods of nitrate in water: a review. *Sensors Actuators A Phys.* 280: 210–221.

55 Moshoeshoe, M.N. and Obuseng, V. (2018). Simultaneous determination of nitrate, nitrite and phosphate in environmental samples by high performance liquid chromatography with UV detection. *S. Afr. J. Chem* 71: 79–85.

56 Drolc, A. and Vrtovšek, J. (2010). Nitrate and nitrite nitrogen determination in wastewater using on-line UV spectrometric method. *Bioresour. Technol.* 101: 4228–4233.

57 Arduino Uno Rev3. https://store.arduino.cc/usa/arduino-uno-rev3 (accessed 28 January 2021).

58 Lora Shield for Arduino – Long Range Transceiver. https://www.iot-store.com. au/products/lora-shield-for-arduino-long-range-transceiver (accessed 28 January 2021).

59 12/24V 10A Dual Battery PWM Solar Charge Controller with LED Indicator. www.jaycar.com.au/12-24v-10a-dual-battery-pwm-solar-charge-controller-with-led-indicator/p/MP3760 (accessed 28 January 2021).

60 12V 10W Solar Panel with Clips. www.jaycar.com.au/12v-10w-solar-panel-with-clips/p/ZM9051?gclid=Cj0KCQiA3smABhCjARIsAKtrg6JvnJioP4MJkkjeCcEs 6E0V7_mhCHOc0f7f-R7TsYnqDVc_OIZX3yIaAsc2EALw_wcB (accessed 28 January 2021).

61 12V 12Ah SLA Battery. www.jaycar.com.au/12v-12ah-sla-battery/p/SB2489 (accessed 28 January 2021).

62 SEAFLO 21 Series Diaphragm Pump 12V. https://12voltpumps.com.au/ product/12v-21-series-diaphragm-pump-sfdp1-010-035-21-seaflo (accessed 28 January 2021).

63 Stepper motor controller module for Arduino projects. www.auselectronicsdirect. com.au/stepper-motor-controller-module-for-arduino-projec?gclid=Cj0KCQiA3s mABhCjARIsAKtrg6LbnXTiUl9Rt4UCoPRYiUTLWGviqctaB_ilo-f3IJpFNeQ6Pp2DgwEaAtsDEALw_wcB (accessed 28 January 2021).

64 Digi-Key Electronics. ANT-916-CW-HWR-SMA. www.digikey.com.au/product-detail/en/linx-technologies-inc/ANT-916-CW-HWR-SMA/ANT-916-CW-HWR-SMA-ND/1139580 (accessed 28 January 2021).

65 Thingspeak. https://thingspeak.com (accessed 28 January 2021).

66 Akhter, F., Siddiquei, H.R., Alahi, M.E.E. et al. (2021). An IoT-enabled portable water quality monitoring system with MWCNT/PDMS multifunctional sensor for agricultural applications. *IEEE Internet Things J.* 9 (16): 14307–14316.

67 Akhter, F., Siddiquei, H.R., Alahi, M.E.E., and Mukhopadhyay, S.C. (2021). Design and development of an IoT-enabled portable phosphate detection system in water for smart agriculture. *Sensors Actuators A Phys.* 330: 1–11.

68 Akhter, F., Alahi, M.E.E., Siddiquei, H.R. et al. (2020). Graphene oxide (GO) coated impedimetric gas sensor for selective detection of carbon dioxide (CO_2) with temperature and humidity compensation. *IEEE Sensors J.* 21: 4241–4249.

Index

Sensing Technologies for Real Time Monitoring of Water Quality, First Edition.
Libu Manjakkal, Leandro Lorenzelli, and Magnus Willander.
© 2023 The Institute of Electrical and Electronics Engineers, Inc.
Published 2023 by John Wiley & Sons, Inc.

IEEE Press Series on Sensors

Series Editor: Vladimir Lumelsky, Professor Emeritus, Mechanical Engineering, University of Wisconsin-Madison

Sensing phenomena and sensing technology is perhaps the most common thread that connects just about all areas of technology, as well as technology with medical and biological sciences. Until the year 2000, IEEE had no journal or transactions or a society or council devoted to the topic of sensors. It is thus no surprise that the IEEE Sensors Journal launched by the newly-minted IEEE Sensors Council in 2000 (with this Series Editor as founding Editor-in-Chief) turned out to be so successful, both in quantity (from 460 to 10,000 pages a year in the span 2001–2016) and quality (today one of the very top in the field). The very existence of the Journal, its owner, IEEE Sensors Council, and its flagship IEEE SENSORS Conference, have stimulated research efforts in the sensing field around the world. The same philosophy that made this happen is brought to bear with the book series.

Magnetic Sensors for Biomedical Applications
Hadi Heidari, Vahid Nabaei

Smart Sensors for Environmental and Medical Applications
Hamida Hallil, Hadi Heidari

Whole-Angle MEMS Gyroscopes: Challenges, and Opportunities
Doruk Senkal and Andrei M. Shkel

Optical Fibre Sensors: Fundamentals for Development of Optimized Devices
Ignacio Del Villar and Ignacio R. Matias.

Pedestrian Inertial Navigation with Self-Contained Aiding
YushengWang and Andrei M. Shkel

Sensing Technologies for Real Time Monitoring of Water Quality
Libu Manjakkal, Leandro Lorenzelli, Magnus Willander

Printed and bound by CPI Group (UK) Ltd, Croydon, CR0 4YY

16/04/2025

14658602-0003